CHILDHOOD OBESITY

Prevalence, Pathophysiology, and Prevention

CHILDHOOD OBESITY

Prevalence, Pathophysiology, and Prevention

Edited by

Rexford S. Ahima, MD, PhD

Apple Academic Press

TORONTO NEW JERSEY

Apple Academic Press Inc.
3333 Mistwell Crescent
Oakville, ON L6L 0A2
Canada

Apple Academic Press Inc.
9 Spinnaker Way
Waretown, NJ 08758
USA

International Standard Book Number-13: 978-1-926895-91-8 (hbk)

Library of Congress Control Number: 2013950925

Library and Archives Canada Cataloguing in Publication

Childhood obesity (2013)
Childhood obesity: prevalence, pathophysiology, and prevention/edited by
Rexford S. Ahima, MD, PhD.

The chapters in this book were previously published in various places and in various formats.
Includes bibliographical references and index.
ISBN 978-1-926895-91-8
1. Obesity in children--Epidemiology. 2. Obesity in children--Genetic aspects. 3. Obesity in children--Pathophysiology. 4. Obesity in children--Prevention. I. Ahima, Rexford S., editor of compilation II. Title.

RJ399.C6C55 2013 618.92'398 C2013-906621-7

Apple Academic Press also publishes its books in a variety of electronic formats. Some content that appears in print may not be available in electronic format. For information about Apple Academic Press products, visit our website at **www.appleacademicpress.com** and the CRC Press website at **www.crcpress.com**

ABOUT THE EDITOR

REXFORD S. AHIMA, MD, PhD

Rexford S. Ahima, MD, PhD, is a Professor of Medicine at the University of Pennsylvania Perelman School of Medicine, Philadelphia, Pennsylvania. He received a BSc from the University of London, an MD from the University of Ghana, and a PhD from Tulane University in New Orleans, Louisiana. After an internship and residency training in internal medicine at the Albert Einstein College of Medicine in New York, Dr. Ahima did his specialty training in endocrinology, diabetes and metabolism, and postdoctoral research at the Beth Israel Deaconess Medical Center and Harvard Medical School in Boston, Massachusetts. Dr. Ahima's research is focused on the central and peripheral actions of adipose hormones on energy homeostasis and glucose and lipid metabolism. He is an attending endocrinologist at the Hospital of the University of Pennsylvania, Director of the Obesity Unit of the Institute for Diabetes, Obesity and Metabolism, and Director of the Diabetes Research Center Mouse Phenotyping Core. He was elected to the American Society for Clinical Investigation and the Association of American Physicians, and he is a Fellow of the American College of Physicians and The Obesity Society.

CONTENTS

Part II: Adverse Health Consequences

Part III: Prevention and Treatment

ACKNOWLEDGMENT AND HOW TO CITE

The chapters in this book were previously published in various places and in various formats. By bringing them together here in one place, we offer the reader a comprehensive perspective on recent investigations of childhood obesity. Each chapter is added to and enriched by being placed within the context of the larger investigative landscape.

We wish to thank the authors who made their research available for this book, whether by granting their permission individually or by releasing their research as open source articles. When citing information contained within this book, please do the authors the courtesy of attributing them by name, referring back to their original articles, using the credits provided at the end of each chapter.

LIST OF CONTRIBUTORS

Alen Agaronov
School of Public Health at Hunter College, City University of New York, 2180 Third Avenue, New York, NY 10035, USA

Noori Akhtar–Danesh
School of Nursing and Department of Clinical Epidemiology and Biostatistics, McMaster University, Hamilton, Canada

Leslie R. Atkinson
Ryerson University, Department of Psychology, 350 Victoria Street, Toronto, ON, M5B 2 K3, Canada

L. Charles Bailey
Children's Hospital of Philadelphia, Philadelphia, Pennsylvania, United States of America, Perelman School of Medicine, University of Pennsylvania, Philadelphia, Pennsylvania, United States of America

V. M. Beauloye
Pediatric Endocrinology Unit, Cliniques Universitaires Saint–Luc, Université Catholique de Louvain, 1200 Bruxelles, Belgium

Rakesh Bhattacharjee
Division of Sleep and Respiratory Medicine, Department of Pediatrics, The Hospital for Sick Children, University of Toronto, Toronto, ON, M5G 1X8, Canada

Michael H Boyle
McMaster University, Department of Psychiatry and Behavioral Neuroscience, Offord Centre for Child Studies, 1280 Main Street West, Chedoke Site, Patterson Building, Hamilton, ON L8S 3K1, Canada

Luc Bruyndonckx
Laboratory of Cellular and Molecular Cardiology, University Hospital Antwerp, Wilrijkstraat 10, 2650 Edegem, Belgium, Cardiovascular Diseases, Department of Translational Pathophysiological Research, Faculty of Medicine and Health Sciences, University of Antwerp, Universiteitsplein 1, 2610 Antwerp, Belgium, Department of Pediatrics, University Hospital Antwerp, Wilrijkstraat 10, 2650 Edegem, Belgium

Oscar Sans Capdevila
Pediatric Sleep Unit, Division of Neurophysiology, Department of Neurology, Hospital Sant Joan de Déu, 08950 Barcelona, Spain

Viviane M. Conraads
Laboratory of Cellular and Molecular Cardiology, University Hospital Antwerp, Wilrijkstraat 10, 2650 Edegem, Belgium, Cardiovascular Diseases, Department of Translational Pathophysiological Research, Faculty of Medicine and Health Sciences, University of Antwerp, Universiteitsplein 1, 2610 Antwerp, Belgium, Department of Cardiology, University Hospital Antwerp, Wilrijkstraat 10, 2650 Edegem, Belgium

Shikta Das
Department of Epidemiology and Biostatistics, School of Public Health, Imperial College, London, United Kingdom

M. M. Dassy
Pediatric Endocrinology Unit, Cliniques Universitaires Saint–Luc, Université Catholique de Louvain, 1200 Bruxelles, Belgium

Mahshid Dehghan
Population Health Research Institute, McMaster University, Hamilton, Canada

Mark Del Beccaro
Seattle Children's Hospital, Seattle, WA

Lise Dubois
Institute of Population Health, University of Ottawa, Ottawa, Ontario, Canada, Department of Epidemiology & Community Medicine, University of Ottawa, Ottawa, Ontario, Canada

A. C. Dubuisson
Pediatric Endocrinology Unit, Cliniques Universitaires Saint–Luc, Université Catholique de Louvain, 1200 Bruxelles, Belgium

Laura Duncan
McMaster University, Department of Psychiatry and Behavioral Neuroscience, Offord Centre for Child Studies, 1280 Main Street West, Chedoke Site, Patterson Building, Hamilton, ON L8S 3K1, Canada

Paul Elliott
Department of Epidemiology and Biostatistics, School of Public Health, Imperial College, London, United Kingdom, Centre for Environment and Health, School of Public Health, Imperial College, London, United Kingdom

Christopher B. Forrest
Children's Hospital of Philadelphia, Philadelphia, Pennsylvania, United States of America, Perelman School of Medicine, University of Pennsylvania, Philadelphia, Pennsylvania, United States of America

Victor L. Fulgoni
Nutrition Impact, LLC, 9725 D Drive North, Battle Creek, MI, 49104, USA

Stefan Gaget
Unité Mixte de Recherche 8199, Centre National de Recherche Scientifique (CNRS) and Pasteur Institute, Lille, France

Katholiki Georgiades
McMaster University, Department of Psychiatry and Behavioral Neuroscience, Offord Centre for Child Studies, 1280 Main Street West, Chedoke Site, Patterson Building, Hamilton, ON L8S 3K1, Canada

Matthew W. Gillman
Obesity Prevention Program, Department of Population Medicine, Harvard Medical School and Harvard Pilgrim Health Care Institute, Boston, Massachusetts, United States of America

Manon Girard
Institute of Population Health, University of Ottawa, Ottawa, Ontario, Canada

Andrea Gonzalez
McMaster University, Department of Psychiatry and Behavioral Neuroscience, Offord Centre for Child Studies, 1280 Main Street West, Chedoke Site, Patterson Building, Hamilton, ON L8S 3K1, Canada

David Gozal
Department of Pediatrics, Comer Children's Hospital, The University of Chicago, Suite K–160, 5721, Maryland Avenue, MC 8000, Chicago, IL 60637, USA

Struan F. A. Grant
Division of Human Genetics, The Children's Hospital of Philadelphia, Philadelphia, PA 19104, USA, Department of Pediatrics, University of Pennsylvania School of Medicine, Philadelphia, PA 19104, USA, Center for Applied Genomics, Abramson Research Center, The Children's Hospital of Philadelphia Research Institute, 34th and Civic Center Boulevard, Philadelphia, PA 19104, USA

Kateryna Grytsenko
School of Public Health at Hunter College, City University of New York, 2180 Third Avenue, New York, NY 10035, USA

Anna–Liisa Hartikainen
Institute of Clinical Medicine/Obstetrics and Gynecology, University of Oulu, Oulu, Finland

Remy A. Hirasing
EMGO Institute for Health and Care Research, VU University Amsterdam, Amsterdam, The Netherlands, Department of Public and Occupational Health, VU University Medical Center, Amsterdam, The Netherlands

Jacob Hjelmborg
Department of Biostatistics, University of Southern Denmark, Odense, Denmark

Vicky Y. Hoymans
Laboratory of Cellular and Molecular Cardiology, University Hospital Antwerp, Wilrijkstraat 10, 2650 Edegem, Belgium, Cardiovascular Diseases, Department of Translational Pathophysiological Research, Faculty of Medicine and Health Sciences, University of Antwerp, Universiteitsplein 1, 2610 Antwerp, Belgium

Marjo–Riitta Järvelin
Institute of Health Sciences and Biocenter, University of Oulu, Oulu, Finland, Department of Epidemiology and Biostatistics, School of Public Health, Imperial College, London, United Kingdom, Centre for Environment and Health, School of Public Health, Imperial College, London, United Kingdom, Department of Life Course and Services, National Institute for Health and Welfare, Oulu, Finland

N. B. Jodogne
Pediatric Endocrinology Unit, Cliniques Universitaires Saint–Luc, Université Catholique de Louvain, 1200 Bruxelles, Belgium

Marika Kaakinen
Institute of Health Sciences and Biocenter, University of Oulu, Oulu, Finland

Michael G. Kahn
Children's Hospital of Colorado, Aurora, Colorado, United States of America

Pekka Kannus
The UKK Institute for Health Promotion Research, Bone Research Group, Tampere, P.O. Box 30, 33501 Tampere, Finland, Medical School, University of Tampere, 33014 University of Tampere, Finland, Division of Orthopaedics and Traumatology, Department of Trauma, Musculoskeletal Surgery and Rehabilitation, Tampere University Hospital, P.O. Box 2000, 33521 Tampere, Finland

Kelly Kelleher
Nationwide Children's Hospital, Columbus, Ohio, United States of America

Abdelnaby Khalyfa
Department of Pediatrics, Comer Children's Hospital, The University of Chicago, Suite K–160, 5721, Maryland Avenue, MC 8000, Chicago, IL 60637, USA

Anokhi Ali Khan
Department of Epidemiology and Biostatistics, School of Public Health, Imperial College, London, United Kingdom

Michael G. Kahn
Children's Hospital of Colorado, Aurora, Colorado, United States of America

Leila Kheirandish–Gozal
Department of Pediatrics, Comer Children's Hospital, The University of Chicago, Suite K–160, 5721, Maryland Avenue, MC 8000, Chicago, IL 60637, USA

Jinkwan Kim
Department of Pediatrics, Comer Children's Hospital, The University of Chicago, Suite K–160, 5721, Maryland Avenue, MC 8000, Chicago, IL 60637, USA

Ken Kleinman
Obesity Prevention Program, Department of Population Medicine, Harvard Medical School and Harvard Pilgrim Health Care Institute, Boston, Massachusetts, United States of America

Kirsten Ohm Kyvik
Institute of Regional Health Services Research, University of Southern Denmark, Odense, Denmark, Odense Patient Data Explorative Network, Odense University Hospital, Odense, Denmark

Jaana Laitinen
Finnish Institute of Occupational Health, Helsinki, Finland

May May Leung
School of Public Health at Hunter College, City University of New York, 2180 Third Avenue, New York, NY 10035, USA

Paul Lichtenstein
Department of Medical Epidemiology and Biostatistics, Karolinska Institute, Stockholm, Sweden

Stéphane Lobbens
Unité Mixte de Recherche 8199, Centre National de Recherche Scientifique (CNRS) and Pasteur Institute, Lille, France

Harriet L MacMillan
McMaster University, Department of Psychiatry and Behavioral Neuroscience, Offord Centre for Child Studies, 1280 Main Street West, Chedoke Site, Patterson Building, Hamilton, ON L8S 3K1, Canada

Claudio Maffeis
Regional Centre for Juvenile Diabetes, Obesity and Clinical Nutrition, University of Verona, Verona, Italy

Nicholas G. Martin
Queensland Institute of Medical Research, Brisbane, Australia

Joseph L. Mathew
Division of Respiratory Medicine, The Hospital For Sick Children, 555 University Avenue, Toronto, ON, M5G 1X8, Canada

Anwar T. Merchant
Department of Clinical Epidemiology and Biostatistics, and Population Health Research Institute, McMaster University, Hamilton, Canada

Filip Mess
Division of Sport Science, University of Konstanz, Konstanz, Germany

David Meyre
Unité Mixte de Recherche 8199, Centre National de Recherche Scientifique (CNRS) and Pasteur Institute, Lille, France, Department of Clinical Epidemiology and Biostatistics, McMaster University, Hamilton, Canada

David E. Milov
Nemours Children's Hospital, Orlando, Florida, United States of America

Anita Morandi
Unité Mixte de Recherche 8199, Centre National de Recherche Scientifique (CNRS) and Pasteur Institute, Lille, France, Regional Centre for Juvenile Diabetes, Obesity and Clinical Nutrition, University of Verona, Verona, Italy

Indra Narang
Division of Respiratory Medicine, The Hospital For Sick Children, 555 University Avenue, Toronto, ON, M5G 1X8, Canada

Matti Pasanen
The UKK Institute for Health Promotion Research, Bone Research Group, Tampere, P.O. Box 30, 33501 Tampere, Finland

Daniel Pérusse
Faculté des Arts et des Sciences, Université de Montréal, Montreal, Quebec, Canada

Anneli Pouta
Department of Children, Young People and Families, National Institute for Health and Welfare, Helsinki, Finland, Institute of Clinical Medicine/Obstetrics and Gynecology, University of Oulu, Oulu, Finland

Erin E. Quann
Dairy Research Institute/National Dairy Council, Rosemont, IL, 60018, USA

Hein Raat
Department of Public Health, Erasmus MC University Medical Center Rotterdam, Rotterdam, The Netherlands

José Ramet
Department of Pediatrics, University Hospital Antwerp, Wilrijkstraat 10, 2650 Edegem, Belgium

Finn Rasmussen
Department of Public Health Sciences, Karolinska Institute, Stockholm, Sweden

Annette Rauner
Division of Sport Science, University of Konstanz, Konstanz, Germany

Carry M. Renders
Department of Health Sciences, Faculty of Earth and Life Sciences, VU University Amsterdam, Amsterdam, The Netherlands, EMGO Institute for Health and Care Research, VU University Amsterdam, Amsterdam, The Netherlands

Thomas Richards
Children's Hospital of Philadelphia, Philadelphia, Pennsylvania, United States of America

Sheryl L. Rifas–Shiman
Obesity Prevention Program, Department of Population Medicine, Harvard Medical School and Harvard Pilgrim Health Care Institute, Boston, Massachusetts, United States of America

Almo Ruokonen
Department of Clinical Sciences and Clinical Chemistry, University of Oulu, Oulu, Finland

Harri Sievänen
The UKK Institute for Health Promotion Research, Bone Research Group, Tampere, P.O. Box 30, 33501 Tampere, Finland

Axel Skytthe
Odense Patient Data Explorative Network, Odense University Hospital, Odense, Denmark

Fabiola Tatone–Tokuda
Institute of Population Health, University of Ottawa, Ottawa, Ontario, Canada

Riva Tauman
Dana Children's Hospital, Tel–Aviv University School of Medicine, Tel–Aviv 64239, Israel

Kirsti Uusi–Rasi
The UKK Institute for Health Promotion Research, Bone Research Group, Tampere, P.O. Box 30, 33501 Tampere, Finland, Research Department, University Hospital, Tampere, P.O. Box 2000, 33521 Tampere, Finland

Amaryllis H. Van Craenenbroeck
Laboratory of Cellular and Molecular Cardiology, University Hospital Antwerp, Wilrijkstraat 10, 2650 Edegem, Belgium, Cardiovascular Diseases, Department of Translational Pathophysiological Research, Faculty of Medicine and Health Sciences, University of Antwerp, Universiteitsplein 1, 2610 Antwerp, Belgium, Department of Nephrology, University Hospital Antwerp, Wilrijkstraat 10, 2650 Edegem, Belgium

Amy van Grieken
Department of Public Health, Erasmus MC University Medical Center Rotterdam, Rotterdam, The Netherlands

Vincent Vatin
Unité Mixte de Recherche 8199, Centre National de Recherche Scientifique (CNRS) and Pasteur Institute, Lille, France

Dirk K. Vissers
Faculty of Medicine and Health Sciences, University of Antwerp, Campus Drie Eiken, Universiteitsplein 1, 2610 Antwerp, Belgium

Christiaan J. Vrints
Laboratory of Cellular and Molecular Cardiology, University Hospital Antwerp, Wilrijkstraat 10, 2650 Edegem, Belgium, Cardiovascular Diseases, Department of Translational Pathophysiological Research, Faculty of Medicine and Health Sciences, University of Antwerp, Universiteitsplein 1, 2610 Antwerp, Belgium, Department of Cardiology, University Hospital Antwerp, Wilrijkstraat 10, 2650 Edegem, Belgium

Anne I. Wijtzes
Department of Public Health, Erasmus MC University Medical Center Rotterdam, Rotterdam, The Netherlands

Alexander Woll
Department of Sport and Sport Science, Karlsruhe Institute of Technology, Karlsruhe, Germany

Margaret J. Wright
Queensland Institute of Medical Research, Brisbane, Australia

Ming–Chin Yeh
School of Public Health at Hunter College, City University of New York, 2180 Third Avenue, New York, NY 10035, USA

Feliciano Yu
St. Louis Children's Hospital, St. Louis, Missouri, United States of America

F. R. Zech
Department of Internal Medicine, Cliniques Universitaires Saint–Luc, Université Catholique de Louvain, 1200 Bruxelles, Belgium

Jianhua Zhao
Division of Human Genetics, The Children's Hospital of Philadelphia, Philadelphia, PA 19104, USA

LIST OF ABBREVIATIONS

AA	African American
AAP	American Academy of Pediatrics
ADMA	asymmetric dimethylarginine
AGE	advanced glycation end-products
AHA	advanced hip analysis
AHI	apnea/hypopnea index
ALSPAC	Avon Longitudinal Study of Parents and Children
ANOVA	analyses of variance
AUROC	area under the receiver operating curve
BF	body fat
BIA	bioelectrical impedance analysis
BM	behavioral modification
BMC	bone mineral content
BMD	bone mineral density
BMI	body mass index
BP	blood pressure
CAC	circulating Angiogenic cells
CATSS	Child and Adolescent Twin Study in Sweden
CCHS	Canadian Community Health Survey
CDC	U.S. Center for Disease Control and Prevention
CETP	cholesteryl ester transfer protein
CEVQ–SF	Childhood Experiences of Violence Questionnaire Short–Form
CFU–EC	Endothelial Cell Colony Forming Unit
CHQ–PF28	Child Health Questionnaire Parent Form 28
CV	cardiovascular
CVD	cardiovascular diseases
DTR	Danish Twin Registry
DXA	dual energy X–ray absorptiometry
DZ	dizygotic twins

ECFC	endothelial colony forming cells
EDC	expanded diagnostic cluster
EDS	excessive daytime sleepiness
EHR	electronic health record
EMP	endothelial microparticles
eNOS	endothelial nitric oxide synthase
EPC	endothelial progenitor cells
FFA	free fatty acid
FFM	fat–free mass
fl oz	fluid ounce
FMD	flow–mediated dilation
FMS	fundamental movement skills
GNPDA2	glucosamine–6–phosphate deaminase 2
GP	general practitioner
GWAS	genome–wide association studies
HDL	high–density lipoprotein
HOMA	homeostasis model assessment
HRV	heart rate variability
ICAM–1	Intercellular Adhesion Molecule–1
ICC	intraclass correlation coefficient
INSIG2	Insulin–induced gene 2
ISAAC	International Study of Asthma and Allergies in Childhood
ISHAGE	International Society of Hematotherapy and Graft Engineering
kcal	kilocalorie
KCTD15	potassium channel tetramerisation domain containing 15
kg	kilogram
LDL	low–density lipoprotein
LVMI	left ventricular mass index
m	meter
MCP	monocyte chemotactic protein
MCP–1	Macrophage Chemo attractant Protein–1
MCS	mental component score
MEND	Mind, Exercise, Nutrition, Do it program
MetS	metabolic syndrome
MICE	multiple imputation by chain equations
MIF	macrophage inhibitory factor

ml	millileter
MMP	matrix metalloproteinase
MONICA	multinational monitoring of trends and determinants in cardiovascular disease
MRI	magnetic resonance imaging
MRP	myeloid–related protein
MTCH2	mitochondrial carrier
MZ	monozygotic twins
NEGR1	neuronal growth regulator 1
NFBC1986	Northern Finland Birth Cohort 1986
NHANES	National Health and Nutrition Examination Survey
NO	nitric oxide
ObA	obese in adulthood
ObC	obese since childhood
OCHS	Ontario Child Health Study
OR	odds ratio
OSA	obstructive sleep apnea
PA	physical abuse
PA	physical activity
PAP	positive airway pressure
PCS	physical component score
PEDSNet	Pediatric EHR Data Sharing Network
POMC	proopiomelanocortin
PSG	polysomnogram
PWA	pulse wave amplitude
QLS	quasi–least–square
QNTS	Québec Newborn Twin Study
QOL	quality of life
RAAS	renin–angiotensin aldosterone system
RBP4	retinol–binding protein 4
RHI	reactive hyperemia index
ROS	reactive oxygen species
SA	sexual abuse
SB	sedentary behavior
SD	standard deviation
SDB	sleep–disordered breathing

SES	socioeconomic status
SHRINE	Shared Health Research Information Network
SNP	single–nucleotide polymorphism
STR	Swedish Twin Registry
T2D	type 2 diabetes
TBW	total body water
TCHAD	Twin Study of Child and Adolescent Development
TG	triglycerides
USDA	United State Department of Agriculture
VCAM	vascular cell adhesion molecule
VEGFR–2	Vascular Endothelial Growth Factor Receptor–2
VSMC	vascular smooth muscle cells
WHO	World Health Organization
YHC	youth health care

INTRODUCTION

The prevalence of childhood overweight and obesity has increased worldwide in recent decades. Obesity in childhood is associated with a wide range of serious health complications and an increased risk of premature illness and death later in life. This book presents childhood obesity trends across multiple demographics. It discusses the contributing genetic and environmental factors of childhood obesity and shows the adverse health consequences of childhood obesity, both as they relate to childhood and as they last into adulthood. The final section presents multiple methods for obesity treatment included community and family–based intervention, pharmacotherapy, and surgical procedures.

The first section of this book focuses on the genetic and environmental factors that lead to childhood obesity. In Chapter 1, by Zhao and Grant, the authors argue that obesity is a major health problem and an immense economic burden on the health care systems both in the United States and the rest of the world. The prevalence of obesity in children and adults in the United States has increased dramatically over the past decade. Besides environmental factors, genetic factors are known to play an important role in the pathogenesis of obesity. Genome–wide association studies (GWAS) have revealed strongly associated genomic variants associated with most common disorders; indeed there is general consensus on these findings from generally positive replication outcomes by independent groups. To date, there have been only a few GWAS–related reports for childhood obesity specifically, with studies primarily uncovering loci in the adult setting instead. It is clear that a number of loci previously reported from GWAS analyses of adult BMI and/or obesity also play a role in childhood obesity.

Dubois et al. examine the genetic and environmental influences on variances in weight, height, and BMI in Chapter 2. Their data is obtained from a total of 23 twin birth–cohorts from four countries: Canada, Sweden, Denmark, and Australia. Participants were monozygotic (MZ) and dizygotic (DZ) (same– and opposite–sex) twin pairs with data available for both height and weight at a given age, from birth through 19 years of age. Approximately 24,036 children were included in the analyses. They found that heritability for body weight, height, and BMI was low at birth (between 6.4 and 8.7% for boys, and between 4.8 and 7.9% for girls) but increased over time, accounting for close to half or more of the variance in body weight and BMI after 5 months of age in both sexes. Common environmental influences on all body measures were high at birth (between 74.1–85.9% in all measures for boys, and between 74.2 and 87.3% in all measures for girls) and markedly reduced over time. For body height, the effect of the common environment remained significant for a longer period during early childhood (up through 12 years of age). Sex–limitation of genetic and shared environmental effects was observed. The authors concluded that genetics appear to play an increasingly important role in explaining the variation in weight, height, and BMI from early childhood to late adolescence, particularly in boys. Common environmental factors exert their strongest and most independent influence specifically in pre–adolescent years and more significantly in girls. These findings emphasize the need to target family and social environmental interventions in early childhood years, especially for females. As gene–environment correlation and interaction is likely, it is also necessary to identify the genetic variants that may predispose individuals to obesity.

Chapter 3, by Morandi et al., recommends that the prevention of obesity should start as early as possible after birth. The authors devised clinically useful equations estimating the risk of later obesity in newborns, as a first step towards focused early prevention against the global obesity epidemic. To this extent, they analyzed the lifetime Northern Finland Birth Cohort 1986 (NFBC1986) (N = 4,032) to draw predictive equations for childhood and adolescent obesity from traditional risk factors (parental BMI, birth weight, maternal gestational weight gain, behaviour and social indicators), and a genetic score built from 39 BMI/obesity–associated polymorphisms. They performed validation analyses in a retrospective cohort of 1,503 Italian children and in a prospective cohort of 1,032 U.S. children. In the NFBC1986, the cumulative accuracy of traditional risk factors predicting childhood obesity, adolescent obesity, and childhood obesity persistent into adolescence was good: AUROC = 0·78[0·74–0.82], 0·75[0·71–0·79] and 0·85[0·80–0·90] respectively (all p<0·001). Adding the genetic score produced discrimination improvements ≤1%. The NFBC1986 equation for childhood obesity remained acceptably accurate when applied to the Italian and the U.S. cohort (AUROC = 0·70[0·63–0·77] and 0·73[0·67–0·80] respectively) and the two additional equations for childhood obesity newly drawn from the Italian and the U.S. datasets showed good accuracy in respective cohorts (AUROC = 0·74[0·69–0·79] and 0·79[0·73–0·84]) (all $p<0·001$). The three equations for childhood obesity were converted into simple Excel risk calculators for potential clinical use. This study provides the first example of handy tools for predicting childhood obesity in newborns by means of easily recorded information, while it shows that currently known genetic variants have very little usefulness for such prediction.

Folgoni and Quann show in Chapter 4 that given the epidemic of childhood obesity, it is crucial to assess food and beverage intake trends. Beverages can provide a large number of calories and since consumption patterns seem to develop at a young age they examined beverage consumption trends over three decades. The objective of this study was to assess the beverage (milk, fruit juice, fruit drinks, tea, soy beverages, and soft drinks) consumption trends in children <1–5 years of age. They found that during the NHANES 1976–1980 and 1988–1994 periods, approximately 84–85% of children were consuming milk, whereas only 77% were consuming milk during NHANES 2001–2006. Flavored milk intake was relatively low, but increased to 14% during the last decade ($p<0.001$). Fruit juice consumption increased dramatically during NHANES 2001–2006 to more than 50% of the population compared to about 30% in the older surveys ($p < 0.001$). No significant changes were observed in fruit drink intake across all three decades with 35–37% of this population consuming fruit drinks. At least 30% of children consumed soft drinks. Milk was the largest beverage calorie contributor in all three decades surveyed and was the primary contributor of calcium (52–62%), phosphorus (37–42%), magnesium (27–28%), and potassium (32–37%). Fruit juice and fruit drinks each provided 8–10% of calories with soft drinks providing 5–6% of calories. Fruit juice was an important provider of potassium (16–19%) and magnesium (11%). Fruit drinks provided less than 5% of nutrients examined and soft drinks provided very little of the nutrients evaluated. Given these results, as well as concerns about childhood obesity and the need to meet nutrition requirements, it is prudent that parents, educators and child caretakers replace some of the nutrient poor beverages young children are currently consuming with more nutrient dense sources like low–fat and fat–free milk.

Chapter 5 examines the relationship between physical fitness and obesity in youth. Rauner et al. report that obesity increases the risk for several diseases in adults as well as children and adolescents. In turn, many factors including genetic variations and environmental influences (e.g. physical activity) increase the risk of obesity. For instance, 25 to 40 percent of people inherit a predisposition for a high body mass index (BMI). The purpose of this systematic review was to summarize current cross–sectional and longitudinal studies on physical activity, fitness and overweight in adolescents and to identify mediator and moderator effects by evaluating the interaction between these three parameters. Twelve cross–sectional and two longitudinal studies were included. Only four studies analyzed the interaction among physical activity, fitness and overweight in adolescents and reported inconsistent results. All other studies analyzed the relationship between either physical activity and overweight, or between fitness and overweight. Overweight—here including obesity—was inversely related to physical activity. Similarly, all studies reported inverse relations between physical fitness and overweight. Mediator and moderator effects were detected in the interrelationship of BMI, fitness and physical activity. Overall, a distinction of excessive body weight as cause or effect of low levels of physical activity and fitness is lacking. They conclude that the small number of studies on the interrelationship of BMI, fitness and physical activity emphasizes the need for longitudinal studies that would reveal 1) the causality between physical activity and overweight / fitness and overweight and 2) the causal interrelationships among overweight, physical activity and fitness. These results must be carefully interpreted given the lack of distinction between self–reported and objective physical activity and that studies analyzing the metabolic syndrome or cardiovascular disease were not considered. The importance of physical activity or fitness in predicting overweight remains unknown.

In chapter 6, Gonzalez et al. examine the effect of family influences on obesity. Overweight and obesity are steadily increasing worldwide with the greatest prevalence occurring in high–income countries. Many factors influence body mass index (BMI); however multiple influences assessed in families and individuals are rarely studied together in a prospective design. Their objective was to model the impact of multiple influences at the child (low birth weight, history of maltreatment, a history of childhood mental and physical conditions, and school difficulties) and family level (parental income and education, parental mental and physical health, and family functioning) on BMI in early adulthood. They found that at the child level, presence of psychiatric disorder and school difficulties were related to higher BMI in early adulthood. At the family level, receipt of social assistance was associated with higher BMI, whereas family functioning, having immigrant parents and higher levels of parental education were associated with lower BMI. They found that gender moderated the effect of two risk factors on BMI: receipt of social assistance and presence of a medical condition in childhood. In females, but not in males, the presence of these risk factors was associated with higher BMI in early adulthood. Overall, these findings indicate that childhood risk factors associated with higher BMI in early adulthood are multi–faceted and long lasting. These findings highlight the need for preventive interventions to be implemented at the family level in childhood.

Section II examines the negative health effects associated with childhood obesity, both physical and pyschological. In Chapter 7, the first chapter in this section, Narang and Mathew look at the connection between sleep apnea and child obesity. The global epidemic of childhood and adolescent obesity and its immediate as well as long–term consequences for obese individuals and society as a whole cannot be overemphasized. Obesity in childhood and adolescence is associated with an increased risk of adult obesity and clinically significant consequences affecting the cardiovascular and metabolic systems. Importantly, obesity is additionally complicated by obstructive sleep apnea (OSA), occurring in up to 60% of obese children. OSA, which is diagnosed using the gold standard polysomnogram (PSG), is characterized by snoring, recurrent partial (hypopneas) or complete (apneas) obstruction of the upper airway. OSA is frequently associated with intermittent oxyhemoglobin desaturations, sleep disruption, and sleep fragmentation. There is emerging data that OSA is associated with cardiovascular burden including systemic hypertension, changes in ventricular structure and function, arterial stiffness, and metabolic syndromes. Thus, OSA in the context of obesity may independently or synergistically magnify the underlying cardiovascular and metabolic burden. This is of importance as early recognition and treatment of OSA in obese children are likely to result in the reduction of cardiometabolic burden in obese children. This paper summarizes the current state of understanding of obesity–related OSA. Specifically, this paper will discuss epidemiology, pathophysiology, cardiometabolic burden, and management of obese children and adolescents with OSA.

Chapter 8, by Bruyndonckx et al., associates obesity with noncommunicable diseases, such as cardiovascular complications and diabetes, all of which are considered a major threat to the management of health care worldwide. Epidemiological findings show that childhood obesity is rapidly rising in Western society, as well as in developing countries. This pandemic is not without consequences and can affect the risk of future cardiovascular disease in these children. Childhood obesity is associated with endothelial dysfunction, the first yet still reversible step towards atherosclerosis. Advanced research techniques have added further insight on how childhood obesity and associated comorbidities lead to endothelial dysfunction. Techniques used to measure endothelial function were further brought to perfection, and novel biomarkers, including endothelial progenitor cells, were discovered. The aim of this chapter is to provide a critical overview on both in vivo as well as in vitro markers for endothelial integrity. Additionally, an in–depth description of the mechanisms that disrupt the delicate balance between endothelial damage and repair will be given. Finally, the effects of lifestyle interventions and pharmacotherapy on endothelial dysfunction will be reviewed.

Chapter 9 connects insulin sensitivity and obesity. Kim et al. argue the impact of obesity as a systemic low–grade inflammatory process has only partially been explored. To this effect, 704 community–based school–aged children (354 obese children and 350 age–, gender–, and ethnicity–matched controls) were recruited and underwent assessment of plasma levels of fasting insulin and glucose, lipids, and a variety of proinflammatory mediators that are associated with cardiometabolic dysfunction. Obese children were at higher risk for abnormal HOMA and cholesterol levels. Furthermore, BMI z score, HOMA, and LDL/HDL ratio strongly correlated with levels of certain inflammatory mediators. Taken together, obesity in children is not only associated with insulin resistance and hyperlipidemia, but is accompanied by increased, yet variable, expression of markers of systemic inflammation. Future community–based intervention and phenotype correlational studies on childhood obesity will require inclusion of expanded panels of inflammatory biomarkers to provide a comprehensive assessment of risk on specific obesity–related morbidities.

Uusi–Rasi et al. examine the connections between childhood obesity and bone density in adulthood in Chapter 10. Associations between childhood obesity and adult bone traits were assessed among 62 obese premenopausal women, of which 12 had been obese since childhood (ObC), and 50 had gained excess weight in adulthood (ObA). Body composition and bone mineral content (BMC) of the total body, spine, and proximal femur were assessed with DXA. Total cross–sectional area and cortical (diaphyseal CoD) and trabecular (epiphyseal TrD) bone density of the radius and tibia were measured with pQCT. Compared to ObA–group, ObC–group was 5.2 cm taller having 2.5 and 3.5 kg more lean and fat mass, respectively. Depending on the statistical adjustment, ObC–group had 5–10% greater TrD both in tibia and in radius. The remaining bone traits did not significantly differ between the groups. Current preliminary observations bring up an interesting question whether childhood obesity can result in denser trabecular bone in adulthood. However, prudence must be exercised in the statistical adjustment.

Chapter 11 examines the psychological effects, rather than the physical effects of child obesity. van Grieken et al. Limited studies have reported on associations between overweight, and physical and psychosocial health outcomes among younger children. This study evaluates associations between overweight, obesity and underweight in 5–year–old children, and parent–reported health outcomes at age 7 years. This study shows that overweight, obesity and underweight at 5 years of age is associated with more parent–reported adverse treatment of the child. Qualitative research examining underlying mechanisms is recommended. Healthcare providers should be aware of the possible adverse effects of childhood overweight and also relative underweight, and provide parents and children with appropriate counseling.

The third section looks at both treatment and prevention. Dubuisson et al. examine the determinants of weight loss in Chapter 12. Efforts are needed to improve the long–term efficiency of childhood obesity treatment. To adapt strategies, the identification of subgroups of patients with a greater weight loss may be useful. The objective was to analyze the results of a chronic care program for childhood obesity and to determine baseline factors (medical, dietary, and psychosocial) associated with successful weight loss. The authors set up a family–targeted and individually adapted interdisciplinary long–term care program and reviewed the medical files of 144 children (59 boys and 85 girls; y; mean BMI–z–score:) who had ≥2 interdisciplinary visits and ≥1–year treatment. Mean treatment length was 2.2 y (1–6.7 y) with visits/year. The duration of treatment did not depend on the initial weight loss, but this was predictive of the weight change over time. Furthermore any additional weight loss was observed with time whatever the initial weight change. High levels of physical activity and daily water intake from baseline conditions were associated with a greater weight loss after 9 months of intervention. In contrast, a high baseline consumption of soft drinks resulted in lower weight loss. Family specific factors such as being a single child or the child's family support were identified as baseline factors which may contribute to better results. Conclusion. The study suggests that the benefit of a chronic weight control program supports the need for its integration into the current concept of treatment. Better prevention policy and parental support may improve the success of the childhood obesity treatment.

Chapter 13 looks at ways to reduce sedentary behaviors among youth. Leung et al. assess the effectiveness of interventions that focus on reducing sedentary behavior (SB) among school–age youth and to identify elements associated with interventions' potential for translation into practice settings. A comprehensive literature search was conducted using 4 databases for peer–reviewed studies published between 1980 and April 2011. Randomized trials, which lasted at least 12 weeks, aimed at decreasing SB among children aged 6 to 19 years were identified. Twelve studies were included; 3 focused only on SB, 1 focused on physical activity (PA), 6 were combined SB and PA interventions, and 2 studies targeted SB, PA, and diet. The majority of the studies were conducted in a school setting, while others were conducted in such settings as clinics, community centers, and libraries. Overall, interventions that focused on decreasing SB were associated with reduction in time spent on SB and/or improvements in anthropometric measurements related to childhood obesity. Several of the studies did consider elements related to the intervention's potential for translation into practice settings.

Bailey et al. argue that institutional sharing of health records has a role to play in assessing childhood obesity in Chapter 14. They conducted a non–concurrent cohort study of 528,340 children with outpatient visits to six pediatric academic medical centers during 2007–08, with sufficient data in the EHR for body mass index (BMI) assessment. EHR data were compared with data from the 2007–08 National Health and Nutrition Examination Survey (NHANES). Among children 2–17 years, BMI was evaluable for 1,398,655 visits (56%). The EHR dataset contained over 6,000 BMI measurements per month of age up to 16 years, yielding precise estimates of BMI. In the EHR dataset, 18% of children were obese versus 18% in NHANES, while 35% were obese or overweight versus 34% in NHANES. BMI for an individual was highly reliable over time (intraclass correlation coefficient 0.90 for obese children and 0.97 for all children). Only 14% of visits with measured obesity (BMI ≥95%) had a diagnosis of obesity recorded, and only 20% of children with measured obesity had the diagnosis documented during the study period. Obese children had higher primary care (4.8 versus 4.0 visits, $p<0.001$) and specialty care (3.7 versus 2.7 visits, $p<0.001$) utilization than non–obese counterparts, and higher prevalence of diverse co–morbidities. The cohort size in the EHR dataset permitted detection of associations with rare diagnoses. Data sharing did not require investment of extensive institutional resources, yet yielded high data quality. They conclude that multi–institutional EHR data sharing is a promising, feasible, and valid approach for population health surveillance. It provides a valuable complement to more resource–intensive national surveys, particularly for iterative surveillance and quality improvement. Low rates of obesity diagnosis present a significant obstacle to surveillance and quality improvement for care of children with obesity.

The final chapter, Chapter 15, by Dehghan et al., reports that childhood obesity has reached epidemic levels in developed countries. Twenty five percent of children in the US are overweight and 11% are obese. Overweight and obesity in childhood are known to have significant impact on both physical and psychological health. The mechanism of obesity development is not fully understood and it is believed to be a disorder with multiple causes. Environmental factors, lifestyle preferences, and cultural environment play pivotal roles in the rising prevalence of obesity worldwide. In general, overweight and obesity are assumed to be the results of an increase in caloric and fat intake. On the other hand, there is supporting evidence that excessive sugar intake by soft drink, increased portion size, and steady decline in physical activity have been playing major roles in the rising rates of obesity all around the world. Consequently, both over–consumption of calories and reduced physical activity are involved in childhood obesity. Almost all researchers agree that prevention could be the key strategy for controlling the current epidemic of obesity. Prevention may include primary prevention of overweight or obesity, secondary prevention or prevention of weight regain following weight loss, and avoidance of more weight increase in obese persons unable to lose weight. Until now, most approaches have focused on changing the behaviour of individuals through diet and exercise. It seems, however, that these strategies have had little impact on the growing increase of the obesity epidemic. While about 50% of the adults are overweight and obese in many countries, it is difficult to reduce excessive weight once it becomes established. Children should therefore be considered the priority population for intervention strategies. Prevention may be achieved through a variety of interventions targeting built environment, physical activity, and diet. Some of these potential strategies for intervention in children can be implemented by targeting preschool institutions, schools or after–school care services as natural setting for influencing the diet and physical activity. Overall, there is an urgent need to initiate prevention and treatment of obesity in children.

PART I

GENETIC AND ENVIRONMENTAL FACTORS

CHAPTER 1

GENETICS OF CHILDHOOD OBESITY

JIANHUA ZHAO and STRUAN F. A. GRANT

1.1 DEFINITION AND EPIDEMIOLOGY OF CHILDHOOD OBESITY

Obesity is a major health problem in modern societies, with a prevalence of up to 25% in Western societies and an increasing incidence in children [1]. Obesity, plus the associated insulin resistance [2, 3], is also considered a contributor to the major causes of death in the United States and is an important risk factor for type 2 diabetes (T2D), cardiovascular diseases (CVD), hypertension, and other chronic diseases.

Approximately 70% of obese adolescents grow up to become obese adults [4–6]. The main direct adverse effects of childhood obesity include orthopedic complications, sleep apnea, and psychosocial disorders [7, 8]. Obesity present in adolescence has been shown to be associated with increased overall mortality in adults [9]; overweight children followed up for 40 [10] and 55 years [11] were more likely to have CVD and digestive diseases, and to die from any cause as compared with those who were lean.

Obesity is a complex disease that involves interactions between environmental and genetic factors. Excess in adipose tissue mass can be seen as a disruption in the balance between energy intake and expenditure. In modern times, this excess in adipose tissue fuel storage is considered a disease; however, a better viewpoint would be that obesity is a survival advantage that has gone astray that is, what is now considered a disease was probably advantageous when food was less available and a high level of energy expenditure through physical activity was a way of life [12].

The true prevalence of childhood obesity is difficult to empirically quantify as there is currently no internationally accepted definition; however, in general terms, childhood obesity is considered to have reached epidemic levels in developed countries.

Approximately 25% of children in the US are overweight and approximately 11% are obese. In the 10–year period between the National Health and Nutrition Examination Survey (NHANES) II (1976–1980) and NHANES III (1988–1991), the prevalence of overweight children in the USA had increased by approximately forty percent [1]. Examination of historical standards for defining overweight in children from many countries tells us that the distribution of BMI is becoming increasingly skewed [13]. The lower part of the distribution has shifted relatively little whereas the upper part has widened substantially. This finding suggests that many children may be more susceptible (genetically or socially) to influence by the changing environment.

Although the definition of obesity and overweight has changed over time [14, 15], it can be defined as an excess of body fat. The definition of childhood obesity continues to be problematic due to the fact that almost all definitions use some variant of BMI (body mass index). A range of other methods are available which allow for accurate estimates of total body fat; however, none of these are widely available and/or are easily applicable to the clinical situation. Body weight is reasonably well correlated with body fat but is also highly correlated with height, and children of the same weight but different heights can have widely differing amounts of adiposity, but in adults BMI correlates more strongly with more specific measurements of body fat, that is, BMI is useful for depicting overweight in the population but is an imperfect approximation of excess adiposity [16].

In addition, the relation between BMI and body fat in children varies widely with age and with pubertal maturation. This in itself makes BMI definitions of overweight for children more complex than definitions for adults, which use a single cutoff value for all ages. Definitions of overweight that use BMI–for–age can be based on a number of different standards that all give slightly different results, and all are essentially statistical not functional definitions. However, useful percentile charts relating BMI to age have now been published in several countries [17]. The Center for

Disease Control and Prevention defined overweight as at or above the 95th percentile of BMI for age and "at risk for overweight" as between 85th to 95th percentile of BMI for age [18, 19]. European researchers classified overweight as at or above 85th percentile and obesity as at or above 95th percentile of BMI [20]. A recent report from the Institute of Medicine has specifically used the term "obesity" to characterize BMI $\geq$ 95th percentile in children and adolescents [21]. By late adolescence, these percentiles approach those used for adult definitions; the 95th percentile is then approximately 30 kg/m2 [8]. These statistical percentile definitions are now general guidelines for clinicians and others [19].

1.2 THERAPEUTIC OPTIONS

Data supporting the use of pharmacological therapy for pediatric overweight are limited and inconclusive [22].

Sibutramine has been studied in a randomized controlled trial of severe obesity [23]. It has been shown to be efficacious as compared with behavior therapy alone, but it may be associated with side effects including increases in heart rate and blood pressure [24]; recent clinical trial studies have concluded that subjects with preexisting cardiovascular conditions who were receiving long–term sibutramine treatment had an increased risk of nonfatal myocardial infarction and nonfatal stroke [25]; indeed, it was recently dropped from further development based on the results from such clinical trials.

Orlistat is approved for use in adolescence but its efficacy has not yet been tested extensively in young patients. Orlistat is associated with gastrointestinal side effects and requires fat–soluble vitamin supplementation and monitoring [26, 27].

Metformin, used to treat T2D, has been used in insulin–resistant children and adolescents who are overweight, but long–term efficacy and safety are unknown [28]. Additionally, surgical approaches to treat severe adolescent obesity are being undertaken by several centers [29].

For rare genetic and metabolic disorders, pharmacological treatment may be useful. For example, recombinant leptin is useful in hereditary leptin deficiency. Octreotide may be useful in hypothalamic obesity [30].

1.3 EVIDENCE FOR A GENETIC COMPONENT IN OBESITY

The rising prevalence of obesity can be partly explained by environmental changes over the last 30 years, in particular the unlimited supply of convenient, highly calorific foods together with a sedentary lifestyle. Despite these changes, there is also strong evidence for a genetic component to the risk of obesity [31, 32]; indeed, obesity is now considered a classic example of a complex multifactorial disease resulting from the interplay between behavioral, environmental and genetic factors which may influence individual responses to diet and physical activity.

A genetic component for obesity is reflected in prevalence differences between racial groups, from 5% or less in Caucasian and Asian populations to 50% or more among Pima Indians and South Sea Island populations [33]. In addition, the familial occurrences of obesity have been long noted with the concordance for fat mass among MZ twins reported to be 70–90%, higher than the 35–45% concordance in DZ twins; as such, the estimated heritability of BMI ranges from 30 to 70% [34–36].

1.4 PREVIOUS GENETIC STUDIES IN OBESITY AND THE NEED FOR GWAS APPROACHES

Genome–wide linkage scans in families with the common form of childhood obesity have yielded several loci, but the genes in these loci have yet to be elucidated. A number of families with rare pleiotropic obesity syndromes have been studied by linkage analysis where chromosomal loci for Prader–Willi syndrome [37], Alström's syndrome [38], and Bardet–Biedl syndrome [39–41] have been mapped but the underlying molecular mechanisms have yet to be determined.

Recent studies of genetic syndromes of obesity in rodents have provided insights in to the underlying mechanisms that may play a role in energy homoeostasis. In recent years, research has begun to identify human disorders of energy balance that arise from defects in these or related genes [42]. These mutations have been shown to result in morbid obesity in children without the developmental features that commonly accompany recognized syndromes of childhood obesity.

The severely obese ob/ob mouse strain [43] inherits its early–onset obesity autosomal recessively and weighs approximately three–times more than normal mice by maturity. Zhang et al. [44] cloned and characterized the ob gene which is expressed primarily in white adipose tissue as the secreted protein, leptin, a mutation of which renders these mice leptin deficient. Administration of recombinant leptin is known to reverse the phenotypic abnormalities in these mice entirely [45–47] while there is no effect in another strain of severely obese mice, db/db, who instead have been characterized to have a mutation in the leptin receptor gene, which is primarily expressed in a different site, namely, the hypothalamus [48]. In human studies, serum leptin concentrations are widely recognized as being positively correlated with obesity–related traits [49].

The behavioral and neuroendocrine effects of leptin could potentially be mediated through its actions at hypothalamic leptin receptors. Proopiomelanocortin (POMC) is produced by the hypothalamus, which is subsequently cleaved by prohormone convertases to yield peptides (including α melanocyte stimulating hormone, αMSH) that play a role in feeding behavior. Forty percent of POMC neurons express mRNA for the long form of the leptin receptor, and POMC expression is positively regulated by leptin [50]. Work in rodents has demonstrated that αMSH acts as a suppressor of feeding behavior; recently, mutations in POMC associated with severe and early–onset obesity have been described in two unrelated German subjects [51]. A single patient with severe early–onset obesity was reported to have compound heterozygote mutations in the prohormone convertase 1 (PC1) gene, a key component in the proteolytic processing of POMC [52, 53].

One form of melanocortin receptor (MC4R) is highly expressed in areas of the hypothalamus involved in feeding; mice with disruption of the MC4R gene are severely obese [54]. More recently in humans, mutations in the MC4R gene have been associated with obesity [55–58]. The MC4R gene is the first locus at which mutations are associated with dominantly inherited morbid human obesity thus making it the commonest genetic cause of human obesity described before the era of GWAS.

1.5 GENOME WIDE ASSOCIATION STUDIES

Overall, linkage analysis studies conducted to date have achieved only limited success in identifying genetic determinants of obesity due to various reasons, importantly including the generic problem that the linkage analysis approach is generally poor in identifying common genetic variants that have modest effects [59, 60]. Comparably, a generic problem with the candidate gene association studies is their general reliance on a suspected disease–causing gene(s) whose identity derives from a particular biological hypothesis on the pathogenesis of obesity. Thus, since the pathophysiological mechanisms underlying obesity are generally unknown, continued use of the hypothesis–driven candidate gene association approach is likely to identify only a relatively small fraction of the genetic risk factors for the disease.

The GWAS approach serves the critical need for a more comprehensive and unbiased strategy to identify causal genes related to obesity. It is also well established that in noncoding regions of the genome there are important regulatory elements, such as enhancers and silencers, and genetic variants that disrupt those elements could equally confer susceptibility to complex disease.

The human genome and International HapMap projects have enabled the development of unprecedented technology and tools to investigate the genetic basis of complex disease. The HapMap project, a large–scale effort aimed at understanding human sequence variation, has yielded new insights into human genetic diversity that is essential for the rigorous study design needed to maximize the likelihood that a genetic association study will be successful [59–61]. Genome–wide genotyping of over 500,000 SNPs can now be readily achieved in an efficient and highly accurate manner [62, 63]. Since much of human diversity is due to single base pair variations together with variations in copy number [64] throughout the genome, current advances in single–base extension (SBE) biochemistry and hybridization/detection to synthetic oligonucleotides now make it possible to accurately genotype and quantitate allelic copy number [63, 65].

There is now a revolution occurring in SNP genotyping technology, with high–throughput genotyping methods allowing large volumes of SNPs (105–106) to be genotyped in large cohorts of patients and controls, therefore enabling large–scale GWAS in complex diseases. Already with this technology compelling evidence for genetic variants involved in type 1 diabetes [66–68], type 2 diabetes [68–72], age–related macular degeneration [73], inflammatory bowel disease [74, 75], heart disease [76, 77], and breast cancer [78] has been described.

1.6 FINDINGS FROM FIRST GWAS ANALYSES OF OBESITY

In the past four years, many genetic loci have been implicated for BMI from the outcomes of GWAS, primarily in adults.

Insulin–induced gene 2 (INSIG2) was the first locus to be reported by this method to have a role in obesity [79] but replication attempts have yielded inconsistent outcomes [80–84]. A common genetic variant with modest relative risk (RR = ~1.2), rs7566605, near the INSIG2 gene has been described to be associated with both adult and childhood obesity from a GWAS employing 100,000 SNPs [79]. This variant, present in 10% of individuals, was subsequently replicated in four separate cohorts in the same study, including individuals who were Caucasian, African American, and children; however, three subsequent technical comments to Science [80–82] disagreed with this observation.

The identification of the second locus, the fat mass– and obesity–associated gene (FTO) [85], which has been more robustly observed by others [86–89], including by us [90]. Interestingly, its role in obesity pathogenesis was actually made indirectly as a consequence of a GWAS of T2D [68, 71], but it became quite clear that its primary influence is on BMI determination which then in turn impairs glycemic control [85]. However, the mechanism by which the variant in FTO influences the risk of obesity is largely unknown.

Studies from both FTO knockout and FTO overexpression mouse model support the fact that FTO is directly involved in the regulation of energy intake and metabolism in mice, where the lack of FTO expression leads to leanness while enhanced expression of FTO leads to obesity [91, 92].

A French sequencing effort in Caucasians (primarily adults) has reported a set of exonic mutations in FTO; however, due to the lack of significant difference in the frequencies of these variants between lean and obese individuals, this study was largely negative [93].

1.7 META–ANALYSES

Subsequent larger studies have uncovered eleven additional genes [94–96], again primarily in adults, firstly melanocortin 4 receptor (MC4R) from a multicenter meta–analysis [94], then the GIANT consortium revealed six more genes (transmembrane protein 18 (TMEM18), potassium channel tetramerisation domain containing 15 (KCTD15), glucosamine–6–phosphate deaminase 2 (GNPDA2), SH2B adaptor protein 1 (SH2B1), mitochondrial carrier 2 (MTCH2), and neuronal growth regulator 1 (NEGR1)) [96], five of which were confirmed in the GWAS reported from Iceland (but not GNPDA2 due to an unavailable proxy SNP), who also uncovered and reported loci on 1q25, 3q27 and 12q13 [95] and verified association with the brain–derived neurotrophic factor (BDNF) gene [97].

The latest GIANT meta–analysis revealed multiple new loci associated with body mass index in a study involving a total of 249,796 individuals [98]. A total of 32 loci reached genome wide significance, which included ten known loci associated with BMI, four known loci associated with weight and/or waist–hip ratio, namely, SEC16B, TFAP2B, FAIM2, NRXN3, and eighteen new BMI loci, namely, RBJ–ADCY3–POMC, GPRC5B–IQCK, MAP2K5–LBXCOR1, QPCTL–GIPR, TNNI3K, SLC39A8, FLJ35779–HMGCR, LRRN6C, TMEM160–ZC3H4, FANCL, CADM2, PRKD1, LRP1B, PTBP2, MTIF3–GTF3A, ZNF608, RPL27A–TUB, and NUDT3–HMGA1. Besides association to SNPs, a correlated copy number variation (CNV), that is, a 21 kb deletion, was identified 50 kb upstream of GPRC5B. This study also leveraged a pediatric cohort to lend further support for their findings.

1.8 TESTING ADULT–DISCOVERED LOCI IN CHILDREN

As described above, a number of genetic determinants of adult BMI have already been established through GWAS. One obvious question is how do these loci operate in childhood with respect to the pathogenesis of obesity? We have an ongoing GWAS of BMI variation in children so we are in position to query these SNPs in our dataset of in excess of 6,000 children with measures of BMI [99]. To date nine such loci have yielded evidence of association to BMI in childhood, of which variants at the FTO locus yielded the strongest association. With a similar magnitude of association to FTO was TMEM18 followed by GNPDA2. The remaining loci with evidence for association were INSIG2, MC4R, NEGR1, 1q25, BDNF, and KCTD15 (Table 1). This is very much in line with the observations made with the pediatric cohort utilized by Willer et al. [96].

TABLE 1: Childhood obesity loci that have been identified to date and the route through which they were implicated.

Category	Loci	Citations
Adult BMI GWAS loci also associated with childhood BMI/obesity in independent studies	*FTO, TMEM18, GNPDA2; INSIG2, MC4R, NEGR1, 1q25, BDNF, KCTD15*	[99]
Adult 2 type diabetes GWAS loci also associated with childhood BMI/obesity	*HHEX–IDE*	[100]
GWAS of extreme childhood obesity—novel loci	*SDCCAG8, TNKS–MSRA*	[101]
CNV analyses—novel loci	*SH2B1, EDIL3, S1PR5, FOXP2, TBCA, ABCB5, ZPLD1, KIF2B, ARL15, EPHA6–UNQ6114, OR4P4–OR4S2–OR4C6*	[102]–[105]

The positive results for FTO and MC4R come as no surprise as we previously reported their association with the CDC–defined 95th percentile of BMI, that is, obesity, in our pediatric cohort, but limited to ages 2–18 years old [90, 106]. One of the more notable results is the positive association with INSIG2. This association with pediatric BMI, albeit at just the nominal level, contributes to the ongoing debate on the relative contribution of INSIG2 in BMI determination.

However, these nine loci only explain 1.12% of the total variation for BMI. In addition, testing pair–wise interactions between the fifteen significant SNPs, none of the interaction effects were significant suggesting that these loci act additively on pediatric BMI [99]. As such, we do observe a cumulative effect but not as striking as reported by the GIANT consortium in their adult cohorts [96].

A number of studies have found that body mass index (BMI) in early life influences the risk of developing type 2 diabetes (T2D) later in life. Indeed, the same variant in IDE–HHEX that increases the risk of developing the disease later in life turns out to be also associated with increased BMI in childhood [100].

1.9 LOCI SPECIFICALLY IDENTIFIED IN CHILDHOOD OBESITY GWAS ANALYSES

Two new loci for body–weight regulation were identified in a joint analysis of GWAS data for early–onset extreme obesity, that is, BMI ≥ 99th, in French and German study groups [101], namely, SDCCAG8 and TNKS/MSRA (Table 1). In the discovery step, association was examined in a combined French and German sample of 1,138 extremely obese children and adolescents and 1120 normal or underweight controls with screening of 2,339,392 genotyped or imputed SNPs and testing ultimately 1,596,878 SNPs. In the replication cohort, all SNPs with strong evidence for association were genotyped in independent samples of 1,181 obese children and adolescents and 1,960 normal or underweight controls and in up to 715 nuclear families with at least one extremely obese offspring. However the two loci were, at most, only marginally associated with adult BMI in the latest GIANT study [98], suggesting their influence may be limited to extreme obesity in children.

The biochemistry employed in the current genome wide SNP arrays allows also for the accurate genotyping and quantitation of allelic CNV genome–wide [62, 63, 65]. Neurological disorders have proven the most challenging complex disease to address using genome wide SNP approaches, primarily as a consequence of the need for strict, uniform phenotyping across very large, multicenter cohorts. However, they have led the way in the uncovering of CNVs in common disorders such as autism [107–110], attention–deficit hyperactivity disorder [111], and schizophrenia [112–114].

Genomic copy number variations (CNVs) have been strongly implicated in subjects with extreme obesity and coexisting developmental delay (Table 1). Two groups in the UK plus collaborators independently reported deletions on chromosome 16p11.2 to be present at much higher in extreme obese cases than normal and obese individuals [102, 103]. These deletions, estimated to range in size from 220 kb to 1.7 Mb, encompass several genes. Bochukova et al. [102] pointed out that the SH2B1 gene is within the deleted region that is common to all five cases studied. SH2B1 may be the culprit as its role in leptin and insulin signaling and energy homeostasis is well described [102], plus common SNPs near SH2B1 locus have already been associated with BMI in GWAS reports [96, 102].

To complement these previous CNV studies on extreme obesity, we addressed CNVs in common childhood obesity by examining children in the upper 5th percentile of BMI but excluding any subject greater than 3 standard deviations from the mean to reduce severe cases in the cohort [104] (Table 1). We performed a whole–genome CNV survey of our cohort of European American (EA) childhood obesity cases (n = ~1,000) and lean controls (n = 2,500) who were genotyped with 550,000 SNP markers. We identified 34 putative CNVR loci (15 deletions and 19 duplications) that were exclusive to EA cases; however, three of the deletions proved to be false positives during the validation process with quantitative PCR (qPCR). Only 17 of these CNVR loci were unique to our cohort that is, not reported in controls by the Database of Genomic Variants. Positive findings were evaluated in an independent African American (AA) cohort (n = ~1,500) of childhood obesity cases and lean controls (n = ~1,500). Surprisingly, eight of these loci, that is, almost half, also replicated exclusively in AA cases (6 deletions and 2 duplications). Replicated deletion loci consisted of EDIL3, S1PR5, FOXP2, TBCA, ABCB5, and ZPLD1 while replicated duplication loci consisted of KIF2B and ARL15. We also observed evidence for a deletion at the EPHA6–UNQ6114 locus when the AA cohort was investigated as a discovery set.

The majority of genes harboring at the loci uncovered in this study have not been implicated in obesity previously. However, the most notable finding is with ARL15, which was recently uncovered in a GWAS of adiponectin levels, with the same risk allele also being associated with a higher risk of CVD and T2D [115].

We also evaluated large rare deletions present in <1% of individuals and >500 kb in size as set previously [104] and did not observe excess of large rare deletions genome–wide. This is not unexpected given that previous reports only found significance when including developmental delay subjects but not when severe early–onset obesity was evaluated alone [102, 103].

More recently, a novel common copy number variation for early–onset extreme obesity was reported on chromosome 11q11, harboring the OR4P4, OR4S2, and OR4C6 genes using a similar approach [105] (Table 1). Indeed, as higher and higher resolution genome wide scans are carried out, one would expect further reports of such findings.

1.10 OTHER ETHNICITIES

Studying populations of different ancestry will also help us to globally identify and understand the genetic and environmental factors associated with estimates of obesity, as variants found in populations of both African and Caucasian ancestry may represent more universally important genes and pathways for subsequent diagnosis, prevention, and treatment of obesity and its complications. In addition, a cohort of African ancestry in many instances can aid in refining the anticipated association(s) made in with the GWAS approach due to lower LD in this ethnicity, for example, the association of T2D with TCF7L2 [116] has been refined utilizing a West African patient cohort [117].

To date, most obesity GWAS reports have come from investigations of populations of European origin. This is partly due to the relatively low haplotypic complexity of Caucasian genomes and partly to get around admixture concerns. Indeed, like many of the other replication efforts, FTO shows the strongest association with BMI in our large European American pediatric cohort [98]. However, the role of the FTO locus in influencing BMI and obesity predisposition in populations of African ancestry has been previously less clear [88, 118], but consensus is emerging from large cohort studies, both in adults [119] and in our own pediatric cohort [90], that SNP rs3751812 captures the FTO association with the trait in both ethnicities; this finding is comparable to similar outcomes working with loci in asthma [120] and T2D [117].

1.11 CONCLUSIONS

While these recently discovered loci unveil several new biomolecular pathways not previously associated with obesity, it is important to note that these well–established genetic associations with obesity explain very little of the genetic risk for this pediatric phenotype, suggesting the existence of additional loci whose number and effect size remain unknown.

These findings have left the genetics community to ponder how we are going to finally uncover the full repertoire of the genetic component of given traits in order to explain the "missing heritability" [121]. Thus, it is clear that in addition to larger and larger cohorts combined in to meta–analyses, new whole genome sequencing technologies will be a large part of the solution. With new advances in sequencing, one would expect further variants to be characterized in this condition so collectively they could build up to a meaningful contribution to the missing heritably for this trait.

Taken together, the unbiased genome wide approach to assess the entire genome has revealed genes that underpin the pathogenesis of childhood obesity. Further functional studies will be needed to fully characterize the function of the genes at these loci in relation to childhood obesity.

REFERENCES

1. R. P. Troiano and K. M. Flegal, "Overweight children and adolescents: description, epidemiology, and demographics," Pediatrics, vol. 101, no. 3, part 2, pp. 497–504, 1998.
2. G. M. Reaven, "Banting Lecture 1988. Role of insulin resistance in human disease. 1988," Nutrition, vol. 13, no. 1, pp. 1595–1607, 1997.

3. R. A. DeFronzo and E. Ferrannini, "Insulin resistance: a multifaceted syndrome responsible for NIDDM, obesity, hypertension, dyslipidemia, and atherosclerotic cardiovascular disease," Diabetes Care, vol. 14, no. 3, pp. 173–194, 1991.
4. T. A. Nicklas, T. Baranowski, K. W. Cullen, and G. Berenson, "Eating patterns, dietary quality and obesity," Journal of the American College of Nutrition, vol. 20, no. 6, pp. 599–608, 2001.
5. R. C. Whitaker, J. A. Wright, M. S. Pepe, K. D. Seidel, and W. H. Dietz, "Predicting obesity in young adulthood from childhood and parental obesity," The New England Journal of Medicine, vol. 337, no. 13, pp. 869–873, 1997.
6. T. J. Parsons, C. Power, S. Logan, and C. D. Summerbell, "Childhood predictors of adult obesity: a systematic review," International Journal of Obesity, vol. 23, supplement 8, pp. S1–S107, 1999.
7. W. H. Dietz, "Health consequences of obesity in youth: childhood predictors of adult disease," Pediatrics, vol. 101, no. 3, pp. 518–525, 1998.
8. S. R. Daniels, D. K. Arnett, R. H. Eckel et al., "Overweight in children and adolescents: pathophysiology, consequences, prevention, and treatment," Circulation, vol. 111, no. 15, pp. 1999–2012, 2005.
9. A. Must, "Does overweight in childhood have an impact on adult health?" Nutrition Reviews, vol. 61, no. 4, pp. 139–142, 2003.
10. H. O. Mossberg, "40–year follow–up of overweight children," The Lancet, vol. 2, no. 8661, pp. 491–493, 1989.
11. A. Must, P. F. Jacques, G. E. Dallal, C. J. Bajema, and W. H. Dietz, "Long–term morbidity and mortality of overweight adolescents—a follow–up of the Harvard Growth Study of 1922 to 1935," The New England Journal of Medicine, vol. 327, no. 19, pp. 1350–1355, 1992.
12. R. H. Eckel, "Obesity: a disease or a physiologic adaptation for survival?" in Obesity Mechanisms and Clinical Management, R. H. Eckel, Ed., pp. 3–30, Lippincott Williams & Wilkins, Philadelphia, Pa, USA, 2003.
13. K. M. Flegal and R. P. Troiano, "Changes in the distribution of body mass index of adults and children in the US population," International Journal of Obesity, vol. 24, no. 7, pp. 807–818, 2000.
14. K. M. Flegal, M. D. Carroll, C. L. Ogden, and C. L. Johnson, "Prevalence and trends in obesity among US adults, 1999–2000," Journal of the American Medical Association, vol. 288, no. 14, pp. 1723–1727, 2002.
15. R. J. Kuczmarski and K. M. Flegal, "Criteria for definition of overweight in transition: background and recommendations for the United States," The American Journal of Clinical Nutrition, vol. 72, no. 5, pp. 1074–1081, 2000.
16. S. R. Daniels, P. R. Khoury, and J. A. Morrison, "The utility of body mass index as a measure of body fatness in children and adolescents: differences by race and gender," Pediatrics, vol. 99, no. 6, pp. 804–807, 1997.
17. T. J. Cole, J. V. Freeman, and M. A. Preece, "Body mass index reference curves for the UK, 1990," Archives of Disease in Childhood, vol. 73, no. 1, pp. 25–29, 1995.
18. K. M. Flegal, R. Wei, and C. Ogden, "Weight–for–stature compared with body mass index–for–age growth charts for the United States from the Centers for Disease Control and Prevention," The American Journal of Clinical Nutrition, vol. 75, no. 4, pp. 761–766, 2002.

19. J. H. Himes and W. H. Dietz, "Guidelines for overweight in adolescent preventive services: recommendations from an expert committee. The expert committee on clinical guidelines for overweight in adolescent preventive services," The American Journal of Clinical Nutrition, vol. 59, no. 2, pp. 307–316, 1994.
20. C.–E. Flodmark, I. Lissau, L. A. Moreno, A. Pietrobelli, and K. Widhalm, "New insights into the field of children and adolescents' obesity: the European perspective," International Journal of Obesity, vol. 28, no. 10, pp. 1189–1196, 2004.
21. J. P. Koplan, C. T. Liverman, and V. I. Kraak, "Preventing childhood obesity: health in the balance: executive summary," Journal of the American Dietetic Association, vol. 105, no. 1, pp. 131–138, 2005.
22. J. A. Yanovski, "Intensive therapies for pediatric obesity," Pediatric Clinics of North America, vol. 48, no. 4, pp. 1041–1053, 2001.
23. R. I. Berkowitz, K. Fujioka, S. R. Daniels et al., "Effects of sibutramine treatment in obese adolescents: a randomized trial," Annals of Internal Medicine, vol. 145, no. 2, pp. 81–90, 2006.
24. R. I. Berkowitz, T. A. Wadden, A. M. Tershakovec, and J. L. Cronquist, "Behavior therapy and sibutramine for the treatment of adolescent obesity: a randomized controlled trial," Journal of the American Medical Association, vol. 289, no. 14, pp. 1805–1812, 2003.
25. W. P. T. James, I. D. Caterson, W. Coutinho et al., "Effect of sibutramine on cardiovascular outcomes in overweight and obese subjects," The New England Journal of Medicine, vol. 363, no. 10, pp. 905–917, 2010.
26. J. R. McDuffie, K. A. Calis, G. I. Uwaifo et al., "Three–month tolerability of orlistat in adolescents with obesity–related comorbid conditions," Obesity Research, vol. 10, no. 7, pp. 642–650, 2002.
27. J. R. McDuffie, K. A. Calis, S. L. Booth, G. I. Uwaifo, and J. A. Yanovski, "Effects of orlistat on fat–soluble vitamins in obese adolescents," Pharmacotherapy, vol. 22, no. 7, pp. 814–822, 2002.
28. M. Freemark and D. Bursey, "The effects of metformin on body mass index and glucose tolerance in obese adolescents with fasting hyperinsulinemia and a family history of type 2 diabetes," Pediatrics, vol. 107, no. 4, p. E55, 2001.
29. T. H. Inge, V. Garcia, S. Daniels et al., "A multidisciplinary approach to the adolescent bariatric surgical patient," Journal of Pediatric Surgery, vol. 39, no. 3, pp. 442–447, 2004.
30. R. H. Lustig, P. S. Hinds, K. Ringwald–Smith et al., "Octreotide therapy of pediatric hypothalamic obesity: a double–blind, placebo–controlled trial," Journal of Clinical Endocrinology and Metabolism, vol. 88, no. 6, pp. 2586–2592, 2003.
31. J. M. Friedman, "Modern science versus the stigma of obesity," Nature Medicine, vol. 10, no. 6, pp. 563–569, 2004.
32. H. N. Lyon and J. N. Hirschhorn, "Genetics of common forms of obesity: a brief overview," The American Journal of Clinical Nutrition, vol. 82, no. 1, supplement, pp. 215S–217S, 2005.
33. W. C. Knowler, D. J. Pettit, M. F. Saad, and P. H. Bennett, "Diabetes mellitus in the pima indians: incidence, risk factors and pathogenesis," Diabetes/Metabolism Reviews, vol. 6, no. 1, pp. 1–27, 1990.

34. J. Hebebrand, S. Friedel, N. Schäuble, F. Geller, and A. Hinney, "Perspectives: molecular genetic research in human obesity," Obesity Reviews, vol. 4, no. 3, pp. 139–146, 2003.
35. I. S. Farooqi and S. O'Rahilly, "New advances in the genetics of early onset obesity," International Journal of Obesity, vol. 29, no. 10, pp. 1149–1152, 2005.
36. C. G. Bell, A. J. Walley, and P. Froguel, "The genetics of human obesity," Nature Reviews Genetics, vol. 6, no. 3, pp. 221–234, 2005.
37. I. Kondo, J. Hamabe, K. Yamamoto, and N. Niikawa, "Exclusion mapping of the Cohen syndrome gene from the Prader–Willi syndrome locus," Clinical Genetics, vol. 38, no. 6, pp. 422–426, 1990.
38. I. M. Russell–Eggitt, P. T. Clayton, R. Coffey, A. Kriss, D. S. I. Taylor, and J. F. N. Taylor, "Alstrom syndrome: report of 22 cases and literature review," Ophthalmology, vol. 105, no. 7, pp. 1274–1280, 1998.
39. P. L. Beales, A. M. Warner, G. A. Hitman, R. Thakker, and F. A. Flinter, "Bardet–Biedl syndrome: a molecular and phenotypic study of 18 families," Journal of Medical Genetics, vol. 34, no. 2, pp. 92–98, 1997.
40. E. A. Bruford, R. Riise, P. W. Teague et al., "Linkage mapping in 29 Bardet–Biedl syndrome families confirms loci in chromosomal regions 11q13, 15q22.3–q23, and 16q21," Genomics, vol. 41, no. 1, pp. 93–99, 1997.
41. T. L. Young, L. Penney, M. O. Woods et al., "A fifth locus for Bardet–Biedl syndrome maps to chromosome 2q31," American Journal of Human Genetics, vol. 64, no. 3, pp. 900–904, 1999.
42. I. S. Farooqi and S. O'Rahilly, "Recent advances in the genetics of severe childhood obesity," Archives of Disease in Childhood, vol. 83, no. 1, pp. 31–34, 2000.
43. A. M. Ingalls, M. M. Dickie, and G. D. Snell, "Obese, a new mutation in the house mouse," The Journal of heredity, vol. 41, no. 12, pp. 317–318, 1950.
44. Y. Zhang, R. Proenca, M. Maffei, M. Barone, L. Leopold, and J. M. Friedman, "Positional cloning of the mouse obese gene and its human homologue," Nature, vol. 372, no. 6505, pp. 425–432, 1994.
45. J. L. Halaas, K. S. Gajiwala, M. Maffei et al., "Weight–reducing effects of the plasma protein encoded by the obese gene," Science, vol. 269, no. 5223, pp. 543–546, 1995.
46. L. A. Campfield, F. J. Smith, Y. Guisez, R. Devos, and P. Burn, "Recombinant mouse OB protein: evidence for a peripheral signal linking adiposity and central neural networks," Science, vol. 269, no. 5223, pp. 546–549, 1995.
47. M. A. Pelleymounter, M. J. Cullen, M. B. Baker et al., "Effects of the obese gene product on body weight regulation in ob/ob mice," Science, vol. 269, no. 5223, pp. 540–543, 1995.
48. S. C. Chua Jr., W. K. Chung, X. S. Wu–Peng et al., "Phenotypes of mouse diabetes and rat fatty due to mutations in the OB (leptin) receptor," Science, vol. 271, no. 5251, pp. 994–996, 1996.
49. R. V. Considine, M. K. Sinha, M. L. Heiman et al., "Serum immunoreactive–leptin concentrations in normal–weight and obese humans," The New England Journal of Medicine, vol. 334, no. 5, pp. 292–295, 1996.
50. C. C. Cheung, D. K. Clifton, and R. A. Steiner, "Proopiomelanocortin neurons are direct targets for leptin in the hypothalamus," Endocrinology, vol. 138, no. 10, pp. 4489–4492, 1997.

51. H. Krude, H. Biebermann, W. Luck, R. Horn, G. Brabant, and A. Grüters, "Severe early–onset obesity, adrenal insufficiency and red hair pigmentation caused by POMC mutations in humans," Nature Genetics, vol. 19, no. 2, pp. 155–157, 1998.
52. R. S. Jackson, J. W. M. Creemers, S. Ohagi et al., "Obesity and impaired prohormone processing associated with mutations in the human prohormone convertase 1 gene," Nature Genetics, vol. 16, no. 3, pp. 303–306, 1997.
53. S. O'Rahilly, H. Gray, P. J. Humphreys et al., "Brief report: impaired processing of prohormones associated with abnormalities of glucose homeostasis and adrenal function," The New England Journal of Medicine, vol. 333, no. 21, pp. 1386–1390, 1995.
54. D. Huszar, C. A. Lynch, V. Fairchild–Huntress et al., "Targeted disruption of the melanocortin–4 receptor results in obesity in mice," Cell, vol. 88, no. 1, pp. 131–141, 1997.
55. G. S. H. Yeo, I. S. Farooqi, S. Aminian, D. J. Halsall, R. G. Stanhope, and S. O'Rahilly, "A frameshift mutation in MC4R associated with dominantly inherited human obesity," Nature Genetics, vol. 20, no. 2, pp. 111–112, 1998.
56. C. Vaisse, K. Clement, B. Guy–Grand, and P. Froguel, "A frameshift mutation in human MC4R is associated with a dominant form of obesity," Nature Genetics, vol. 20, no. 2, pp. 113–114, 1998.
57. W. Gu, Z. Tu, P. W. Kleyn et al., "Identification and functional analysis of novel human melanocortin–4 receptor variants," Diabetes, vol. 48, no. 3, pp. 635–639, 1999.
58. A. Hinney, A. Schmidt, K. Nottebom et al., "Several mutations in the melanocortin–4 receptor gene including a nonsense and a frameshift mutation associated with dominantly inherited obesity in humans," Journal of Clinical Endocrinology and Metabolism, vol. 84, no. 4, pp. 1483–1486, 1999.
59. J. N. Hirschhorn and M. J. Daly, "Genome–wide association studies for common diseases and complex traits," Nature Reviews Genetics, vol. 6, no. 2, pp. 95–108, 2005.
60. C. S. Carlson, M. A. Eberle, L. Kruglyak, and D. A. Nickerson, "Mapping complex disease loci in whole–genome association studies," Nature, vol. 429, no. 6990, pp. 446–452, 2004.
61. The International HapMap Consortium, "A haplotype map of the human genome," Nature, vol. 437, no. 7063, pp. 1299–1320, 2005.
62. D. Reich, N. Patterson, P. L. De Jager, et al., "A whole–genome admixture scan finds a candidate locus for multiple sclerosis susceptibility," Nature Genetics, vol. 37, no. 10, pp. 1113–1118, 2005.
63. F. J. Steemers, W. Chang, G. Lee, D. L. Barker, R. Shen, and K. L. Gunderson, "Whole–genome genotyping with the single–base extension assay," Nature Methods, vol. 3, no. 1, pp. 31–33, 2006.
64. R. Redon, S. Ishikawa, K. R. Fitch et al., "Global variation in copy number in the human genome," Nature, vol. 444, no. 7118, pp. 444–454, 2006.
65. K. L. Gunderson, F. J. Steemers, G. Lee, L. G. Mendoza, and M. S. Chee, "A genome–wide scalable SNP genotyping assay using microarray technology," Nature Genetics, vol. 37, no. 5, pp. 549–554, 2005.

66. H. Hakonarson, S. F. A. Grant, J. P. Bradfield et al., "A genome–wide association study identifies KIAA0350 as a type 1 diabetes gene," Nature, vol. 448, no. 7153, pp. 591–594, 2007.
67. J. A. Todd, N. M. Walker, J. D. Cooper et al., "Robust associations of four new chromosome regions from genome–wide analyses of type 1 diabetes," Nature Genetics, vol. 39, no. 7, pp. 857–864, 2007.
68. P. R. Burton, D. G. Clayton, L. R. Cardon et al., "Genome–wide association study of 14,000 cases of seven common diseases and 3,000 shared controls," Nature, vol. 447, no. 7145, pp. 661–678, 2007.
69. R. Sladek, G. Rocheleau, J. Rung et al., "A genome–wide association study identifies novel risk loci for type 2 diabetes," Nature, vol. 445, no. 7130, pp. 881–885, 2007.
70. R. Saxena, B. F. Voight, V. Lyssenko et al., "Genome–wide association analysis identifies loci for type 2 diabetes and triglyceride levels," Science, vol. 316, no. 5829, pp. 1331–1336, 2007.
71. E. Zeggini, M. N. Weedon, C. M. Lindgren et al., "Replication of genome–wide association signals in UK samples reveals risk loci for type 2 diabetes," Science, vol. 316, no. 5829, pp. 1336–1341, 2007.
72. L. J. Scott, K. L. Mohlke, L. L. Bonnycastle et al., "A genome–wide association study of type 2 diabetes in finns detects multiple susceptibility variants," Science, vol. 316, no. 5829, pp. 1341–1345, 2007.
73. R. J. Klein, C. Zeiss, E. Y. Chew et al., "Complement factor H polymorphism in age–related macular degeneration," Science, vol. 308, no. 5720, pp. 385–389, 2005.
74. R. H. Duerr, K. D. Taylor, S. R. Brant et al., "A genome–wide association study identifies IL23R as an inflammatory bowel disease gene," Science, vol. 314, no. 5804, pp. 1461–1463, 2006.
75. M. Imielinski, R. N. Baldassano, A. Griffiths et al., "Common variants at five new loci associated with early–onset inflammatory bowel disease," Nature Genetics, vol. 41, no. 12, pp. 1335–1340, 2009.
76. A. Helgadottir, G. Thorleifsson, A. Manolescu et al., "A common variant on chromosome 9p21 affects the risk of myocardial infarction," Science, vol. 316, no. 5830, pp. 1491–1493, 2007.
77. R. McPherson, A. Pertsemlidis, N. Kavaslar et al., "A common allele on chromosome 9 associated with coronary heart disease," Science, vol. 316, no. 5830, pp. 1488–1491, 2007.
78. D. F. Easton, K. A. Pooley, A. M. Dunning et al., "Genome–wide association study identifies novel breast cancer susceptibility loci," Nature, vol. 447, no. 7148, pp. 1087–1093, 2007.
79. A. Herbert, N. P. Gerry, M. B. McQueen et al., "A common genetic variant is associated with adult and childhood obesity," Science, vol. 312, no. 5771, pp. 279–283, 2006.
80. R. J. Loos, I. Barroso, S. O'rahilly, and N. J. Wareham, "Comment on "A common genetic variant is associated with adult and childhood obesity"," Science, vol. 315, no. 5809, p. 187, 2007.
81. C. Dina, D. Meyre, C. Samson et al., "Comment on "A common genetic variant is associated with adult and childhood obesity"," Science, vol. 315, no. 5809, p. 187, 2007.

82. D. Rosskopf, A. Bornhorst, C. Rimmbach et al., "Comment on "A common genetic variant is associated with adult and childhood obesity"," Science, vol. 315, no. 5809, p. 187, 2007.
83. H. N. Lyon, V. Emilsson, A. Hinney, et al., "The association of a SNP upstream of INSIG2 with body mass index is reproduced in several but not all cohorts," PLoS Genetics, vol. 3, no. 4, article e61, 2007.
84. K. Hotta, M. Nakamura, Y. Nakata et al., "INSIG2 gene rs7566605 polymorphism is associated with severe obesity in Japanese," Journal of Human Genetics, vol. 53, no. 9, pp. 857–862, 2008.
85. T. M. Frayling, N. J. Timpson, M. N. Weedon et al., "A common variant in the FTO gene is associated with body mass index and predisposes to childhood and adult obesity," Science, vol. 316, no. 5826, pp. 889–894, 2007.
86. A. Hinney, T. T. Nguyen, A. Scherag et al., "Genome Wide Association (GWA) study for early onset extreme obesity supports the role of fat mass and obesity associated gene (FTO) variants," PLoS One, vol. 2, no. 12, Article ID e1361, 2007.
87. C. Dina, D. Meyre, S. Gallina, et al., "Variation in FTO contributes to childhood obesity and severe adult obesity," Nature Genetics, vol. 39, no. 6, pp. 724–726, 2007.
88. A. Scuteri, S. Sanna, W. M. Chen, et al., "Genome–wide association scan shows genetic variants in the FTO gene are associated with obesity–related traits," PLoS Genetics, vol. 3, no. 7, article e115, 2007.
89. K. A. Fawcett and I. Barroso, "The genetics of obesity: FTO leads the way," Trends in Genetics, vol. 26, no. 6, pp. 266–274, 2010.
90. S. F. A. Grant, M. Li, J. P. Bradfield et al., "Association analysis of the FTO gene with obesity in children of Caucasian and African ancestry reveals a common tagging SNP," PLoS One, vol. 3, no. 3, Article ID e1746, 2008.
91. C. Church, L. Moir, F. McMurray et al., "Overexpression of Fto leads to increased food intake and results in obesity," Nature Genetics, vol. 42, no. 12, pp. 1086–1092, 2010.
92. J. Fischer, L. Koch, C. Emmerling et al., "Inactivation of the Fto gene protects from obesity," Nature, vol. 458, no. 7240, pp. 894–898, 2009.
93. D. Meyre, K. Proulx, H. Kawagoe–Takaki et al., "Prevalence of loss–of–function FTO mutations in lean and obese individuals," Diabetes, vol. 59, no. 1, pp. 311–318, 2010.
94. R. J. F. Loos, C. M. Lindgren, S. Li et al., "Common variants near MC4R are associated with fat mass, weight and risk of obesity," Nature Genetics, vol. 40, no. 6, pp. 768–775, 2008.
95. G. Thorleifsson, G. B. Walters, D. F. Gudbjartsson et al., "Genome–wide association yields new sequence variants at seven loci that associate with measures of obesity," Nature Genetics, vol. 41, no. 1, pp. 18–24, 2009.
96. C. J. Willer, E. K. Speliotes, R. J. F. Loos et al., "Six new loci associated with body mass index highlight a neuronal influence on body weight regulation," Nature Genetics, vol. 41, no. 1, pp. 25–34, 2009.
97. J. Gunstad, P. Schofield, R. H. Paul et al., "BDNF Val66Met polymorphism is associated with body mass index in healthy adults," Neuropsychobiology, vol. 53, no. 3, pp. 153–156, 2006.

98. E. K. Speliotes, C. J. Willer, S. I. Berndt, et al., "Association analyses of 249,796 individuals reveal 18 new loci associated with body mass index," Nature Genetics, vol. 42, no. 11, pp. 937–948, 2010.
99. J. Zhao, J. P. Bradfield, M. Li et al., "The role of obesity–associated loci identified in genome–wide association studies in the determination of pediatric BMI," Obesity, vol. 17, no. 12, pp. 2254–2257, 2009.
100. J. Zhao, J. P. Bradfield, H. Zhang et al., "Examination of all type 2 diabetes GWAS loci reveals HHEX–IDE as a locus influencing pediatric BMI," Diabetes, vol. 59, no. 3, pp. 751–755, 2010.
101. A. Scherag, C. Dina, A. Hinney et al., "Two new loci for body–weight regulation identified in a joint analysis of genome–wide association studies for early–onset extreme obesity in French and German study groups," PLoS Genetics, vol. 6, no. 4, Article ID e1000916, 2010.
102. E. G. Bochukova, N. I. Huang, J. Keogh et al., "Large, rare chromosomal deletions associated with severe early–onset obesity," Nature, vol. 463, no. 7281, pp. 666–670, 2010.
103. R. G. Walters, S. Jacquemont, A. Valsesia et al., "A new highly penetrant form of obesity due to deletions on chromosome 16p11.2," Nature, vol. 463, no. 7281, pp. 671–675, 2010.
104. J. T. Glessner, J. P. Bradfield, K. Wang et al., "A genome–wide study reveals copy number variants exclusive to childhood obesity cases," American Journal of Human Genetics, vol. 87, no. 5, pp. 661–666, 2010.
105. I. Jarick, C. I. Vogel, S. Scherag et al., "Novel common copy number variation for early onset extreme obesity on chromosome 11q11 identified by a genome–wide analysis," Human Molecular Genetics, vol. 20, no. 4, pp. 840–852, 2011.
106. S. F. A. Grant, J. P. Bradfield, H. Zhang et al., "Investigation of the locus near MC4R with childhood obesity in Americans of European and African ancestry," Obesity, vol. 17, no. 7, pp. 1461–1465, 2009.
107. J. Sebat, B. Lakshmi, D. Malhotra et al., "Strong association of de novo copy number mutations with autism," Science, vol. 316, no. 5823, pp. 445–449, 2007.
108. C. R. Marshall, A. Noor, J. B. Vincent et al., "Structural variation of chromosomes in autism spectrum disorder," American Journal of Human Genetics, vol. 82, no. 2, pp. 477–488, 2008.
109. L. A. Weiss, Y. Shen, J. M. Korn et al., "Association between microdeletion and microduplication at 16p11.2 and autism," The New England Journal of Medicine, vol. 358, no. 7, pp. 667–675, 2008.
110. J. T. Glessner, K. Wang, G. Cai et al., "Autism genome–wide copy number variation reveals ubiquitin and neuronal genes," Nature, vol. 459, no. 7246, pp. 569–573, 2009.
111. J. Elia, X. Gai, H. M. Xie et al., "Rare structural variants found in attention–deficit hyperactivity disorder are preferentially associated with neurodevelopmental genes," Molecular Psychiatry, vol. 15, no. 6, pp. 637–646, 2010.
112. H. Stefansson, D. Rujescu, S. Cichon, et al., "Large recurrent microdeletions associated with schizophrenia," Nature, vol. 455, no. 7210, pp. 232–236, 2008.
113. T. Walsh, J. M. McClellan, S. E. McCarthy et al., "Rare structural variants disrupt multiple genes in neurodevelopmental pathways in schizophrenia," Science, vol. 320, no. 5875, pp. 539–543, 2008.

114. J. T. Glessner, M. P. Reilly, C. E. Kim et al., "Strong synaptic transmission impact by copy number variations in schizophrenia," Proceedings of the National Academy of Sciences of the United States of America, vol. 107, no. 23, pp. 10584–10589, 2010.
115. J. B. Richards, D. Waterworth, S. O'Rahilly et al., "A genome–wide association study reveals variants in ARL15 that influence adiponectin levels," PLoS Genetics, vol. 5, no. 12, Article ID e1000768, 2009.
116. S. F. A. Grant, G. Thorleifsson, I. Reynisdottir et al., "Variant of transcription factor 7–like 2 (TCF7L2) gene confers risk of type 2 diabetes," Nature Genetics, vol. 38, no. 3, pp. 320–323, 2006.
117. A. Helgason, S. Pálsson, G. Thorleifsson et al., "Refining the impact of TCF7L2 gene variants on type 2 diabetes and adaptive evolution," Nature Genetics, vol. 39, no. 2, pp. 218–225, 2007.
118. A. Adeyemo, G. Chen, J. Zhou et al., "FTO genetic variation and association with obesity in West Africans and African Americans," Diabetes, vol. 59, no. 6, pp. 1549–1554, 2010.
119. M. T. Hassanein, H. N. Lyon, T. T. Nguyen, et al., "Fine mapping of the association with obesity at the FTO locus in African–derived populations," Human Molecular Genetics, vol. 19, no. 14, pp. 2907–2916, 2010.
120. P. M. A. Sleiman, J. Flory, M. Imielinski et al., "Variants of DENND1B associated with asthma in children," The New England Journal of Medicine, vol. 362, no. 1, pp. 36–44, 2010.
121. T. A. Manolio, F. S. Collins, N. J. Cox, et al., "Finding the missing heritability of complex diseases," Nature, vol. 461, no. 7265, pp. 747–753, 2009.

CHAPTER 2

GENETIC AND ENVIRONMENTAL CONTRIBUTIONS TO WEIGHT, HEIGHT, AND BMI FROM BIRTH TO 19 YEARS OF AGE: AN INTERNATIONAL STUDY OF OVER 12,000 TWIN PAIRS

LISE DUBOIS, KIRSTEN OHM KYVIK, MANON GIRARD, FABIOLA TATONE–TOKUDA, DANIEL PÉRUSSE, JACOB HJELMBORG, AXEL SKYTTHE, FINN RASMUSSEN, MARGARET J. WRIGHT, PAUL LICHTENSTEIN, and NICHOLAS G. MARTIN

2.1 INTRODUCTION

The global obesity epidemic is accelerating [1] and has affected virtually all ages, races, and sexes in developed and developing countries [2], [3]. The obesity increase in childhood is especially troubling as overweight/obesity is shown to track into later adolescent and adult years [4] and is associated with numerous immediate and long–term health risks that lead to morbidity and premature mortality (e.g. asthma, type 2 diabetes, cardiovascular diseases, and cancer) [1].

Overweight/obesity has a multifactorial aetiology; moreover, there has not been a substantial change in mankind's genetic makeup to explain the obesity epidemic that has ravaged the world over the last three decades [5].This epidemic is mainly attributed to a global shift in the consumption of calorie–dense diets and reduced physical activity, a trend that has accompanied globalization and is further exacerbated by various individual, societal, and socioeconomic factors [3], [6], [7]. Nonetheless, not all individuals exposed to obesogenic environments become obese. A genetic propensity for weight gain and obesity must be present for the environment to precipitate an overweight/obese phenotype. Twin, family, and adoption studies provide strong evidence for large genetic influences on variations in body mass index (BMI), with heritability estimates ranging from 50% to over 90%, leaving the remaining variance attributed to environmental influences, whether common to family members/siblings or unique to the individual [8]–[10]. However, these estimates have varied widely across studies due to differences in study types, populations, and ages targeted.

Twin studies generally provide higher heritability estimates in comparison to adoption and family studies, and they are considered to provide the most precise estimates of the genetic and environmental influences on behavioural and physical phenotypes [8]. Most large–scale twin studies involve adult populations, and these show a very small to no effect of the common environment on variations in BMI [9], [11]; rather, it is the unique environment that generally influences the remaining variance in BMI in adulthood. The role of the common environment may be more significant in childhood, however, as there are more frequent opportunities for twins to be exposed to the same environmental influences while living together with parents and other siblings. In fact, a recent systematic review and meta–analysis of twin and adoption studies found that common environmental factors showed a substantial influence on variations in BMI in mid–childhood, although this influence vanished in adolescence between the ages of 14 and 17 years [10]. A review of twin and family studies also observed that, while there are strong genetic influences on the tracking of BMI from early childhood to the beginning of adulthood, there is also evidence that common environmental influences are important throughout childhood [12]. However, much of research on child and adolescent twins is limited to specific local populations, fairly small samples, or only examines a portion of childhood and adolescence. The critical years where interventions can be made to target common and unique environmental influences on these body measures have not been identified as, to date, no large–scale twin study has yet examined the genetic and environmental influences on variances in body weight, height, and BMI over each year of age, from birth to adulthood.

Little is also known about sex–differences in the heritability of body weight, height, and BMI from birth to late adolescence. It is well known that girls in general mature more rapidly than boys, but only a few studies have examined sex–limitation in genetic and environmental influences on variations in these body measures over childhood and adolescence; these show inconsistent results. One study in twins aged 8–11 years found no evidence of sex–limitation in the heritability of BMI or waist circumference [13]; whereas, other studies report age–specific sex–differences in twins at 5 months [14], between 16 and 17 years, [15] and between 18 and 25 years [16]. Furthermore, where sex–differences are observed, it is unclear which sex is more strongly influenced by genetic factors for the variability of various anthropometric measures. Some studies report higher heritability estimates in body weight, height, or BMI for females [14], [16]–[18], whereas others report stronger genetic influences in males [19]. Further large–scale studies on child and adolescent monozygotic (MZ) and dizygotic (DZ) twin pairs are needed to clarify these inconsistencies.

Using international, population–based data obtained from large twin birth–cohorts in three different continents, the present study aims to describe the distribution of weight, height, and BMI in MZ and DZ same– and opposite–sex twin pairs, from birth through 19 years of age, and examine the genetic and environmental influences on variances in these body measures over each year of age during the first 19 years of life; sex–limitation of genetic and environmental effects will also be explored at each age examined.

2.2 METHODS

2.2.1 ETHICS STATEMENT

Ethics approval was obtained for each of the respective cohorts and participants gave informed consent.

2.2.2 DATA SOURCES

The present study analyzed data obtained from a total of 23 twin birth–cohorts from four different countries: Canada, Sweden, Denmark, and Australia.

Analyses included data from MZ and DZ (same– and opposite–sex) twin pairs with available measures for both height and weight at a given age, from birth through 19 years of age. From an initial sample of approximately 30,500 children, 24,036 children provided data for the analyses. A brief overview of the cohorts from which the study samples were drawn is provided in the following subsections.

2.2.2.1 FROM CANADA: 'QUÉBEC NEWBORN TWIN STUDY' (QNTS).

The Québec Newborn Twin Study (QNTS) [20] is a population–based birth cohort of twin births occurring between April 1st, 1995 and December 31st, 1998 in the seven health districts of the greater Montreal area in the province of Québec, Canada. Out of a total of 989 families contacted, 672 agreed to participate (68%). Twins with chronic diseases and those who died prior to the age of 5 months were excluded from the cohort. The twins were first seen between the ages of 59 and 61 weeks (or 5 months, corrected for gestational age); these were followed annually thereafter. Each year, parents provided their informed consent.

Zygosity was determined when twins were 5 and 18 months old through the aggregation of independent tester ratings based on live assessments of physical similarity between twins; this was accomplished using a shortened version of the Zygosity Questionnaire for Young Twins [21]. At ages 5 and 18 months, mouth swabs were also collected from a random subsample of same–sex twins; DNA was extracted from the cells, amplified by polymerase chain reaction, and typed using 8 to 10 highly polymorphic micro–satellite markers. A comparison between physical assessments and genotyping yielded a 91.9% concordance among a random subsample of 123 pairs assessed at 5 months of age, and a 93.8% concordance among a subsample of 113 pairs assessed at 18 months [22]. With consideration for chorionicity data obtained from the twins' medical files, 96% twin pairs were thus classified correctly.

Anthropometric measures of children's weights (in grams; g) and heights (in centimeters; cm) were taken at birth (drawn from medical records), at 5 months, and at 5 and 8 years through laboratory measures. In order to eliminate potential biases related to perceived zygosity, different research assistants took the measures for each child within a twin pair.

2.2.2.2 FROM SWEDEN: 'CHILD AND ADOLESCENT TWIN STUDY IN SWEDEN' (CATSS) AND 'TWIN STUDY OF CHILD AND ADOLESCENT DEVELOPMENT' (TCHAD).

From Sweden, three different cohorts of children and adolescents were used for the analyses: two from the Child and Adolescent Twin Study in Sweden (CATSS) and one from the Twin Study of Child and Adolescent Development (TCHAD) [23]. Both studies are based on twins included in the Swedish Twin Registry (STR) [24].

The Child and Adolescent Twin Study in Sweden (CATSS) [24] started in September 2004 and it is an ongoing study that includes twins born between 1992 and 2001. Zygosity determination for 571 pairs of twins in whom DNA from both twins was available was based on a panel of 48 single–nucleotide polymorphisms derived for zygosity analyses [23]. For the remaining twins, an algorithm based on 5 items concerning twin similarity and confusion [24] derived from the twins with known zygosity was used. Only twins with more than 95% probability of being correctly classified were assigned a zygosity. In this study, parents of all Swedish twins turning 9 or 12 years were approached to complete a telephone interview regarding various health and behavioural issues about their twin children. Certain families were followed to complete additional questionnaires, genotyping, and further clinical interviews. To date, the survey holds an 80% response frequency, with 7,408 interviews completed by November 2008. Twins' birth–weights, lengths, and heights and weights at age 9 and 12 years are parent–reported and obtained through telephone interviews.

The Twin Study of Child and Adolescent Development (TCHAD) [23] follows 1,480 twin pairs from ages 8 to 20 years. Twins included in the study were those born in Sweden between May 1985 and December 1986. Zygosity was determined by using discriminant analyses on 385 twin pairs with known zygosity which were confirmed by 47 polymorphic DNA–markers [23]. This algorithm is restricted to classify monozygotic twins (MZ) and dizygotic twins (DZ) with 95% accuracy [25]. A questionnaire of four items covering the twins' physical similarities were answered at age 8–9 (via parent–reports) and at age 13–14 and 16–17 years (via both parent– and self–reports). Zygosity classification was made for each response separately through discriminant analysis. A final zygosity assignment was set if there were no disagreements between the five separate assignments. In cases of any contradictions between the assignments, the zygosity score was set to 'unknown'.

The study was conducted in four waves, starting in 1994 (when the twins were 8–9 years old), then again in 1999 (at age 13–14 years), in 2002 (at age 16–17 years), and in 2006 (at age 19–20 years). Questionnaires were administered to the parents and twins over the telephone. Twins' birthweights were obtained from the Swedish Medical Birth Register. Measures for twins' heights and weights at later ages were parent–reported in the first questionnaires at age 8–9 years, and were both parent– and self–reported by each twin in the following study waves. The response rates for the four study waves were: 91% (n = 1339 parents) in Wave 1 for the parent–questionnaires; 73% (n = 1063 parents) and 78% (n = 2263 twins) in Wave 2 for the parent– and twin– questionnaires, respectively; and 74% (n = 1067 parents) and 87% (n = 2369 twins) in Wave 3.

2.2.2.3 FROM DENMARK: 'DANISH TWIN REGISTRY' (DTR).

For the present study, data from a total of 18 birth cohorts of twins born in 1983–2000 were obtained from the Danish Twin Registry [26]. The DTR is an ongoing population–based twin registry that initiated in 1954 and, by the end of 2005, included over 75,000 twin pairs born between 1870 and 2004 [27]. Zygosity in the DTR is determined through questions examining the degree of similarity between same–sex co–twins, which has been validated by DNA finger printing and found to be correct in more than 97% of cases [28]. The present cohort was approached with the Danish Twin Child Survey in 2003: a short questionnaire was administered to the parents if the twins were born in 1988–2000 and to the twins themselves if they were born in 1983–87. The questionnaire included questions on weight and height at birth and at the age of the twins when answering in 2003 (i.e. from ages 3 to 19 inclusively). A total of 29,711 twin individuals were approached and 19,782 (66.6%) provided answers.

2.2.2.4 FROM AUSTRALIA: BRISBANE LONGITUDINAL TWIN STUDY (BTLS).

The data from this sample were collected through the ongoing Brisbane Longitudinal Twin Study (1992–2010) where twins are evaluated for melanoma risk factors at ages twelve and fourteen [29], [30], and for cognitive variables at age sixteen [31]. Participants are ascertained from schools in south–east Queensland and are of mainly European extraction, most with Anglo–Celtic ancestry. Blood samples are obtained for zygosity confirmation and DNA extraction. At each visit, height is measured with a stadiometer and weight is measured using frequently recalibrated scales.

2.2.3 MEASURES

Data on weight (in kilograms, Kg), height (in meters, m), and BMI (kg/m^2) from birth through 19 years of age were standardized to z–scores with a mean of zero and a standard deviation of one. The test for normality was employed to ensure that data were normally distributed (i.e., no transformation was needed). Furthermore, data were adjusted for repeated measurement (in SAS, the REPEATED statement controls the covariance structure imposed on the residuals or errors). Where data were available from more than one cohort or more than one country for a given age, they were pooled (data from 6 datasets were used in multivariate analyses). The proportions that twin pairs represent are given by dataset, age, zygosity, and sex after confirming that cohort distributions were similar (data not shown). No family outliers (i.e., bivariate outliers) exceeding three standard deviation from the mean were identified using the mahalanobis distance for each family represented as a Z–score.

2.2.4 STATISTICAL ANALYSES

Intra–class correlations were computed for five zygosity–by–sex groups (MZ–boys, MZ–girls, DZ–boys, DZ–girls, DZ–opposite–sex) and for the total number of MZ and DZ twin pairs included in the study. MZ twins, being genetically identical, share 100% of their segregating genes, whereas non–identical DZ twins share on average 50%. For this reason, if phenotypic variation in a specific trait is due to genetic effects, more resemblance in that trait will be found within MZ twins in comparison to DZ twins. However, two important assumptions must be made: Firstly, the environment to which each twin in a MZ and DZ pair is exposed is assumed to be similar; and secondly, results for the genetic and environmental influences on phenotypic variation in twins is assumed to be generalizable to singletons in the rest of the population. Several publications have discussed these two assumptions [32]–[34].

Classical model–fitting techniques [17] were used to test for different models and to quantify the magnitude of the genetic and environmental influences on variations in the body measures. As twins form a natural two–level hierarchy, a hierarchical random–effect multilevel model of twin data [35] that allows for a full likelihood estimation of all parameters [36] was used. A model was first built by specifying means, between–pair, and within–pair variances separately for MZ and DZ twins [14]; doing so equates the predicted means, variances, and covariances of the model to their observed values in both twin groups. The conditions of equal means and variances of MZ and DZ twins were imposed (as well as for twin A and twin B). In a twin study, the random part of this model can be specified to reflect four components of phenotypic variation in a specific trait: 1) additive genetic (A) variation, the sum of the effect of all alleles on a specific trait over all loci; 2) non–additive genetic (D) variation, the non–additive effect of alleles in the same locus with the inclusion of dominance genetic effects, caused by interactions between alleles in the same locus, and epistasis (interactions between alleles at different loci); 3) common environmental (C) variation, which consists of environmental factors shared by twins; and, 4) unique environmental (E) variation, which consists of environmental factors that are unique to each individual and includes measurement error. When analyzing information on twins reared together, the

C and D components cannot be estimated simultaneously [37], [38]. Thus, one can estimate four parameters in a resulting ACE (or ADE) model with two degrees of freedom: A phenotypic mean, additive genetic variance, common environmental variance (or non–additive genetic variance), and unique environmental variance. Specifications can also be formulated to examine submodels, including: 1) a CE model that removes all genetic components (suggesting no genetic effect); and, 2) an AE model that suggests no effect of the common/family environment. These are all considered to be univariate models. The square of path coefficients (i.e. a^2, c^2 or d^2, and e^2) or variance components are typically used to express the expected variances and covariances between individuals in twin pairs. These values are calculated using matrix algebra to identify the A, C (or D), and E components, respectively.

Generally, factors that constitute C, 'common environmental influences', in childhood and early adolescence that relate to body weight and BMI include: family's socioeconomic status [39] parenting style and parental modeling of healthy eating and activity behaviours [40], the home, school, and community food environment [41], [42], and neighbourhood characteristics [43]. Examples of E, 'unique environmental influences', include exposure to a virus or an injury/accident, among others. Given that data at several ages were pooled from two or more of the countries included in the present study sample, we have chosen to focus primarily on ACE (instead of ADE) models (i.e., to examine C rather than D) because the magnitude of the MZ and DZ same–sex intraclass correlation ratios tend to satisfy inequality [$2r_{DZ}>r_{MZ}>r_{DZ}$] at different ages, thus evidencing common (shared) genetic influences. Therefore, examining components A, C, and E yields a broader portrait of the genetic and environmental influences on weight, height, and BMI in this international population.

Using these classical methods, the twin design can also be extended to examine sex–limitation in the genetic and/or common environmental influences on the variability in a specific trait. This is accomplished by testing two models (one model per sex) simultaneously, while controlling for the covariance between opposite–sex DZ twins in A and C components. Including opposite–sex DZ twins in these analyses increases power and permits one to examine an additional male or female additive genetic (A'_F or A'_M) or common environmental (C'_F or C'_M) component that does

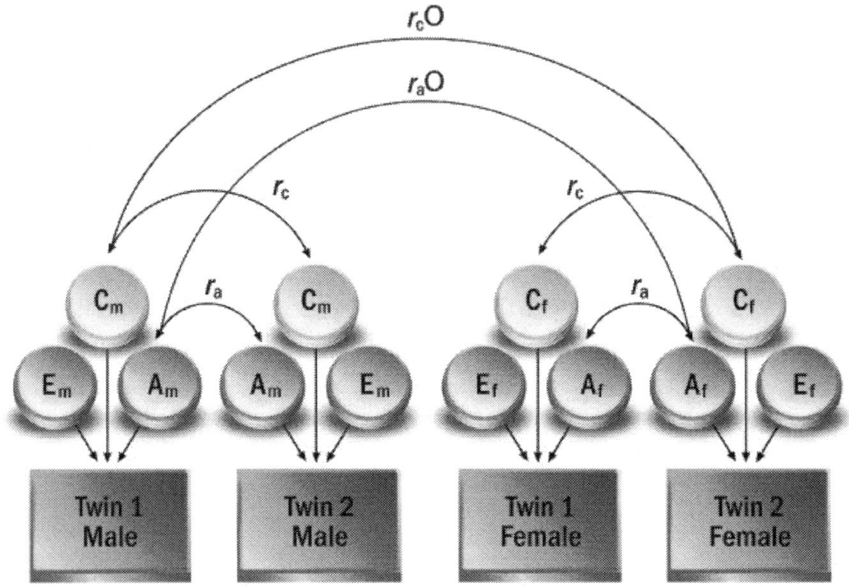

FIGURE 1: General sex–limited model: The m and f subscripts refer to males and females, respectively. ra and rd are additive genetic and common environmental correlations between same sex twins; raO and rcO are additive genetic and common environmental correlations between opposite sex twins.

not correlate with the genetic or environmental influences observed in the phenotype displayed on the female or male counterpart (Figure 1). Thus, ACE, CE, and AE models can be examined all with common effects, correlated effects, and uncorrelated effects. Observing significant estimates in a sex–limitation model provides evidence that the genetic or environmental factors that influence variability in a trait are not identical across sexes. Further detail pertaining to sex–limited modeling techniques is available elsewhere [17].

For each dependent variable and for each age examined in the present study, a sex–limited saturated model and a univariate saturated model were first fitted to examine a sex–effect in the genetic component. Using the likelihood ratio test (−2Log), the resulting two saturated models were then compared; as this test was shown to be significant at the 0.05 level for almost all models at every age (except for weight and height at age 4 and 7 years, and BMI at 9 years), the sex–limited models were used in the analyses at all ages. Subsequent nested models were examined, beginning from saturated to reduced models (ACE, AE, CE, and E). Nested models were compared to the full saturated models using a likelihood ratio test (–2Log) and Akaike's Information Criterion (AIC: chi–square–2df) which also considers both goodness–of–fit and parsimony in a model's explanatory value. Selecting a model based on the AIC tends to produce more power. All statistical analyses and model–fitting were conducted using SAS/NLMIXED 8.2 and statistical significance was set at 0.05.

2.3 RESULTS

Data on weight, height, and BMI were available from all 23 cohorts at birth, from two cohorts (from two different countries) at ages 5, 8, 9, 13, and 14 years, and from three cohorts (from three different countries) at ages 12 and 16 years; data were obtained from single cohorts for all remaining ages (at 5 months, and at ages 3, 4, 6, 7, 10, 11, 15, and 17 to 19 years, inclusively). Data at ages 1 and 2 years were not available from any of the cohorts included in the present study.

The mean birthweight for all children included in analyses, irrespective of zygosity and sex, was 2.6 kg (SEM = 0.00). Mean weight, height, and BMI values from birth through age 19 years for MZ and DZ twins from all cohorts combined are presented in Figure 2. Over all ages, MZ and DZ twins maintained similar patterns of growth in mean weight, height and BMI, with a sharp increase in growth from birth to age 3 years, and then a steady increase in both weight and height through 19 years of age. When intra–class correlations between MZ and DZ twins were examined (Figure 3), irrespective of sex, MZ twin correlations were consistently greater than those of DZ twins for weight, height and BMI, and the gap between MZ and DZ correlations increased over time. While MZ twins maintained a correlation of approximately 0.8 or greater in weight, height, and BMI from birth through age 19 years, DZ twin correlations in weight reduced from around 0.7 to close to 0.2 by 19 years of age, and from around 0.8 to approximately 0.3 for height over those same years, indicating the presence

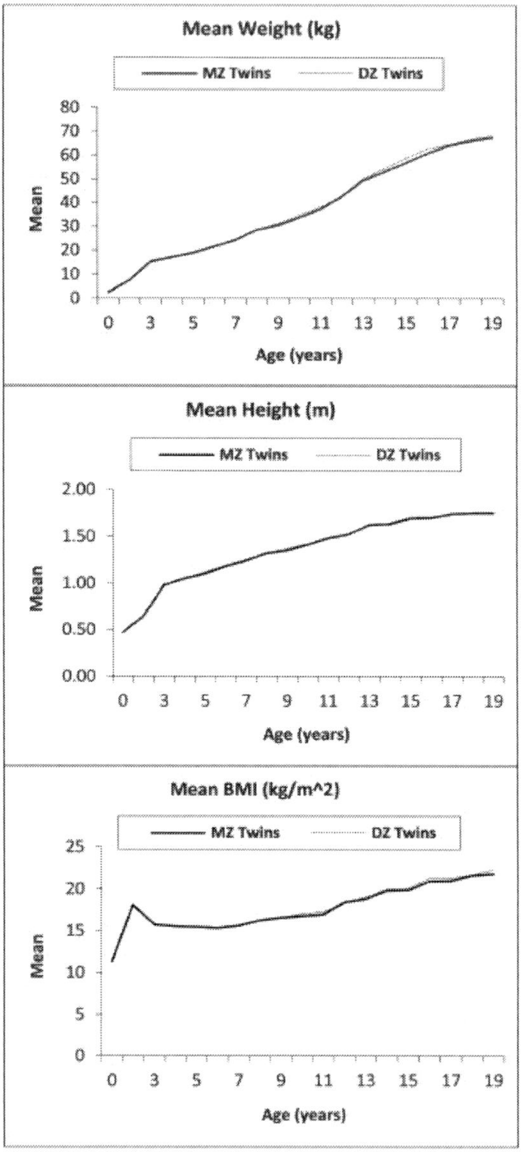

FIGURE 2: Mean of weight (kg), height (m), and BMI (kg/m2) in MZ and DZ twins of four countries, from birth through 19 years of age.

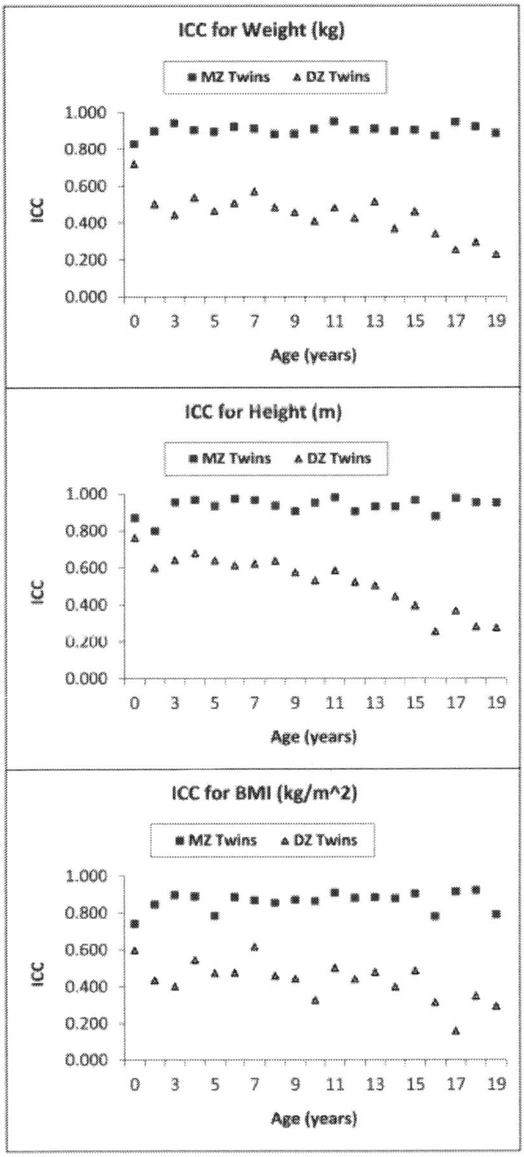

FIGURE 3: Intra–class correlations (ICC) between MZ and DZ twin pairs for weight (kg), height (m), and BMI (kg/m2), from birth through 19 years of age.

of strong genetic effects. Sex–specific correlations between MZ and DZ twins were also examined for weight, height and BMI. At all ages examined, and for all measures (except for height at age 5 months), intraclass correlations for MZ twins by sex differed significantly from DZ twins of the same sex, suggesting possible sex–limitation in the heritability of these body measures.

The proportion of variance in weight, height, and BMI explained by additive genetic (a^2), common environmental (c^2), and unique environmental (e^2) factors, according to full ACE and nested AE sex–limitation models from birth through 19 years of age, is presented for boys and girls in Tables 1 and 2, respectively; best fitting and most parsimonious models are displayed in bold in the tables. No sex–limitation was observed in either body weight or height at 4 and 7 years of age, nor in BMI at 9 years of age (data not shown); however, as significant sex–limitation was observed in all variables at every other age, all modeling results are presented in sex–limited form for consistency in Tables 1 and 2. The proportion of the phenotypic variance in weight, height and BMI explained by a^2 and c^2 according to the full ACE non–sex–limited model (with 95% confidence intervals), from birth through 19 years, are presented in Figure 4.

In both sexes, weight and BMI shared similar aetiologies; however, the aetiology of body height differed from other body measures. ACE was the best–fitting model for weight only at birth and at ages 4, 7, 8, 13, and 16 years. Similarly, for BMI, ACE was the best–fitting model at birth and at ages 4, 5, 7, 8, and 16 years. For all other ages an AE model provided best fit for weight and BMI (Tables 1 and 2). With regard to height, the effect of the common environment played a more important role, with the ACE model being consistently chosen as the best–fitting model from birth through 12 years of age, and again at 16 years of age. Only from 13 through 15 years of age, and from 17 through 19 years was AE the best–fitting model for height.

Figure 5 presents the proportion of phenotypic variance in weight, height, and BMI explained by a^2 and c^2 (according to the full ACE sex–limitation model) from birth through 19 years of age, in boys and girls separately. The proportion of variance in body weight and BMI explained by genetic influence was greater in boys than in girls, with the gap between the sexes increasing consistently from 4 through 19 years of age. This sex

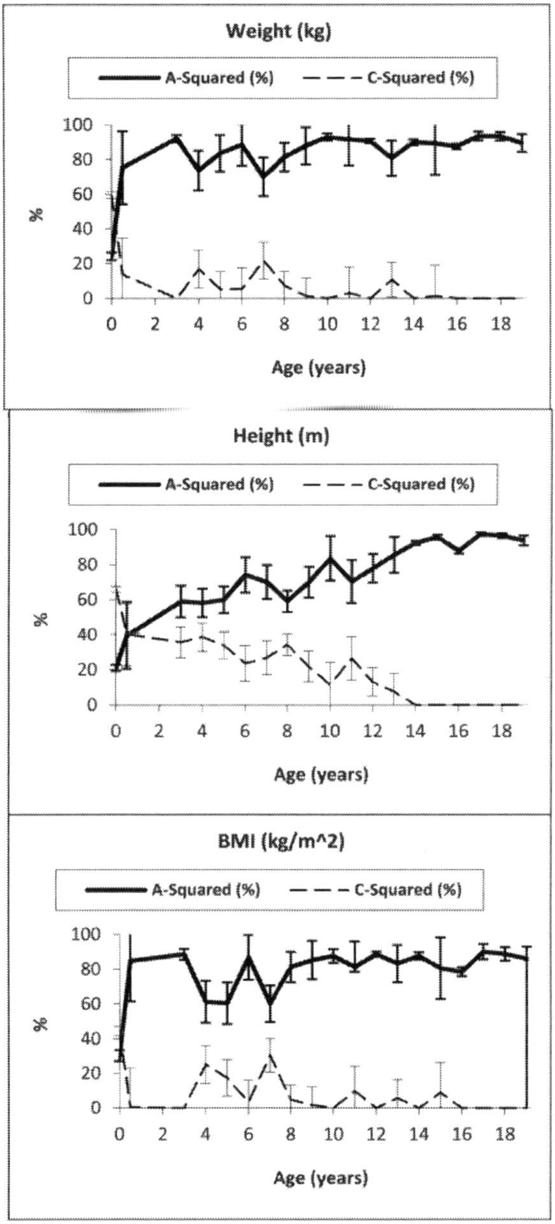

FIGURE 4: Proportion of the variance in weight (kg), height (m), and BMI (kg/m2) explained by A–squared and C–squared (with 95% confidence interval), in boys and girls (combined), from birth through 19 years of age – ACE models assumed.

TABLE 1: Best fitting model (in bold) for weight, height, and BMI, from birth through age 19 years, and the proportion of variance explained by additive genetic (a^2), common environmental (c^2), and unique environmental (e^2) influences: ACE–AE sex–limited model results for boys only.

	Weight					Height					BMI				
	ACE			AE		ACE			AE		ACE			AE	
Age *(Cohorts included)*	A^2	C^2	E^2	A^2	E^2	A^2	C^2	E^2	A^2	E^2	A^2	C^2	E^2	A^2	E^2
Birth *(All cohorts)*	**8.7**	**81.2**	**10.1**	83.9	16.1	**6.4**	**85.9**	**7.81**	86.7	13.3	**8.2**	**74.1**	**17.6**	77.8	22.2
5 mos *(QNTS)*	58.2	29.8	11.9	**86.3**	**13.7**	**18.5**	**65.4**	**16.2**	78.0	22.0	65.2	20.4	14.4	**85.8**	**14.2**
3y *(DTR)*	47.3	45.7	7.0	**92.6**	**7.4**	**40.6**	**57.2**	**2.2**	96.7	3.3	41.3	48.9	9.8	**90.9**	**9.1**
4y *(DTR)*	**60.5**	**31.4**	**8.1**	90.8	9.2	**38.7**	**58.4**	**2.9**	95.7	4.3	**47.9**	**42.2**	**9.9**	88.1	11.9
5y *(DTR & QNTS)*	82.7	9.2	8.0	**91.6**	**8.4**	**42.5**	**52.5**	**4.0**	94.3	5.7	**64.5**	**24.6**	**10.9**	87.5	12.5
6y *(DTR)*	62.6	31.1	6.3	**93.2**	**6.8**	**55.9**	**42.4**	**1.8**	97.6	2.4	70.6	19.1	10.3	**89.3**	**10.7**
7y *(DTR)*	**58.2**	**33.9**	**7.9**	92.1	7.9	**51.7**	**45.9**	**2.4**	96.9	3.1	48.5	43.0	8.5	90.8	9.2
8y *(DTR, QNTS, & TCHAD)*	**76.7**	**11.9**	**11.4**	88.8	11.2	**43.1**	**51.6**	**5.3**	92.7	7.3	**75.9**	**9.6**	**14.6**	86.5	13.5
9y *(CATSS & DTR)*	78.7	10.3	11.0	**89.6**	**10.4**	**51.9**	**40.6**	**7.5**	90.9	9.1	78.6	9.1	12.3	**87.6**	**12.4**
10y *(DTR)*	92.2	0.0	7.8	**92.2**	**7.8**	**66.7**	**29.6**	**3.6**	95.7	4.3	87.0	0.0	13.0	**87.3**	**12.7**
11y *(DTR)*	93.9	0.8	5.3	**95.5**	**4.5**	**60.4**	**38.0**	**1.6**	98.2	1.8	73.0	20.6	6.4	**94.2**	**5.8**
12y *(CATTS, DTR, & BTLS)*	88.7	1.1	10.2	**90.2**	**9.8**	**68.1**	**23.0**	**9.0**	90.0	10.0	86.0	0.9	13.1	**88.1**	**11.9**
13y *(DTR & TCHAD)*	**56.0**	**37.0**	**7.0**	92.0	8.0	63.5	30.6	5.9	**93.0**	**7.0**	70.3	19.7	9.9	**89.6**	**10.4**
14y *(DTR & BTLS)*	88.8	0.0	11.2	**89.0**	**11.0**	77.7	15.5	6.9	**92.4**	**7.6**	86.0	0.9	13.1	**88.2**	**11.8**
15y *(DTR)*	67.2	25.5	7.4	**92.2**	**7.8**	87.1	7.8	5.2	**94.8**	**5.2**	76.9	14.4	8.7	**90.5**	**9.5**
16y *(DTR, BTLS, & TCHAD)*	**67.8**	**15.9**	**16.3**	84.2	15.8	**71.8**	**9.4**	**18.8**	80.2	19.8	**73.6**	**0.0**	**26.4**	73.7	26.3
17y *(DTR)*	92.0	0.0	8.0	**92.0**	**8.0**	72.7	21.9	5.4	**93.9**	**6.1**	90.6	0.0	9.4	**91.0**	**9.0**
18y *(DTR)*	90.8	3.3	5.9	**94.2**	**5.8**	79.8	8.9	11.3	**90.8**	**9.2**	86.6	1.1	12.3	**91.4**	**8.6**
19y *(DTR)*	82.8	0.0	17.2	**84.3**	**15.7**	71.8	12.5	15.7	**82.9**	**17.1**	89.1	0.0	10.9	**90.1**	**9.9**

Note: CE sex–limited model excluded as it never provided the best fit.

difference was significant, but less apparent for height. For girls, the effect of the common environment played a more important role, particularly in explaining the variability in BMI. For both sexes, heritability in variances for body weight, height, and BMI was low at birth, between 6.4 and 8.7% in all measures for boys and between 4.8 and 7.9% in all measures for girls, but increased over time. Genetic effects accounted for close to half or more of the variance in weight and BMI after 5 months of age in both sexes, while the effect of the common environment in all body measures was high at birth, between 74.1 and 85.9% in all measures for boys and between 74.2 and 87.3% in all measures for girls, and markedly reduced over time. For body height, however, the effect of the common environment maintained a greater influence over a longer period during early childhood (from birth up to 12 years of age), in comparison to its influence on body weight or BMI. The effect of the unique environment generally remained

TABLE 2: Best fitting model (in bold) for weight, height, and BMI, from birth through age 19 years, and the proportion of variance explained by additive genetic (a^2), common environmental (c^2), and unique environmental (e^2) influences: ACE–AE sex–limited model results for girls only.

	Weight					Height					BMI				
	ACE			AE		ACE			AE		ACE			AE	
Age ***(Cohorts included)***	A^2	C^2	E^2	A^2	E^2	A^2	C^2	E^2	A^2	E^2	A^2	C^2	E^2	A^2	E^2
Birth *(All cohorts)*	**4.9**	**84.9**	**10.2**	85.0	15.0	**4.8**	**87.3**	**7.91**	87.4	12.6	**7.9**	**74.2**	**17.8**	76.5	23.5
5 mos *(QNTS)*	70.9	19.5	9.7	**89.5**	**10.5**	**18.1**	**68.2**	**13.7**	80.8	19.2	76.9	7.0	16.2	**83.7**	**16.3**
3y *(DTR)*	54.4	38.0	7.6	**91.6**	**8.4**	**31.0**	**64.0**	**5.0**	93.0	7.0	48.0	39.8	12.1	**86.7**	**13.3**
4y *(DTR)*	**59.4**	**32.7**	**7.9**	90.8	9.2	**40.2**	**58.2**	**1.7**	97.5	2.5	**45.9**	**41.6**	**12.5**	86.1	13.9
5y *(DTR & QNTS)*	78.6	8.3	13.1	**86.4**	**13.6**	**43.3**	**51.4**	**5.2**	93.1	6.9	**50.6**	**25.0**	**24.3**	73.8	26.2
6y *(DTR)*	47.6	48.5	3.8	**95.4**	**4.6**	**54.5**	**43.7**	**1.8**	97.7	2.3	29.6	64.0	6.4	**91.9**	**8.1**
7y *(DTR)*	**48.2**	**46.0**	**5.8**	91.8	8.2	**50.2**	**47.4**	**2.4**	96.8	3.2	**40.7**	**53.3**	**5.9**	91.6	8.4
8y *(DTR, QNTS, & TCHAD)*	**66.6**	**23.7**	**9.6**	88.6	11.4	**42.9**	**52.7**	**4.4**	93.8	6.2	**61.8**	**26.4**	**11.0**	85.6	14.4
9y *(CATSS & DTR)*	73.1	16.2	10.7	**88.8**	**11.2**	**49.0**	**45.2**	**5.8**	92.9	7.1	76.1	10.7	13.2	**86.4**	**13.6**
10y *(DTR)*	94.0	0.0	6.0	**94.0**	**6.0**	**76.8**	**17.8**	**5.4**	94.1	5.9	81.5	7.1	11.4	**88.5**	**11.5**
11y *(DTR)*	71.6	22.8	5.6	**93.5**	**6.5**	**56.5**	**50.4**	**3.1**	95.4	4.6	52.7	36.0	11.4	**86.3**	**13.7**
12y *(CATTS, DTR, & BTLS)*	83.8	7.7	8.4	**91.1**	**8.9**	**62.5**	**30.4**	**7.1**	91.6	8.4	85.2	4.3	10.5	**89.2**	**10.8**
13y *(DTR & TCHAD)*	**60.3**	**31.6**	**8.1**	90.6	9.4	63.9	30.3	5.8	**92.9**	**7.1**	65.5	23.8	10.6	**88.5**	**11.5**
14y *(DTR & BTLS)*	88.0	2.8	9.2	**90.8**	**9.2**	56.8	35.7	7.5	**91.3**	**8.7**	73.1	15.2	11.7	**86.9**	**13.1**
15y *(DTR)*	49.4	39.6	11.0	**87.9**	**12.1**	92.8	1.1	6.2	**93.8**	**6.2**	42.0	48.0	10.0	**88.8**	**11.2**
16y *(DTR, BTLS, & TCHAD)*	**87.8**	**0.0**	**12.2**	87.7	12.3	**65.2**	**17.1**	**17.7**	81.0	19.0	**82.9**	**1.0**	**16.4**	83.9	16.1
17y *(DTR)*	84.2	6.4	9.5	**90.6**	**9.4**	72.7	23.1	4.3	**95.2**	**4.8**	73.8	15.1	11.1	**89.4**	**10.6**
18y *(DTR)*	39.8	49.6	10.6	**88.4**	**11.6**	96.2	0.0	3.8	**96.5**	**3.5**	51.6	37.7	10.7	**86.1**	**13.9**
19y *(DTR)*	73.9	12.1	14.0	**85.9**	**14.1**	60.4	32.8	6.8	**92.4**	**7.6**	65.1	17.8	17.1	**80.7**	**19.3**

Note: CE sex–limited model excluded as it never provided the best fit.

stable for both sexes, across all body measures, and at all ages, accounting for less than 19% (ACE models) of the variance in weight, height, and BMI from birth through age 19 years, with the exception for BMI at age 16 years in boys (26.4%) and at 5 years in girls (24.3%), where the effect was slightly higher.

2.4 DISCUSSION

While others have conducted international twin studies to compare the heritability of body height and/or BMI during certain years in adolescence [44] and over broader age ranges in adulthood [9], [45], the present study is unique for its description and examination of the genetic and environmental influences on body weight, height, and BMI over virtually every

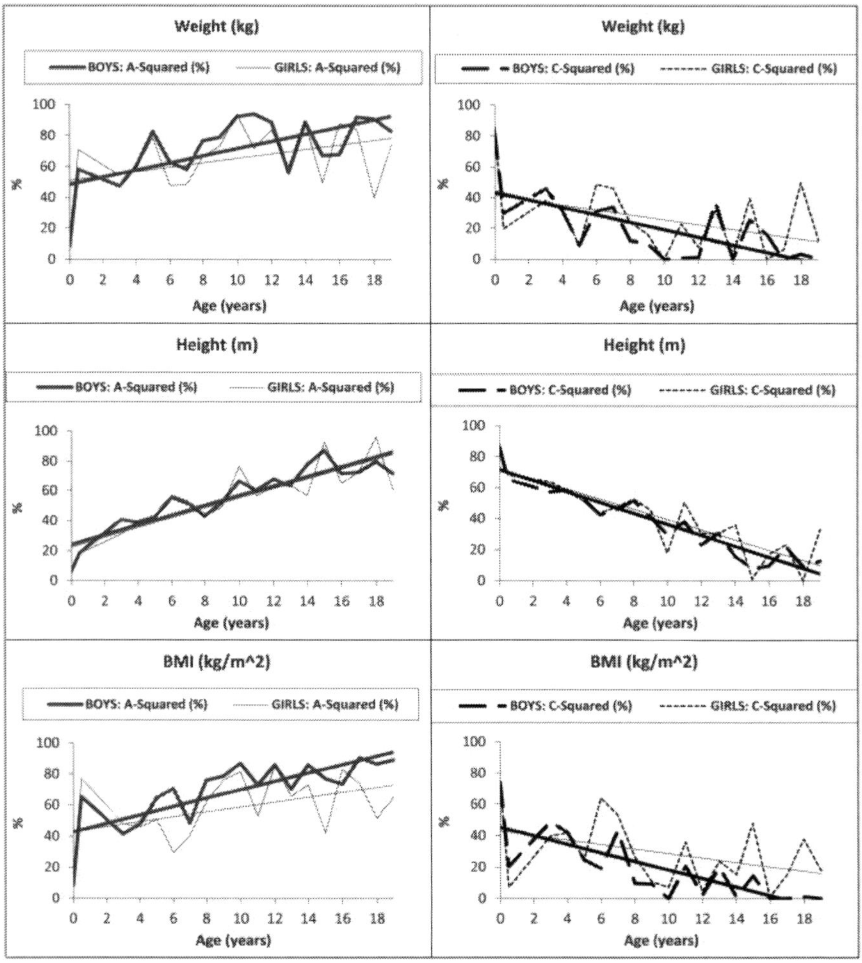

FIGURE 5: Proportion of the variance (with linear trend) in weight (kg), height (m), and BMI (kg/m2) explained by A–squared and C–squared, in boys and girls separately, from birth through 19 years of age – ACE models assumed.

year of age, from birth through 19 years, in a large sample of MZ and DZ same– and opposite–sex twins from three continents. The findings indicate that variability in weight, height, and BMI amongst twins from four developed countries is strongly influenced by genetic factors in both sexes as early as from 5 months of age, and increasingly so through late adolescence. The increasing heritability in these body measures over time, which was found to explain at times more than 80–90% of the variance in mid– to late adolescence, was observed along with a decreased influence of the common environment in early childhood years. Common environmental influences were found to play an important role in influencing variability in body weight, height, and BMI in early childhood in both sexes, particularly for body height where the influence of the common environment remained significant for a longer period, up through 12 years of age; however, common environmental influences were, for the most part, no longer significant by early– to mid–adolescence. On the other hand, the influence of the unique environment on all body measures was found to be small, but significant, at all ages and in both sexes.

2.4.1 INCREASING HERITABILITY

The increasing heritability observed in the present study is in agreement with results from other large twin–studies which found similar results for BMI in twins followed longitudinally from 4 to 7 years of age [46] and from 11–12 to 14 and 17 years of age [19]. A large longitudinal study of Dutch twin children followed from ages 3 to 12 years also found strong evidence for the role of genetic influences on body height and BMI in males and females, in addition to moderate non–significant increases in heritability specifically for height [47]. In correspondence with the present study's findings, these increasing heritability estimates for height resulted from the decreased influence of the common environment; however, common environmental influences remained important for both height and BMI at all ages examined [47].

Increasing heritability estimates have also been observed in other traits over childhood, such as in cognitive development [48], and this is generally attributed in part to changing gene expression, but also to gene–

environment interaction (G×E) [19], [49] and gene–environment correlations [50]–[52]. In the case of gene–environment interactions, individuals with differing genotypes may react to specific environmental stimuli in different ways. If a gene interacts with a factor within the common/shared environment, but the phenotypic expression associated with this interaction is only activated in adolescent years, this may explain the observed diminished effect of the common environment in early– to mid–adolescence in the present study. Alternatively, with gene–environment correlations, a genetic factor may influence an environmental exposure such that individuals may seek environments that correlate with the same phenotype. For example, an individual with a genetic predisposition for weight gain who is inclined to maintain a sedentary lifestyle or poor dietary practices may seek the company of other individuals who share similar attributes and practices, possibly increasing exposure to obesogenic environments that would lead to furthered weight gain. Such gene–environment correlations could increase heritability over time, firstly, due to increased independence to choose one's personal environment in adolescence, with MZ twins choosing more similar environments in comparison to DZ twins due to their identical genes, and, secondly, due to the genetic influences becoming more reinforced and direct over time by means of the respective correlated and phenotype–proliferating environmental influences. In the present study, the genetic modeling technique used includes "gene–environment interaction" within the component of heritability if the environmental component of the interaction is shared within a twin pair, and within the unique environment component if not shared [49]. Although effects of interaction and genotype–environment correlation were not quantified in the present study, these are not sufficient to fully explain the dramatic decrease in the independent effect of the common environment by early adolescence.

2.4.2 DECREASING EFFECT OF THE COMMON ENVIRONMENT

The present study's finding that the effect of the common environment was highest at birth, with heritability thus being the lowest at that point, is consistent with results from other twin studies [14], [47], [53]–[55]. This observation reiterates the special intrauterine situation characterized by twinning whereby, due to shared placental membranes and environmental/nutritional constraints in the uterus, monochorionic MZ twins may compete more intensely for prenatal resources than dichorionic DZ twins, making them less similar at birth than their genetic potential would allow [22], [56]. However, the influence of the common environment decreased rapidly through later childhood years, particularly for body weight and BMI. In correspondence, a longitudinal study of Finnish twins found persistent effects of the common environment on variations in BMI up through 14 years of age, but a disappearance of this influence by 17 years of age [19]. A recent systematic review on twin and adoption studies also reported a substantial effect of the common environment on variations in BMI in mid–childhood, but a disappearance of this effect in adolescence [10]. Several studies have reported significant effects of the common environment on BMI in children aged 12 years or less, but not over the age of 12 years [13], [47], [57], [58]. However, not all studies support this finding [54].

Our results suggest that potential common environmental factors exert their strongest and most independent influence on variations in weight, height, and BMI specifically in pre–adolescent years. However, this does not signify that environmental factors are irrelevant as targets for intervention once a child reaches adolescence; rather, this may signify a lack of common environmental influences that are independent of genetic predisposition in later adolescent years [14]. Furthermore, given the situation of gene–environment correlation and gene–environment interaction, it is all the more necessary to continue investigating potential environmental interventions that counter the obesity epidemic in those most genetically predisposed, as early in life as possible.

2.4.3 PRESENCE OF SEX–LIMITATION

The sex–difference observed in the present study revealed that the proportion of the variance explained by genetic influence was greater in boys than in girls over the majority of years examined; this was most pronounced for weight and BMI. This sex–difference appears to have resulted from a difference in the magnitude of the influence of the common environment across the sexes, which seems to have played a more important role in girls than in boys. The influence of the unique environment on all body measures was similar in both sexes across the ages examined. These sex–differences concur with findings from a longitudinal study of Finnish twins, which found that boys had slightly higher heritability estimates at all ages examined in comparison to girls [19], and studies that report higher heritability estimates for BMI in men than in women [8], [9]. However, some studies report higher heritability estimates in girls in comparison to boys [16]–[18].

The reasons behind these sex–differences are intriguing. From early infancy and through pre–puberty, sex–differences are observed not only in fat mass and pattern of fat distribution [59], [60], but also in hormone levels that are implicated in feeding behaviours, metabolic processes and body composition, e.g. insulin & leptin [61]–[63]. However, given the present study's findings, it is important to consider what common environmental factors may influence variances in body weight and BMI in girls, more so than boys. Some studies suggest sex–differences in childhood obesity, demonstrating that boys and girls differ in their susceptibility to various social and ethnic environmental influences [64]. Furthermore, environmental influences, such as the availability of unhealthy foods in the home or exposure to family conflicts, are seen to associate with obesity–promoting dietary practices, such as the consumption of sweet snacks or take–away, in girls more so than in boys [65]. Similarly, a meta–analysis found that in girls, but not in boys, increasing parental food restriction (i.e. the degree to which parents attempt to restrict their child's eating during meals) was associated with an increased tendency to eat in the absence of hunger [66]. An association between parental level of education and BMI has also been observed in adolescent females, but not in males [67]. Another study showed an association between girls' weight concerns and mother's gender attitudes, with no such association observed in boys [68]. Research is needed to examine potential sex–differences in gene–environment interaction that may occur as a result of underlying sex–differences in hormonal regulation (e.g. leptin) and its interaction with the family and social environmental influences mentioned above [64].

Finally, epigenetic factors relating to body weight, height, and BMI, must also be considered. The epigenetic model of obesity theorizes that maternal weight gain/obesity, and/or nutrition prior to, and during pregnancy may create permanent changes in the gene regulation of the foetus, promoting an obese phenotype [69]. Much remains to be understood about the role of epigenetics and how its effects may differ across the sexes. However, the twin modeling techniques used model epigenetic effects as part of the variation accounted for by the unique environment, and no notable differences were observed across the sexes for this component.

2.4.4 STRENGTHS & LIMITATIONS

The present study has important strengths. Firstly, it analyzes a large sample of MZ and DZ same– and opposite–sex twins from birth cohorts obtained from four different countries in three continents; this not only provides high power and high confidence in the study results, but also allows for a strong assessment of sex–limitation. This study also provides a comprehensive overview of the genetic and environmental influences on body weight, height, and BMI over the entire span of childhood and adolescence, with data available from birth, at 5 months of age, and yearly from 3 years to 19 years of age, inclusively.

A limitation of this study lies in the use of self– or parent–reported weights and heights for some cohorts, which may have caused some bias due to the common under–reporting of weight known to occur especially in individuals with true values in the upper end of the BMI distribution [70], [71]. However, there is no reason to suspect a difference in the degree of bias across a twin pair, with an exception perhaps for dizygotic twins of the opposite sex, since girls have been shown to under–report their weight to a much larger extent in comparison to boys [72]. It is also reported that mothers overestimate their children's weights more than their heights, especially for boys, such that this was found to overestimate the prevalence of overweight by over 3% in 4–year–old children, and by 5% specifically for boys [14]. However, another study reports that parents generally tend to underestimate the prevalence of overweight in young children, typically by underreporting body weight and over–reporting height in children with a high BMI, and over–reporting body weight in children with a low BMI [73]. Such a potential bias may lead to decreased heritability estimates and increased estimates for the effect of the unique environment. A bias specifically across the sexes, though, would mainly affect the results obtained through the sex–limitation analyses, leaving the general analyses unaffected. Nonetheless, it is important to note that, several twin studies from various populations (Australian, Finnish, Danish, and British) have reported a good agreement between self–reported and measured weight and height [9], [74].

Due to the lack of information about the weight, height and BMI of parents, another limitation arises with the techniques used in twin analyses. Without such information, one must make the assumption of random mating. Several studies, however, have shown that this is an unrealistic assumption as assortative mating, the increased preference to marry someone with similar traits, occurs with several phenotypic, physical, and psychological characteristics, including BMI [75], [76], [76], [77]. Assortative mating may potentially inflate heritability estimates in twin studies since it increases the genetic similarities between DZ twins above the 0.5 correlation assumed in the twin modeling techniques. Furthermore, even if twin modeling techniques reveal that an AE model best fits the data (i.e. no effect of the common environment observed), this does not mean that assortative mating and non–additivity are not acting [78]. On the other hand, assortative mating may also inflate the variance estimates obtained as part of the common environmental component when not accounted for in the modeling, since this estimate is derived from comparing correlations between MZ and DZ twin pairs with an assumption of a correlation of 0.5 between DZ twin pairs. However, given that the effect of the common environment became insignificant after early adolescence, it is very unlikely that assortative mating would have significantly inflated the estimates obtained for that component. We are thus confident that the effect of the common environment observed in early childhood years is in fact a true influence and is not simply an artificial result of the twin–modeling techniques used [10]. Finally, the present study's findings may be limited in their generalizability to other ethnicities due to the main inclusion of Caucasian populations.

2.4.5 CONCLUSIONS AND FUTURE DIRECTIONS

Genetics appear to play an increasingly important role in explaining the variation in weight, height, and BMI from early childhood to late adolescence, with boys being more significantly affected by these effects. This finding emphasizes the need for future studies to continue identifying common genetic variants that may predispose individuals to obesity. It is hoped that with the identification of such variants and a furthered understanding of potential gene–environment interactions, interventions may be tailored to an individual's personal genetic predisposition so that greater success can be attained in the battle against obesity. The findings also emphasize the need to target family and social environmental interventions in early childhood years, particularly for females, as the effect of the common environment was particularly influential until early adolescence.

REFERENCES

1. Lobstein T, Baur L, Uauy R, IASO International Obesity TaskForce (2004) Obesity in children and young people: A crisis in public health. Obes Rev 5: Suppl 14–104. doi: 10.1111/j.1467–789X.2004.00133.x.
2. Monteiro CA, Moura EC, Conde WL, Popkin BM (2004) Socioeconomic status and obesity in adult populations of developing countries: A review. Bull World Health Organ 82(12): 940–946.
3. World Health Organization (WHO). (2006) Obesity and overweight: Fact sheet [no 311]. Accessed: 2010 April 19.
4. Johannsson E, Arngrimsson SA, Thorsdottir I, Sveinsson T (2006) Tracking of overweight from early childhood to adolescence in cohorts born 1988 and 1994: Overweight in a high birth weight population. Int J Obes (Lond) 30(8): 1265–1271. doi: 10.1038/sj.ijo.0803253.
5. Sorensen TI, Echwald SM (2001) Obesity genes. BMJ 322(7287): 630–631. doi: 10.1111/j.1600–0447.1993.tb05363.x.
6. Kumanyika S, Jeffery RW, Morabia A, Ritenbaugh C, Antipatis VJ, et al. (2002) Obesity prevention: The case for action. Int J Obes Relat Metab Disord 26(3): 425–436. doi: 10.1038/sj.ijo.0801938.

7. Vandenbroeck P, Goossens J, Clemens M (2007) Obesity system atlas. in: Foresight – tackling obesities: Future choices.
8. Maes HHM, Neale MC, Eaves LJ (1997) Genetic and environmental factors in relative body weight and human adiposity. Behav Genet 27(6): 600.
9. Schousboe K, Willemsen G, Kyvik KO, Mortensen J, Boomsma DI, et al. (2003) Sex differences in heritability of BMI: A comparative study of results from twin studies in eight countries. Twin Res 6(5): 409–421. doi: 10.1375/136905203770326411.
10. Silventoinen K, Rokholm B, Kaprio J, Sorensen TI (2010) The genetic and environmental influences on childhood obesity: A systematic review of twin and adoption studies. Int J Obes (Lond) 34(1): 29–40. doi: 10.1038/ijo.2009.177.
11. Hjelmborg JB, Fagnani C, Silventoinen K, McGue M, Korkeila M, et al. (2008) Genetic influences on growth traits of BMI: A longitudinal study of adult twins. Obesity (Silver Spring) 16(4): 847–852. doi: 10.1038/oby.2007.135.
12. Silventoinen K, Kaprio J (2009) Genetics of tracking of body mass index from birth to late middle age: Evidence from twin and family studies. Obesity Facts 2(3): 196–202. doi: 10.1159/000219675.
13. Wardle J, Carnell S, Haworth CM, Plomin R (2008) Evidence for a strong genetic influence on childhood adiposity despite the force of the obesogenic environment. Am J Clin Nutr 87(2): 398–404.
14. Dubois L, Girard M, Girard A, Tremblay R, Boivin M, et al. (2007) Genetic and environmental influences on body size in early childhood: A twin birth–cohort study. Twin Research and Human Genetics 10(3): 479–485. doi: 10.1375/twin.10.3.479.
15. Pietilainen KH, Kaprio J, Rissanen A, Winter T, Rimpela A, et al. (1999) Distribution and heritability of BMI in finnish adolescents aged 16 y and 17 y: A study of 4884 twins and 2509 singletons. Int J Obes Relat Metab Disord 23(2): 107–115. doi: 10.1038/sj.ijo.0800767.
16. Harris JR, Tambs K, Magnus P (1995) Sex–specific effects for body mass index in the new norwegian twin panel. Genet Epidemiol 12(3): 251–265. doi: 10.1002/gepi.1370120303.
17. Neale MC, Cardon LR (1992) Methodology for genetic studies of twins and families. Dordecht: Kluwer Academic Publishers.
18. Hur YM (2007) Sex difference in heritability of BMI in south korean adolescent twins. Obesity (Silver Spring) 15(12): 2908–2911. doi: 10.1038/oby.2007.346.
19. Lajunen H–, Kaprio J, Keski–Rahkonen A, Rose RJ, Pulkkinen L, et al. (2009) Genetic and environmental effects on body mass index during adolescence: A prospective study among finnish twins. Int J Obes 33(5): 559–567. doi: 10.1038/ijo.2009.51.
20. Pérusse D (1995) The quebec longitudinal twin study of infant temperament. New Orleans: American Academy of Child and Adolescent Psychiatry.
21. Goldsmith HH (1991) A zygosity questionnaire for young twins: A research note. Behav Genet 21(3): 257–269. doi: 10.1007/BF01065819.
22. Forget–Dubois N, Perusse D, Turecki G, Girard A, Billette JM, et al. (2003) Diagnosing zygosity in infant twins: Physical similarity, genotyping, and chorionicity. Twin Res 6(6): 479–485. doi: 10.1375/136905203322686464.
23. Lichtenstein P, Tuvblad C, Larsson H, Carlstrom E (2007) The swedish twin study of Child and adolescent development: The TCHAD–study. Twin Res Hum Genet 10(1): 67–73. doi: 10.1375/twin.10.1.67.

24. Lichtenstein P, Sullivan PF, Cnattingius S, Gatz M, Johansson S, et al. (2006) The swedish twin registry in the third millennium: An update. Twin Res Hum Genet 9(6): 875–882. doi: 10.1375/183242706779462444.
25. Hannelius U, Gherman L, Mäkelä VV, Lindstedt A, Zucchelli M, et al. (2007) Large–scale zygosity testing using single nucleotide polymorphisms. Twin Research & Human Genetics: The Official Journal of the International Society for Twin Studies 10(4): 604–625. doi: 10.1375/twin.10.4.604.
26. Skytthe A, Kyvik K, Holm NV, Vaupel JW, Christensen K (2002) The danish twin registry: 127 birth cohorts of twins. Twin Res 5(5): 352–357. doi: 10.1375/136905202320906084.
27. Skytthe A, Kyvik K, Bathum L, Holm N, Vaupel JW, et al. (2006) The danish twin registry in the new millennium. Twin Res Hum Genet 9(6): 763–771. doi: 10.1375/183242706779462732.
28. Christiansen L, Frederiksen H, Schousboe K, Skytthe A, von Wurmb–Schwark N, et al. (2003) Age– and sex–differences in the validity of questionnaire–based zygosity in twins. Twin Res 6(4): 275–278. doi: 10.1375/136905203322296610.
29. Zhu G, Duffy DL, Eldridge A, Grace M, Mayne C, et al. (1999) A major quantitative–trait locus for mole density is linked to the familial melanoma gene CDKN2A: A maximum–likelihood combined linkage and association analysis in twins and their sibs. Am J Hum Genet 65(2): 483–492. doi: 10.1086/302494.
30. Zhu G, Montgomery GW, James MR, Trent JM, Hayward NK, et al. (2007) A genome–wide scan for naevus count: Linkage to CDKN2A and to other chromosome regions. Eur J Hum Genet 15(1): 94–102. doi: 10.1038/sj.ejhg.5201729.
31. Wright MJ, Martin NG (2004) Brisbane adolescent twin study: Outline of study methods and research projects. Aust J Psychol 56(2): 65–78. doi: 10.1080/00049530410001734865.
32. Boomsma D, Busjahn A, Peltonen L (2002) Classical twin studies and beyond. Nature Reviews Genetics 3(11): 872–882. doi: 10.1038/nrg932.
33. Derks EM, Dolan CV, Boomsma DI (2006) A test of the equal environment assumption (EEA) in multivariate twin studies. Twin Research and Human Genetics 9(3): 403–411. doi: 10.1375/183242706777591290.
34. Visscher PM, Hill WG, Wray NR (2008) Heritability in the genomics era – concepts and misconceptions. Nature Reviews Genetics 9(4): 255–266. doi: 10.1038/nrg2322.
35. Guo G, Wang J (2002) The mixed or multilevel model for behavior genetic analysis. Behav Genet 32(1): 37–49.
36. Goldstein H (1995) Multilevel statistical models. London/New York: Arnold/Halstead.
37. Grayson DA (1989) Twins reared together: Minimizing shared environmental effects. Behav Genet 19(4): 593–604. doi: 10.1007/BF01066256.
38. Hewitt JK (1989) Of biases and more in the study of twins reared together: A reply to grayson. Behav Genet 19(4): 605–608. doi: 10.1007/BF01066257.
39. Danielzik S, Czerwinski–Mast M, Langnase K, Dilba B, Muller MJ (2004) Parental overweight, socioeconomic status and high birth weight are the major determinants of overweight and obesity in 5–7 y–old children: Baseline data of the kiel obesity prevention study (KOPS). Int J Obes Relat Metab Disord 28(11): 1494–1502. doi: 10.1038/sj.ijo.0802756.

40. Rhee K (2008) Childhood overweight and the relationship between parent behaviors, parenting style, and family functioning. Ann Am Acad Pol Soc Sci 615(1): 12–37. doi: 10.1177/0002716207308400.
41. Fox MK, Dodd AH, Wilson A, Gleason PM (2009) Association between school food environment and practices and body mass index of US public school children. J Am Diet Assoc 109(2 Suppl): S108–17. doi: 10.1016/j.jada.2008.10.065.
42. Haire–Joshu D, Nanney MS (2002) Prevention of overweight and obesity in children: Influences on the food environment. Diabetes Educ 28(3): 415–423. doi: 10.1177/014572170202800311.
43. Nelson MC, Gordon–Larsen P, Song Y, Popkin BM (2006) Built and social environments. associations with adolescent overweight and activity. Am J Prev Med 31(2): 109–117.
44. Hur Y–, Kaprio J, Iacono WG, Boomsma DI, McGue M, et al. (2008) Genetic influences on the difference in variability of height, weight and body mass index between caucasian and east asian adolescent twins. Int J Obes 32(10): 1455–1467. doi: 10.1038/ijo.2008.144.
45. Silventoinen K, Sammalisto S, Perola M, Boomsma DI, Cornes BK, et al. (2003) Heritability of adult body height: A comparative study of twin cohorts in eight countries. Twin Res 6(5): 399–408. doi: 10.1375/136905203770326402.
46. Haworth CM, Carnell S, Meaburn EL, Davis OS, Plomin R, et al. (2008) Increasing heritability of BMI and stronger associations with the FTO gene over childhood. Obesity (Silver Spring) 16(12): 2663–2668. doi: 10.1038/oby.2008.434.
47. Silventoinen K, Bartels M, Posthuma D, Estourgie–van Burk GF, Willemsen G, et al. (2007) Genetic regulation of growth in height and weight from 3 to 12 years of age: A longitudinal study of dutch twin children. Twin Res Hum Genet 10(2): 354–363. doi: 10.1375/twin.10.2.354.
48. Davis OSP, Haworth CMA, Plomin R (2009) Dramatic increase in heritability of cognitive development from early to middle childhood: An 8–year longitudinal study of 8,700 pairs of twins: Research article. Psychological Science 20(10): 1301–1308. doi: 10.1111/j.1467–9280.2009.02433.x.
49. Purcell S (2002) Variance components models for gene–environment interaction in twin analysis. Twin Research 5(6): 554–571. doi: 10.1375/136905202762342026.
50. Bergen SE, Gardner CO, Kendler KS (2007) Age–related changes in heritability of behavioral phenotypes over adolescence and young adulthood: A meta–analysis. Twin Research and Human Genetics 10(3): 423–433. doi: 10.1375/twin.10.3.423.
51. Jaffee SR, Price TS (2007) Gene–environment correlations: A review of the evidence and implications for prevention of mental illness. Mol Psychiatry 12(5): 432–442. doi: 10.1038/sj.mp.4001950.
52. Plomin R (1994) Genetics and experience: The interplay between nature and nurture. (1994).Genetics and Experience: The Interplay between Nature and Nurture. Xvi. Thousand Oaks, CA, US: Sage Publications, Inc; US. 189 p.
53. Pietilainen KH, Kaprio J, Rasanen M, Rissanen A, Rose RJ (2002) Genetic and environmental influences on the tracking of body size from birth to early adulthood. Obes Res 10(9): 875–884. doi: 10.1038/oby.2002.120.

54. Silventoinen K, Pietilainen KH, Tynelius P, Sorensen TIA, Kaprio J, et al. (2007) Genetic and environmental factors in relative weight from birth to age 18: The swedish young male twins study. Int J Obes 31(4): 615–621. doi: 10.1038/sj.ijo.0803577.
55. Whitfield JB, Treloar SA, Zhu G, Martin NG (2001) Genetic and non–genetic factors affecting birth–weight and adult body mass index. Twin Res 4(5): 365–370. doi: 10.1375/1369052012533.
56. Vlietinck R, Derom R, Neale MC, Maes H, van Loon H, et al. (1989) Genetic and environmental variation in the birth weight of twins. Behav Genet 19(1): 151–161. doi: 10.1007/BF01065890.
57. Haworth CM, Plomin R, Carnell S, Wardle J (2008) Childhood obesity: Genetic and environmental overlap with normal–range BMI. Obesity (Silver Spring) 16(7): 1585–1590. doi: 10.1038/oby.2008.240.
58. Cornes BK, Zhu G, Martin NG (2007) Sex differences in genetic variation in weight: A longitudinal study of body mass index in adolescent twins. Behav Genet 37(5): 648–660. doi: 10.1007/s10519–007–9165–0.
59. Dunger DB, Salgin B, Ong KK (2007) Session 7: Early nutrition and later health early developmental pathways of obesity and diabetes risk. Proc Nutr Soc 66(3): 451–457. doi: 10.1017/S0029665107005721.
60. He Q, Horlick M, Thornton J, Wang J, Pierson J, et al. (2002) Sex and race differences in fat distribution among asian, african–american, and caucasian prepubertal children. J Clin Endocrinol Metab 87(5): 2164–2170. doi: 10.1210/jc.87.5.2164.
61. Mann DR, Johnson AOK, Gimpel T, Castracane VD (2003) Changes in circulating leptin, leptin receptor, and gonadal hormones from infancy until advanced age in humans. Journal of Clinical Endocrinology & Metabolism 88(7): 3339–3345. doi: 10.1210/jc.2002–022030.
62. Murphy MJ, Metcalf BS, Voss LD, Jeffery AN, Kirkby J, et al. (2004) Girls at five are intrinsically more insulin resistant than boys: The programming hypotheses revisited – the EarlyBird study (EarlyBird 6). Pediatrics 113(1 I): 82–86. doi: 10.1542/peds.113.1.82.
63. Petridou E, Mantzoros CS, Belechri M, Skalkidou A, Dessypris N, et al. (2005) Neonatal leptin levels are strongly associated with female gender, birth length, IGF–I levels and formula feeding. Clin Endocrinol (Oxf) 62(3): 366–371. doi: 10.1111/j.1365–2265.2005.02225.x.
64. Wisniewski AB, Chernausek SD (2009) Gender in childhood obesity: Family environment, hormones, and genes. Gender Medicine 6(SUPPL. 1): 76–85. doi: 10.1016/j.genm.2008.12.001.
65. Campbell KJ, Crawford DA, Salmon J, Carver A, Garnett SP, et al. (2007) Associations between the home food environment and obesity–promoting eating behaviors in adolescence. Obesity 15(3): 719–730. doi: 10.1038/oby.2007.553.
66. Faith MS, Berkowitz RI, Stallings VA, Kerns J, Storey M, et al. (2004) Parental feeding attitudes and styles and child body mass index: Prospective analysis of a gene–environment interaction. Pediatrics 114(4): e429–e436. doi: 10.1542/peds.2003–1075–L.
67. Tschumper A, Nagele C, Alsaker FD (2006) Gender, type of education, family background and overweight in adolescents. Int J Pediatr Obes 1(3): 153–160. doi: 10.1080/17477160600881767.

68. McHale SM, Corneal DA, Crouter AC, Birch LL (2001) Gender and weight concerns in early and middle adolescence: Links with well–being and family characteristics. J Clin Child Psychol 30(3): 338–348. doi: 10.1207/S15374424JCCP3003_6.
69. Waterland RA (2008) Epigenetic epidemiology of obesity: Application of epigenomic technology. Nutr Rev 66Suppl 1(Suppl 1): S21–3. doi: 10.1111/j.1753–4887.2008.00060.x.
70. Crawley H, Portides G (1995) Self–reported versus measured height, weight and body mass index amongst 16–17 year old british teenagers. Int J Obes 19(8): 579–584.
71. Elgar FJ, Moore L, Roberts C, TudorSmith C (2004) Validity of Self–Reported Height and Weight, and Predictors of Bias in Adolescents. Psychol Health 19(Supplement): 49. doi: 10.1016/j.jadohealth.2004.07.014.
72. Betz NE, Mintz L, Speakmon G (1994) Gender differences in the accuracy of self–reported weight. Sex Roles 30(7–8): 543–552. doi: 10.1007/BF01420801.
73. Scholtens S, Brunekreef B, Visscher TLS, Smit HA, Kerkhof M, et al. (2007) Reported versus measured body weight and height of 4–year–old children and the prevalence of overweight. Eur J Public Health 17(4): 369–374. doi: 10.1093/eurpub/ckl253.
74. Stunkard AJ, Harris JR, Pedersen NL, McClearn GE (1990) The body–mass index of twins who have been reared apart. N Engl J Med 322(21): 1483–1487. doi: 10.1056/NEJM199005243222102.
75. Mascie–Taylor CGN (1987) Assortative mating in a contemporary british population. Ann Hum Biol 14(1): 59–68. doi: 10.1080/03014468700008841.
76. Silventoinen K, Kaprio J, Lahelma E, Viken RJ, Rose RJ (2003) Assortative mating by body height and BMI: Finnish twins and their spouses. Am J Hum Biol 15(5): 620–627. doi: 10.1002/ajhb.10183.
77. Spuhler JN (1982) Assortative mating with respect to physical characteristics. Soc Biol 29(1–2): 53–66. doi: 10.1080/19485565.1982.9988478.
78. Eaves LJ, Heath AC, Martin NG, Neale MC, Meyer JM, et al. (1999) Biological and cultural inheritance of stature and attitudes. In: Cloninger CR, editor. Personality and psychopathology. Washington, DC, US: American Psychiatric Association; US. pp. 269–308.

CHAPTER 3

ESTIMATION OF NEWBORN RISK FOR CHILD OR ADOLESCENT OBESITY: LESSONS FROM LONGITUDINAL BIRTH COHORTS

ANITA MORANDI, DAVID MEYRE, STÉPHANE LOBBENS, KEN KLEINMAN, MARIKA KAAKINEN, SHERYL L. RIFAS-SHIMAN, VINCENT VATIN, STEFAN GAGET, ANNELI POUTA, ANNA-LIISA HARTIKAINEN, JAANA LAITINEN, AIMO RUOKONEN, SHIKTA DAS, ANOKHI ALI KHAN, PAUL ELLIOTT, CLAUDIO MAFFEIS, MATTHEW W. GILLMAN, MARJO-RIITTA JÄRVELIN, and PHILIPPE FROGUEL

3.1 INTRODUCTION

Childhood and adolescent overweight and obesity, which are leading causes of early type 2 diabetes and cardiovascular disease, have become major public health problems both in westernized and more recently in developing countries [1]. Traditional approaches for the management of overweight and obesity have had poor long term efficacy and therefore prevention is currently the most promising strategy for controlling the obesity epidemic [1].

Prevention of obesity should start as early as possible after birth. Longitudinal studies have shown a strong association between early infancy weight gain rate or adiposity and childhood and even adult body weight, fat mass and body mass index (BMI) [2]–[3]. Moreover, the efficacy of preventive behavioural and nutrition interventions targeting school children, either in primary schools or at home, is very limited [4]–[5]. Finally, in many countries pre–school and school children are already burdened by a high prevalence of overweight or obesity [4].

Assessing the risk for future overweight or obesity in newborns may be a basis for focused preventive interventions for at–risk individuals during the very first months of their life. Even though several sociodemographic and anthropometric predictors, as well as several common genetic variants, have been associated with childhood overweight/obesity, no longitudinal study has attempted to explore the cumulative predictive properties of these known early life risk factors, or to propose possible tools to predict childhood obesity at birth [6]–[20].

We aimed to build such predictive algorithms for the early identification of newborns at an increased risk for childhood and adolescent overweight/obesity. For this purpose, we estimated the ability of clinical, socio–demographic, and genetic risk factors to predict childhood and adolescent overweight/obesity in a large Finnish birth cohort. We then confirmed the promising usefulness of socio–demographic and anthropometric factors in predicting childhood obesity in two independent paediatric cohorts.

3.2 METHODS

3.2.1 ETHICS STATEMENT

The study conducted on the NFBC1986 cohort was approved by the Ethical Committee of Northern Ostrobothnia Hospital District. The retrospective study of the Veneto cohort was approved by the Ethical Committee of the University of Verona and Project Viva was approved by the Human subjects Committees of Harvard Pilgrim Health Care, Brigham and Women's Hospital, and Beth Israel Deaconess Medical Center.

Written informed consent was obtained from parents or guardians of all participants and all clinical investigations were conducted according to the principles expressed in the Declaration of Helsinki.

3.2.2 SUBJECTS

3.2.2.1 DEVELOPMENT SAMPLE.

The Northern Finland Birth Cohort 1986 (NFBC1986) (http://kelo.oulu.fi/NFBC) was followed prospectively from 12th gestational week and several well known early risk factors for childhood obesity were recorded systematically. Participants who had their weight and height recorded at seven and sixteen years of age and met data completeness criteria (see below, N = 4,032) were used to build the models. We separately predicted childhood obesity (obesity at 7 years of age), childhood overweight/obesity (overweight or obesity at 7 years of age), adolescent obesity (obesity at 16 years of age), adolescent overweight/obesity (overweight or obesity at 16 years of age), and the severe sub–phenotypes of childhood obesity persistent into adolescence (obesity at 7 and 16 years of age) and childhood overweight/obesity persistent into adolescence (overweight or obesity at 7 and 16 years of age) (Table 1). Overweight and obesity were defined by the IOTF BMI cut–offs [21].

The traditional predictors used for building the predictive models (gender, pre–pregnancy parental BMI, parental professional category, single parenthood, gestational weight gain, pre–pregnancy maternal smoking, gestational smoking, number of household members, birth weight) were a–priori selected among all available baseline NFBC1986 variables according to their association with early obesity in previous literature (Table 1) [2], [6]–[11]. Forty–four obesity predisposing single–nucleotide polymorphisms (SNPs) were selected according to the following criterion: genome–wide significant level of association ($P<5\times10^{-8}$) for BMI and/or obesity reported in a population of European ancestry [12]–[20]. Genotyping was performed by TaqMan (Applied Biosystems, Foster City, CA): the average genotyping success was of 99.4% (95.1–100) and the average consensus rate from 255 duplicates was 99.8% (99.2–100).

Five SNPs were discarded during the genotyping procedure, since they did not pass the genotyping quality control criteria, leaving 39 SNPs. All 39 SNPs were in Hardy–Weinberg equilibrium (P>0.05). We assumed an additive model and constructed a cumulative genotype score by summing the number of risk alleles (0–78).

3.2.2.2 VALIDATION SAMPLES.

We used a school–based retrospective sample of 1,503 children aged 4–12 from Veneto, Italy, as one of the two validation samples to explore whether results from the NFBC1986 could be applied to a European paediatric cohort contemporary to the NFBC1986, with similar obesity prevalence (4%) but different cultural background [22]. The second validation set used was a prospective sample of 1032 children (7 years) from Massachusetts (United States) from the Project Viva (http://www.dacp.org/viva/index.html) to explore whether results would remain valid when applied to a very recent U.S. child cohort, with higher obesity prevalence (8%) and very different cultural background. Genetic variants were not available for the validation analyses. All children meeting the international criterion for obesity definition at the time of recruitment in the Italian sample and at 7 years of age in the U.S. sample were classified as affected by childhood obesity [21].

3.2.3 STATISTICAL ANALYSIS

3.2.3.1 DEVELOPMENT PHASE.

Predictive models were fitted by stepwise logistic regression analysis (criterion for variable entry: p<0.05, for variable removal: p>0.10) using traditional risk factors only, genetic score only and traditional risk factors plus genetic score for each obesity outcome. Each risk factor entering the analysis as continuous or ordinal scale variable showed a linear relationship

TABLE 1: Characteristics of the NFBC1986 cohort.

Baseline	
Males	1,917 (47.5)
Mother's age (years)	28.5 (16.9–50.8)
Father's age (years)	30.8 (17.9–59.8)
Single parenthood	113 (2.9)
Mothers smoking before pregnancy	994 (24.7)
Mothers smoking during pregnancy	737 (18.3)
Maternal BMI before pregnancy	22.3 (13.2–48.2)
Paternal BMI	24.0 (16.9–41.3)
Maternal professional category	
4 Professional/entrepreneur	277 (6.9)
3 Skilled–non–manual	866 (21.5)
2 Skilled–manual	1,625 (40.3)
1 Unskilled/apprentice/unemployed	1,264 (31.3)
Paternal professional category	
4 Professional/entrepreneur	545 (13.5)
3 Skilled–non–manual	856 (21.2)
2 Skilled–manual	1,934 (48)
1 Unskilled/apprentice/unemployed	691 (17.1)
Household members	3.6 (1–18)
Maternal percentage weight gain during pregnancy	23.3 (–12.0–11.6)
Gestational age	39.3 (27–43)
Birth weight (kg)	3.560 (0.740–5.560)
Genetic score	37.2 (25–50)
Outcome	
Childhood obesity (at 7 years of age)	121 (3)
Childhood overweight/obesity (at 7 years of age)	645 (16)
Adolescent obesity (at 16 years of age)	163 (4)
Adoescent overweight/obesity (at 16 years of age)	678 (17)
Persistent childhood obesity (at both 7 and 16 years of age)	47 (1)
Persistent childhood overweight/obesity (at both 7 and 16 years of age)	331 (8)

Note: Data are given as MEAN (range) or as N (percentage)

with the logit–risk of childhood obesity in a preliminary linear regression analysis. For persistent childhood obesity, not all the a priori selected traditional predictors were used for the stepwise analysis but only the five with the strongest association with persistent childhood obesity in a preliminary univariate analysis, in order to avoid possible model over–fitting due to the relatively small number (forty–seven) of outcome events.

The discrimination accuracy of each model was evaluated by the area under the receiver operating curve (AUROC) of the modeled risk [23]. Models with AUROCs larger than 0.7 were considered potentially clinically useful and those with AUROCs larger than 0.8 were considered to have excellent accuracy [23]. The model calibration, that is the "precision" or correlation between the predicted and observed event rate, was assessed by the Hosmer–Lemeshow test [23]. The possible accuracy improvement associated with adding the genetic score to the traditional risk factors was evaluated by calculating the integrated discrimination improvement (IDI) compared to the traditional risk factors alone [24].

For each model a risk threshold was arbitrarily adopted at the 75th percentile of the modeled risk, identifying the top 25% as being at increased risk and the thresholds' predictive properties (sensitivity, specificity and predictive values) were calculated.

An average of 1.67% (0–11.4%) of data was missing for each traditional risk factor, while an average of 0.72% (0–4.95%) of genotypes was missing for each SNP. We included participants with zero or one missing traditional baseline variable and three or fewer missing SNPs. Multiple imputation was performed for the remaining missing values, in order to avoid possible bias associated with missing potentially important information [25]. Win MICE (Multiple Imputation by Chain Equations) V0.1. was used for multiple imputation [25]. By the MICE procedure, imputed values for missing data are drawn from modelling them on the basis of the other considered variables, with logistic regression if the variable to impute is dichotomous, polytomous logistic regression if it is categorical with three or more categories and with linear regression if it is continuous [25]. So each missing value is replaced by an estimated value modelled on the other variables. Indeed, the method estimates a distribution of each missing variable, taking all aspects of uncertainty in the imputations into account. From this distribution, values are sampled and filled in for the missing data. So every imputation cycle produces, for each missing data, one estimated value sampled among several possible ones, giving rise to a unique dataset which can not be reproduced by following imputation cycles [25].

Five imputation cycles were run so that five values were imputed for each missing datum to get variation in the imputed values, thus reflecting the uncertainty introduced by imputation itself. Inference was based on the five resulting datasets [25]: areas under AUROCs were obtained by averaging the five single data sets coefficients, while 95% confidence intervals were delimited by the two overall most extreme boundaries, the lowest and the highest [25]. All the coefficients, the AUROCs and the 95% C.I. boundaries were identical up to the first or second decimal for any considered variable across the five datasets.

3.2.3.2 VALIDATION AND REPLICATION PHASE.

Only the model developed for childhood obesity was used for validation because the model for prediction of childhood overweight/obesity was not considered accurate enough to be clinically useful and the models concerning adolescent phenotypes required older cohorts than Veneto and Project Viva. The NFBC1986 equation was applied to the validation cohorts after recalculation of the intercept according to the cohort–specific phenotype prevalence and mean values of predictors. In the Veneto sample, number of household members and gestational smoking were not available.

A replication analysis was also performed in which the model for childhood obesity was re–built in the two validation samples by stepwise logistic regression using the available traditional risk factors.

Statistics were performed with R 2.11.0 (www.r–project.org), SPSS.18 (IBM Company, Chicago, Illinois) and SAS 9.3 (SAS Institute, Cary, North Carolina).

3.3 RESULTS

Parental BMI, birth weight, maternal gestational weight gain, number of household members, maternal professional category and smoking habits were independent predictors of all or most of the six obesity outcomes (Table 2–3).

The equations to estimate the risk for the obesity outcomes from these traditional risk factors are represented in supporting information.

TABLE 2: Stepwise multiple logistic models for prediction of overweight phenotypes: ORs and p values associated with predictors, AUROC and P of Hosmer–Lemeshow test in the final models (bold characters) and AUROCs and P of Hosmer–Lemeshow of each step (italic characters).

	OR in the final cumulative model	P	AUROC when term is added	P of H–L test when term is added
Childhood Overweight–Obesity				
Maternal BMI	**1.13 (1.10–1.16)**	**<0.001**	*0.63 (0.60–0.65)*	*<0.001*
Paternal BMI	**1.11 (1.08–1.15)**	**<0.001**	*0.65 (0.62–0.67)*	*0.042*
Gestational weight gain	**0.88 (0.84–0.93)**	**<0.001**	*0.66 (0.64–0.68)*	*0.023*
N of household members	**1.02 (1.01–1.03)**	**<0.001**	*0.66 (0.64–0.69)*	*0.015*
Birth weight	**1.45 (1.22–1.73)**	**<0.001**	*0.67 (0.65–0.69)*	*0.29*
Maternal smoking	**1.28 (1.05–1.57)**	**0.013**	**0.67 (0.65–0.69)**	**0.46**
Adolescent Overweight–Obesity				
Maternal BMI	**1.17 (1.14–1.20)**	**<0.001**	*0.66 (0.63–0.67)*	*0.05*
Paternal BMI	**1.12 (1.09–1.15)**	**<0.001**	*0.68 (0.66–0.70)*	*0.13*
N of household members	**1.02 (1.09–1.15)**	**0.001**	*0.70 (0.68–0.72)*	*<0.001*
Gestational weight gain	**0.90 (0.86–0.95)**	**<0.001**	*0.70 (0.68–0.72)*	*<0.001*
Birth weight	**1.31 (1.12–1.53)**	**<0.001**	*0.71 (0.69–0.72)*	*0.07*
Maternal occupation	**0.75 (0.60–0.93)**	**0.009**	*0.71 (0.69–0.73)*	*0.20*
Maternal smoking	**1.28 (1.06–1.54)**	**0.009**	**0.71 (0.69–0.73)**	**0.09**
Persistent Childhood Overweight–Obesity				
Maternal BMI	**1.18 (1.14–1.22)**	**<0.001**	*0.69 (0.66–0.72)*	*0.001*
Paternal BMI	**1.14 (1.10–1.19)**	**<0.001**	*0.72 (0.69–0.75)*	*0.002*
Gestational weight gain	**1.03 (1.02–1.04)**	**<0.001**	*0.73 (0.70–0.75)*	*0.009*
N of household members	**0.88 (0.82–0.95)**	**<0.001**	*0.73 (0.71–0.75)*	*0.001*
Maternal occupation	**0.57 (0.42–0.77)**	**<0.001**	*0.74 (0.72–0.77)*	*0.01*
Birth weight	**1.41 (1.12–1.77)**	**0.003**	*0.74 (0.72–0.77)*	*0.06*
Gestational smoking	**1.45 (1.09–1.94)**	**0.11**	**0.75 (0.73–0.78)**	**0.07**

TABLE 3: Stepwise multiple logistic models for prediction of obesity phenotypes: ORs and p values associated with predictors, AUROC and P of Hosmer–Lemeshow test in the final models (bold characters) and AUROCs and P of Hosmer–Lemeshow of each step (italic characters).

	OR in the final cumulative model	P	AUROC when term is added	P of H–L test when term is added
Childhood Obesity				
Paternal BMI	**1.19 (1.13–1.27)**	**<0.001**	*0.68 (0.64–0.73)*	*0.39*
Maternal BMI	**1.13 (1.08–1.17)**	**<0.001**	*0.74 (0.70–0.78)*	*0.06*
N of household members	**0.73 (0.63–0.84)**	**<0.001**	*0.77 (0.73–0.80)*	*0.007*
Birth weight (kg)	**2.12 (1.48–3.04)**	**<0.001**	*0.77 (0.73–0.80)*	*0.47*
Maternal occupation	**0.50 (0.31–0.79)**	**0.003**	*0.77 (0.73–0.81)*	*0.57*
Gestational smoking	**1.84 (1.20–2.81)**	**0.005**	**0.78 (0.74–0.82)**	**0.52**
Adolescent Obesity				
Maternal BMI	**1.18 (1.13–1.23)**	**<0.001**	*0.67 (0.63–0.71)*	*0.13*
Paternal BMI	**1.16 (1.10–1.22)**	**<0.001**	*0.70 (0.66–0.74)*	*0.29*
N of household members	**0.83 (0.74–0.92)**	**0.001**	*0.73 (0.69–0.76)*	*0.29*
Maternal occupation	**0.47 (0.32–0.69)**	**<0.001**	*0.74 (0.71–0.78)*	*0.81*
Gestatational weight gain (%)	**1.03 (1.01–1.05)**	**0.001**	**0.75 (0.71–0.79)**	**0.69**
Persistent Childhood Obesity				
Paternal BMI	**1.23 (1.13–1.34)**	**<0.001**	*0.69 (0.61–0.76)*	*0.93*
Maternal BMI	**1.14 (1.07–1.21)**	**<0.001**	*0.81 (0.76–0.87)*	*0.32*
Birth weight	**2.30 (1.29–4.08)**	**0.005**	*0.82 (0.76–0.88)*	*0.06*
Maternal occupation	**0.31 (0.16–0.57)**	**<0.001**	*0.84 (0.79–0.89)*	*0.55*
Single parenthood	**4.27 (1.39–13.12)**	**0.011**	**0.85 (0.80–0.90)**	**0.33**

Discrimination accuracy of the risk calculation from traditional risk factors was excellent for persistent childhood obesity (AUROC = 0.85[0.80–0.90], p<0.001), clinically meaningful for persistent childhood overweight/obesity (AUROC = 0.75[0.73–0.78], p<0.001), childhood obesity (AUROC = 0.78 [0.74–0.82], p<0.001), adolescent obesity (AUROC = 0.75[0.71–0.79], p<0.001) and adolescent overweight/obesity (AUROC =

A

Paternal BMI		Maternal BMI (kg/m²) 20	25	30	35
	20	0.13	0.24	0.44	0.79
	25	0.32	0.59	1.07	1.93
	30	0.79	1.43	2.58	4.61
	35	1.93	3.46	6.12	10.62

No gestational smoking
Birth weight = 3 kg
Maternal profession = professional
Number of household members = 5

B

Paternal BMI		Maternal BMI (kg/m²) 20	25	30	35
	20	0.51	0.93	1.69	3.03
	25	1.26	2.26	4.05	7.14
	30	3.03	5.39	9.41	15.91
	35	7.14	12.29	20.34	31.76

No gestational smoking
Birth weight = 3.5 kg
Maternal profession = skilled non manual
Number of household members = 4

C

Paternal BMI		Maternal BMI (kg/m²) 20	25	30	35
	20	3.59	6.36	11.01	18.39
	25	8.39	14.31	23.33	35.66
	30	18.39	29.11	42.80	57.69
	35	35.66	50.25	64.79	77.03

History of gestational smoking
Birth weight = 4 kg
Maternal profession = skilled manual
Number of household members = 3

FIGURE 1: Estimates of risk percentages for childhood obesity for given pairs of parental BMIs according to the NFBC1986 equation. Estimates are provided for three different combinations of birth weight, maternal professional category, number of household members and maternal gestational smoking, corresponding to three progressively higher risk backgrounds. Grey cells correspond to risk estimates within the highest risk quartile in the overall population.

0.71[0.69–0.73], p<0.001), and below the threshold for clinical usefulness for childhood overweight/obesity (AUROC = 0.67[0.65–0.69], p<0.001) (Figure 1 and Table 2–3) (23). All of the six models developed from traditional risk factors were adequately calibrated (all p for Hosmer–Lemeshow test >0.05).

Parental BMI was the main contributor to discrimination accuracy while other predictors contributed moderately to the model discrimination effectiveness but increased the overall model calibration (Table 2–3).

For any given pair of parental BMIs, estimation of the probability of childhood obesity varied greatly, depending on the combination of other predictors (Figure 1).

Genetic score was an independent predictor of all of the six considered outcomes, with ORs associated with unitary score increase ranging from 1.05[1.03–1.08] to 1.09[1.03–1.14] (0.05> all P>4 × 10^{-8}) but its discrimination accuracy was poor, with AUROCs ranging from 0.56[0.54–0.58] to 0.59[0.54–0.64]. Adding the genetic score to the traditional risk factors did not produce better AUROCs than using traditional risk factors alone and was associated with modest IDIs not larger than 1%. The genetic score composed of only the twenty SNPs identified for childhood obesity traits exhibited similar associations with early obesity phenotypes. Then only the models developed from traditional risk factors were taken into consideration for further analyses. Predictive properties of the risk thresholds corresponding to the highest risk quartile for each obesity phenotype are represented in Table 4. Positive predictive values were low, due to the low prevalence of predicted conditions, while negative predictive values were high (Table 4).

The version of the NFBC1986 equation for childhood obesity lacking gestational smoking and number of household members (AUROC = 0.73[0.69–0.77] in the NFBC1986) had an AUROC = 0.70[0.63–0.77] (p<0.001) when applied to the Veneto cohort, with acceptable calibration accuracy (p for Hosmer–Lemeshow test = 0.12).

The NFBC1986 equation for childhood obesity had an acceptable AUROC = 0.73[0.67–0.80] (p<0.001) when applied to the project Viva children. However, calibration in the Project Viva sample was not satisfactory (p for Hosmer–Lemeshow test = 0.02).

The VENETO equation, i.e. the equation to predict childhood obesity issued from the Italian sample (model replication), included parental BMIs and gender, had an AUROC of 0.74[0.69–0.79] ($p<0.001$) in the Veneto sample and was adequately calibrated (p for Hosmer–Lemeshow test = 0.11).

The Project Viva equation, i.e., the equation to predict childhood obesity issued from the U.S. sample (model replication), included parental BMI, race, gestational smoking and gestational weight gain, had an AUROC of 0.79[0.73–0.84] ($p<0.001$) in the Project Viva sample and was adequately calibrated (p for Hosmer–Lemeshow test = 0.91).

The three equations predicting childhood obesity in the three studied cohorts were converted in an electronic automatic risk calculator for potential clinical use.

3.4 DISCUSSION

Our study provides the first example of predictive tool for assessing the risk of developing early obesity phenotypes, based on readily available traditional risk factors about newborns. The potential inclusion of genetic variants was explored, but due to their modest contribution to predictive accuracy, they were not included in the final models.

TABLE 4: Risk threshold and predictive properties corresponding to the 75° percentile of calculated risk for the obesity phenotypes in the NFBC1986.

	Risk threshold	Sensitivity %	Specificity %	Positive Predictive value %	Negative Predictive value %
Childhood obesity	0.036	72 [65–79]	76.5 [75–78]	9 [7–11]	99 [98.5–99.5]
Adolescent obesity	0.048	66 [59–73]	77 [75.5–78.5]	11 [9–13]	98 [97–99]
Persistent childhood obesity	0.011	79 [69–89]	75.5 [74–77]	4 [3–5]	99.5 [98–100]
Childhood overweight/ obesity	0.194	45 [37–53]	79 [77–81]	29 [26–32]	88 [97–89]
Adolescent overweight/ obesity	0.210	49 [45–53]	80 [78.5–81.5]	33 [30–36]	88.5 [87–90]
Persistent childhood overweight/obesity	0.097	63 [58–68]	78 [76.5–79.5]	21 [18–24]	96 [95–97]

Analysis of the NFBC1986 showed that traditional risk factors performed better in prediction of severe rather than mild obesity phenotypes. Importantly, the predictive accuracy of the models did not decline from childhood to adolescence, suggesting that the association between the traditional risk factors and obesity is stable until early adulthood. This is consistent with recent evidence about the relationship between single early risk factors and adolescent and adult obesity [6], [9], [10]. The risk of childhood obesity was largely driven by parental BMI. However, other predictors moderately improved the discrimination accuracy and increased the exactitude of risk estimation. They also produced large ranges of possible risk estimates for any given parental BMI, significantly improving risk classification at any level of parental BMI (Table 2–3, Figure 1).

Predictive tools need to satisfy important requisites before they can be applied in clinical settings. First, significant preventive advantages should derive from prediction. Although medical societies have been called on to provide reasonable guidance on prevention based on available data and the American Academy of Paediatrics has recently underlined the emergent need of finding effective clinical tools to enable primary care providers to contribute to obesity prevention [26]–[27], there is no compelling evidence of any efficient obesity preventive strategy involving infancy. Then, robust trials proving the effectiveness of strategies of early prevention are still needed to justify the adoption of early obesity prediction in the everyday clinical practice. Should trials prove the efficacy of preventive strategies implying special interventions going beyond paediatric counselling and public health campaigns routinely provided to the general population, a predictive tool like that proposed here would offer the important advantage to exclude a large proportion of infants from such interventions, thanks to its good negative predictive value. This would improve the cost/effectiveness ratio of preventive actions.

However few available controlled prevention trials suggest that interventions directly involving parents of pre–school children outside education settings are more effective than school or community–based interventions targeting later ages, supporting the hypothesis that involving parents in the prevention of their offspring's obesity as early as possible is likely to be a good strategy [1]. In this view, it has been suggested that «Let's Move» against child obesity campaign, which is a U.S. government–sponsored obesity prevention program targeting children aged 2–10, might be more effective if children under 2 could be identified as prevention targets [4].

Parents of newborns are particularly sensitive to information given about their child's health. Once informed of their baby's increased risk for obesity, they might be more receptive to routine advice provided from birth during the first two years of life within population–wide prevention: breastfeeding, feeding on demand, weaning no earlier than the sixth month with recommended meal patterns and food portions, avoiding of television and sugar–sweetened beverages [28]. Moreover, families of newborns at risk could be enrolled in more intensive schedules of growth monitoring and nutritional counselling than those offered to general population, in order to avoid excessive weight gain in infancy. Encouraging strategies aiming at significantly decreasing energy intake in infants should be avoided however, both because of the well known difficulties encountered by parents in doing it and because of potential, unknown harmful effects of an early caloric restriction. In contrast, recent evidence suggests that some preventive strategies prevention of obesity based on educating mothers could be useful to limit excessive infant weight gain promoting appropriate maternal responses to satiety cues and decreasing non–responsive feeding behaviours which over–ride satiety cues, such as food rewards, non food rewards to encourage infant to eat, etc… [29]. Such strategies do not imply a direct food restriction, but rather a limitation of "passive" (not hunger–driven) infant over–eating.

Obviously, even in case of proved efficacy of early obesity prevention, the targeted approach should also be carefully assessed by means of trials with a "focused intervention" design, before any dissemination of the early obesity prediction into broad clinical practice. In fact, targeted approach might also imply deleterious effects, among which, for example, stigmatization of families of infants classified as "at risk" or false reassurance of other families. Indeed, early prediction should not mean a "diagnostic" attitude towards any of the two categories of families. In particular, the assessment of age and BMI at adiposity rebound, which are good predictors of childhood and adult obesity, should be carried on in young children in order to optimize the overall detection rate of those likely to become obese and possibly sensitize families previously "missed" by the neonatal score [30].

Accuracy is another important requisite for a predictive tool. The model predicting persistent obesity had excellent accuracy (AUROC = 0.85) while the models predicting obesity and persistent overweight had clinically useful discrimination accuracy (AUROCs = 0.75 to 0.78) [23], similar to that of widely used tools for predicting multifactor medical conditions, such as the Framingham risk score for coronary heart disease (AUROC = 0.74 to 0.77 depending on gender and type of scoring adopted) [31]. Due to low prevalence of the obesity phenotypes in the NFCB1986, the fourth quartiles of predicted risk had low to moderate prevalence of cases even if they "captured" most or a high percentage of cases (low positive predictive value despite good sensitivity) (Table 4). This represents a possible drawback of preventive strategies based on risk assessment [32]. Nevertheless, risk thresholds conceived for prediction and focused prevention are not required to be "diagnostic" but rather cost–effective. Thus, the criteria we propose to select newborns at risk for obesity, could have a strong impact on public health, despite their low specificity/positive predictive value, because they could justify cost–effective preventive strategies on a subsection of the general population, similarly to several sensitive though little specific selective criteria used for widespread preventive interventions, such as: age higher than 30 years as criterion to recommend pap test against cervical cancer, age higher than 50 years as criterion to recommend the faecal occult blood test against colon cancer, etc…[33]–[34]. The adequate discrimination and calibration accuracy achieved by the equations presented in the manuscript imply that a high percentage of future obese children (more than two–thirds), is included in the highest quartile of calculated risk. Thus, using the highest risk quartile of calculated risk as selective criterion would allow focused preventive strategies to reach 70–75% of potential future cases though involving only 25% of newborns. Should these strategies have just about 50% effectiveness, the number of future obese children would have a 35–38% decrease, which would represent much greater success compared with results obtained to date by large scale preventive strategies involving later infancy and childhood [5]

The models using traditional risk factors had good calibration, which suggests that it may be possible to use the newborns' calculated risks in addition to the two risk categories. This would add precision to prediction and potential further effectiveness to related prevention

Finally, the equations we present use easily accessible information, do not incur additional costs to clinical care, and only require minimal time to calculate, if converted into simple automatic calculators like those we propose in the Supporting Information. Such electronic risk calculators could be part of an electronic medical record system and/or be housed within computer–assisted standardised programs of obesity prevention, which are promising tools for the prevention and care of paediatric obesity [35].

The results of the validation/replication analyses allow for important considerations. First of all, traditional risk factors have a good cumulative accuracy (AUROC = 0.79) in the recent U.S. paediatric cohort, which has a significantly higher prevalence of childhood obesity than the NFBC1986. This demonstrates that the environmental pressure towards obesity has not weakened the role of early risk factors. Moreover, it supports the hypothesis that, at the current phase of the obesity pandemic, the use of "familial and personal" risk factors for early prediction may be useful, in addition to population wide interventions, in those regions, like Massachusetts, where the prevalence of obesity is still moderate and characterised by ethnic and social disparities rather than influenced by country–related risk factors [36]. In these regions, focused preventive strategies based on personal risk stratification may effectively integrate large scale interventions based on nation wide characteristics [32], [36]. Interestingly, since 2010 the U.S Government has been supporting a preventive strategy against childhood obesity involving low–income children from Boston (http://www.cdc.gov/CommunitiesPuttingPreventiontoWork/communities/profiles/both–ma_boston.htm), indicating efforts towards focused prevention. Employing focused strategies involving newborns whose risk is high according to diverse factors beyond social parameters, could lead to earlier, more effective prevention of overweight/obesity in children.

The NFBC1986 equation for childhood obesity proved to keep acceptably discriminative when applied to both the validation cohorts, but showed a lost of calibration when applied to the Viva cohort, suggesting that its adoption in the U.S. would have acceptable validity to discriminate newborns at risk for early obesity but not to perform exact risk estimations. This is probably due to inconsistency of some predictors, such as maternal professional category and number of household members. Accordingly, the Project Viva equation lacks these variables while it includes race, which is not present among obesity predictors in the NFBC1986 equation, because of the high ethnical homogeneity of the NFBC1986. Inconsistency of the role of SES variables across different populations is expected and it is the main reason why it would be very difficult to build a highly accurate and calibrated score that also has complete widespread validity [36].

Overall, the validation analysis suggests that "local" equations, including parental BMI but also other locally important early predictors, may have good accuracy in predicting childhood obesity at birth, even in countries like the U.S., with high environmental pressure towards early weight excess, and should be preferred, whenever possible, to the universal adoption of the NFBC1986 equation. Interestingly, parental BMI, which partly reflects the degree of familial genetic predisposition to obesity, had very similar effect size and accuracy in the three studied cohorts, consistently with the evidence that the growing obesity epidemic has not lowered the heritability of childhood adiposity [37].

Our study also explored, with the largest list of obesity–SNPs ever used, the performance of genetics in predicting early obesity phenotypes, showing very modest predictive accuracy of the assessed genetic variants, consistently with previous evidence on adult obesity [20]. Even if a modest predictive accuracy of the studied genetic variants was expected, the accuracy estimates obtained in this study rule out, for the first time, the hypothesis that genetics may perform a little better in predicting early obesity than adult obesity, due to presumed lower impact of environmental determinants during childhood than later in life. This result is consistent with recent evidence that polygenic risk and BMI show substantially similar correlation coefficients between childhood and adulthood and further contributes to the growing evidence that common genetic variants are not yet "ready for use" for the prediction of several complex diseases, due to the still small proportion of heritability explained by the newly discovered variants [30], [38]. It is possible that next–generation sequencing techniques will reduce significantly the gap of "missing heritability" of obesity, identifying rare causative variants and clarifying the role of epigenetics by the genome–wide characterisation of DNA methylation patterns in foetuses or infants developing later obesity or not [39].

Finally, the most important evidence obtained by including currently known SNPs in our analyses is that not only common genetic variants have very low accuracy in predicting early obesity but also they produce a very little improvement of the prediction when combined with clinical factors. This is particularly important because although the notion that genetic variants have poor value in predicting common diseases is quite well established, the possible utility of including polygenic risk scoring within management strategies for complex diseases is a topical subject of current research and genetic testing services including obesity are being offered to consumers by private companies [39]–[40].

The main limitations of our manuscript are the lack of external validation for the equations predicting adolescent and persistent obesity, due to the young age of our validation cohorts and the use, in one of the validation analyses, of a retrospective paediatric cohort with some variables lacking and an age of assessment not perfectly corresponding to that of the original cohort (4–12 years versus 7 years).

The main strengths include: the novelty and the potential strong public health impact of multivariate obesity predicting tools valid for newborns; the optimization of results reliability and robustness by the adoption of several recommended methods shown recently to be lacking in several recent high impact prediction studies [41]: external geographical and temporal validation (for the model predicting childhood obesity), use of multiple imputation for missing values, avoidance of predictor dichotomisation, assessment of models calibration accuracy, avoidance of model over–fitting.

In summary, our study provides the first example of at birth prediction of early obesity by means of traditional, routinely available risk factors and should guide future efforts towards randomized trials of very early preventive approaches for identified high risk individuals to help combat the obesity epidemic.

REFERENCES

1. Waters E, de Silva–Sanigorski A, Hall BJ, Brown T, Campbell KJ, et al.. (2011) Interventions for preventing obesity in children. Cochrane Database of Systematic Reviews 12: DOI: 10.1002/14651858.
2. Stocks T, Renders CM, Bulk–Bunschoten AM, Hirasing RA, van Buuren S, et al. (2011) Body size and growth in 0– to 4–year–old children and the relation to body size in primary school age. Obes Rev 12(8): 637–52. doi: 10.1111/j.1467–789X.2011.00869.x.
3. Druet C, Stettler N, Sharp S, Simmons RK, Cooper C, et al. (2012) Prediction of childhood obesity by infancy weight gain: an individual–level meta–

analysis. Paediatr Perinat Epidemiol 26(1): 19–26. doi: 10.1111/j.1365–3016.2011.01213.x.

4. Wojcicki JM, Heyman MB (2010) Let's move – Childhood Obesity Prevention from Pregnancy and Infancy Onward. N Eng J Med 362: 1457–1459. doi: 10.1056/NEJMp1001857.
5. Summerbell CD, Waters E, Edmunds LD, Kelly S, Brown T, et al. (2011) Interventions for preventing obesity in children. Cochrane Database Syst Rev 12: CD001871.
6. Whitaker RC, Wright JA, Pepe MS, Seidel KD, et al. (1997) Predicting obesity in young adulthood from childhood and parental obesity. N Engl J Med 337: 869. doi: 10.1056/NEJM199709253371301.
7. Yu ZB, Han SP, Zhu GZ, Zhu C, Wang XJ, et al. (2011) Birth weight and subsequent risk of obesity: a systematic review and meta–analysis. Obes Rev 12(7): 525–42. doi: 10.1111/j.1467–789X.2011.00867.x.
8. Ino T (2010) Maternal smoking during pregnancy and offspring obesity: meta–analysis. Pediatr Int 52(1): 94–9. doi: 10.1111/j.1442–200X.2009.02883.x.
9. Plachta–Danielzik S, Landsberg B, Johannsen M, Lange D, Müller MJ (2010) Determinants of the prevalence and incidence of overweight in children and adolescents. Public Health Nutr 13(11): 1870–81. doi: 10.1017/S1368980010000583.
10. Mamun AA, O'Callaghan M, Callaway L, Williams G, Najman J, et al. (2009) Associations of gestational weight gain with body mass index and blood pressure at 21 years of age: evidence from a birth cohort study. Circulation 119: 1720. doi: 10.1161/CIRCULATIONAHA.108.813436.
11. Smith GD, Steer C, Leary S, Ness A (2007) Is there an intrauterine influence on obesity? Evidence from parent–child associations in the Avon Longitudinal Study of parents and children (ALSPAC). Arch Dis Child 92: 876. doi: 10.1136/adc.2006.104869.
12. Benzinou M, Creemers JWM, Choquet H, Lobbens S, Dina C, et al. (2008) Common nonsynonymous variants in PCSK1 confer risk of obesity. Nat Genet 40: 943–945. doi: 10.1038/ng.177.
13. Dina C, Meyre D, Gallina S, Durand E, Körner A, et al. (2007) Variation in FTO contributes to childhood obesity and severe adult obesity. Nat Genet 39(6): 724–6. doi: 10.1038/ng2048.
14. Loos RJF, Lindgren CM, Li S, Wheeler E, Zhao JH, et al. (2008) Common variants near MC4R are associated with fat mass, weight and risk of obesity. Nat Genet 40: 768–775. doi: 10.1038/ng.140.
15. Willer CJ, Speliotes EK, Loos RJF, Li S, Lindgren CM, et al. (2009) Six new loci associated with body mass index highlight a neuronal influence on body weight regulation. Nat Genet 41: 25–34. doi: 10.1038/ng.287.
16. Rhorleifsson G, Walters GB, Gudbjartsson DF, Steinthorsdottir V, Sulem P, et al. (2009) Genome–wide association yields new sequence variants at seven loci that associate with measures of obesity. Nat Genet 41: 18–24. doi: 10.1038/ng.274.
17. Meyre D, Delplanque J, Chèvre J–C, Locoeur C, Lobbens S, et al. (2009) Genome–wide association study for early–onset and morbid adult obesity identifies three new risk loci in European populations. Nat Genet 41: 157–159. doi: 10.1038/ng.301.
18. Liu YJ, Liu XG, Wang L, Dina C, Yan H, et al. (2008) Genome–wide association scans identified CTNNBLI as a novel gene for obesity. Hum Mol Genet 17: 1803–1813. doi: 10.1093/hmg/ddn072.

19. Scherag A, Dina C, Hinney A, Vatin V, Scherag S, et al.. (2010) Two new loci for body–weight regulation identified in a joint analysis of genome–wide association studies for early onset extreme obesity in French and German study groups. Plos Genet e1000916.
20. Speliotes EK, Willer CJ, Berndt SI, Monda K, Thorleifsson G, et al. (2010) Association analyses of 249,796 individuals reveal 18 new loci associated with body mass index. Nature Genet 42(11): 937–48. doi: 10.1038/ng.686.
21. Cole TJ, Bellizzi MC, Flegal KM, Dietz WH (2000) Establishing a standard definition for child overweight and obesity worldwide : international survey. BMJ 320: 1–6. doi: 10.1136/bmj.320.7244.1240.
22. Maffeis C, Shutz Y, Piccoli R, Gonfiantini E, Pinelli L (1993) Prevalence of obesity in children in north–east Italy. Int J Obesity 17: 287–294.
23. Hosmer DW, Lemeshow S (2000) Applied logistic regression. Wiley–Interscience Publication. II EDITION.
24. Pencina MJ, D'Agostino RB, D'Agostino RB, Vasan RS (2008) Evaluating the added predictive ability of a new marker: From area under the ROC curve to reclassification and beyond. Statistics in Medicine 27: 157–172. doi: 10.1002/sim.2929.
25. Langkamp DL, Lehman A, Lemeshow S (2010) Techniques for Handling Missing Data in Secondary Analyses of Large Surveys. Acad Pediatr 10(3) 205–10. doi: 10.1016/j.acap.2010.01.005.
26. Steering Committee on Quality Improvement and Management (2008) Towards Transparent Clinical Policies. Pediatrics 121: 643–646. doi: 10.1542/peds.2007–3624.
27. Haemer M, Cluett S, Hassink SG, Liu L, Mangarelli C, et al.. (2011) Building Capacity for Childhood Obesity Prevention and Treatment: Call to Action. Pediatrics 128: S 71.
28. US Department of Health and Human Services, US Department of Agriculture (2005) Dietary Guidelines for Americans, 2005. 6th ed. Washington, DC: Government Printing Office.
29. Daniels LA, Mallan KM, Battistutta D, Nicholson JM, Perry R, et al. (2012) Evaluation of an intervention to promote protective infant feeding practices to prevent childhood obesity: outcomes of the NOURISH RCT at 14 months of age and 6 months post the first of two intervention modules. IJO 36(10): 1292–8. doi: 10.1038/ijo.2012.96.
30. Belsky DW, Moffitt TE, Houts R, Bennett GG, Biddle AK, et al. (2012) Polygenic Risk, Rapid Childhood Growth, and the Development of Obesity. Arch Pediatr Adolesc Med. 166(6) 515–521.
31. Wilson PWF, D'Agostino RB, Levy D, Belanger AM, Silbershatz H, et al. (1998) Prediction of Coronary Heart Disease Using Risk Factor Categories. Circulation 97: 1837–1847. doi: 10.1161/01.CIR.97.18.1837.
32. Rose G (1985) Sick individuals and sick populations. International Journal of Epidemiology 14: 32. doi: 10.1093/ije/14.1.32.
33. Saslow D, Runowicz CD, Solomon D, Killackey M, Kulasingam SL, et al. (2002) American Cancer Society. American Cancer Society guideline for the early detection of cervical neoplasia and cancer. CA Cancer J Clin 52(6): 342–362. doi: 10.3322/canjclin.52.6.342.

34. Levin B, Lieberman DA, McFarland B, Smith RA, Brooks D, et al. (2008) Screening and Surveillance for the Early Detection of Colorectal Cancer and Adenomatous Polyps, 2008: A Joint Guideline from the American Cancer Society, the US Multi–Society Task Force on Colorectal Cancer, and the American College of Radiology. CA Cancer J Clin 58: 130–160. doi: 10.3322/CA.2007.0018.
35. Rattay KT, Ramakrishnan M, Atkinson A, Gilson M, Drayton V (2009) Use of an Electronic Medical Record System to Support Primary Care Recommendations to Prevent, Identify, and Manage Childhood Obesity. Pediatrics 123: S100–S107. doi: 10.1542/peds.2008–1755J.
36. Bethell C, Read D, Goodman E, Johnson J, Besl J, et al. (2009) Consistently Inconsistent: A Snapshot Across and Within State Disparities in the Prevalence of Childhood Overweight and Obesity. Pediatrics 123: S277–S286. doi: 10.1542/peds.2008–2780F.
37. Wardle J, Carnell S, Haworth CM, Plomin R (2008) Evidence for a strong genetic influence on childhood adiposity despite the force of the obesogenic environment. Am J Clin Nutr 87: 398–404.
38. Kraft P, Hunter DJ (2009) Genetic risk prediction – Are we there yet? N Engl J Med 360: 1701. doi: 10.1056/NEJMp0810107.
39. Manco M, Dallapiccola B (2012) Genetics of pediatric obesity. Pediatrics 130(1): 123–33. doi: 10.1542/peds.2011–2717.
40. Waxler JL, O'Brien KE, Delahanty LM, Meigs JB, Florez JC, et al.. (2012) Genetic Counseling as a Tool for Type 2 Diabetes Prevention: A Genetic Counseling Framework for Common Polygenetic Disorders. J Genet Couns [Epub ahead of print]
41. Bouwmeester W, Zuithoff NP, Mallett S, Geerlings MI, Vergouwe Y, et al. (2012) Reporting and methods in clinical prediction research: a systematic review. PLoS Med 9(5): e1001221. doi: 10.1371/journal.pmed.1001221.

CHAPTER 4

NATIONAL TRENDS IN BEVERAGE CONSUMPTION IN CHILDREN FROM BIRTH TO 5 YEARS: ANALYSIS OF NHANES ACROSS THREE DECADES

VICTOR L. FULGONI and ERIN E. QUANN

4.1 INTRODUCTION

A child's first 5 years of life is characterized by rapid growth and development and is the period during which food preferences and behaviors develop that often serve as the foundational basis for future eating habits [1,2]. Beverage consumption patterns developed during the early years can have lasting implications throughout childhood and adolescence and into adult years. The American Academy of Pediatrics (AAP) provides recommendations for the introduction of beverages to children's diets. They discourage fruit juice intake before the infant is at least 4–6 months of age and fruit drinks, sports drinks and energy drinks are generally discouraged all together [3–5]. The introduction of cow's milk is recommended after 1 year of age and low–fat and fat–free varieties are encouraged once children turn 2 years old [3]. The 2010 Dietary Guidelines for Americans acknowledge the important link between childhood and adulthood beverage consumption by reinforcing the important habit of drinking milk during childhood because "those who consume milk at an early age are more likely to do so as adults" [6].

In the nationwide sample of infants and toddlers aged 4–24 months who participated in the Feeding Infants and Toddlers Study, beverages provided 36% of the total daily energy [7]. According to 2005–2006 National Health and Nutrition Examination Survey (NHANES) data, total beverage intake among children ages 2–6 years accounts for an average of

367 kcal/day [8]. Although milk intake accounted for the greatest amount of fluids consumed in this young age group (331 ml/day, on average), sodas and fruit drinks provided 206 ml of average daily fluids and by the time children were 7 years of age and older, sodas accounted for the most fluid intake, aside from water, while milk intake continued to decline [8]. The type of beverages consumed can have a significant impact on diet quality and nutrient adequacy.

Lower milk intake makes it more difficult for young children to meet daily milk and milk product group and nutrient intake recommendations, and it could have negative impacts on development since milk and the nutrients it provides are associated with growth and bone health in preschool children [9,10]. Health professionals are concerned that high calorie, high sugar, nutrient poor beverages like soft drinks and fruit–flavored drinks displace nutrient rich beverages such as milk [3]. Sugar–sweetened beverage consumption, primarily referring to carbonated beverages and fruit drinks, begins during the preschool years and generally increases with age [11–13]. The consumption of milk by preschoolers has been positively correlated with diet quality and meeting nutrient recommendations, including vitamin A, folate, vitamin B12, calcium, and magnesium, whereas sugar–sweetened beverage consumption, including soda and fruit drinks, by young children has been linked to reduced milk intake and negatively correlated with achieving intakes of vitamin A, vitamin C, and calcium [14–16].

Research indicates a link between beverage consumption and body weight as well. Childhood overweight and obesity has increased at alarming rates in the United States. Even among children aged 2–5 years of age, obesity defined as body mass index (BMI) greater than the 95th percentile for age, increased from 5 to 10.4% between 1976–1980 and 2007–2008 [17]. There is evidence that a higher intake of sugar–sweetened beverages, such as soda, is associated with increased energy intake [18] and body weight and/or adiposity in children [6,18]. Several studies have observed this phenomenon in preschool children [12,19,20].

In light of the importance of beverage consumption to diet quality and obesity among preschoolers, the purpose of this study was to identify beverage trends in young children over the past three decades. Particular emphasis was placed on evaluating the role of milk and its possible displacement with the rising consumption of empty calorie beverages at this early age.

4.2 METHODS

Our analysis utilized beverage intake data available from NHANES 1976–1980, 1988–1994, and 2000–2006 for children less than 1 year of age to 5 years of age. NHANES is an ongoing cross–sectional nationally representative health and nutrition examination survey of a stratified, multistage probability sample of the civilian, non–institutionalized U.S. population [21]. Beverage consumption data was based on 24–hour dietary recalls collected in each NHANES cycle in which proxy respondents reported on all foods and beverages consumed by the children in a 24–hour period. While multiple days of recall were obtained in some NHANES surveys, only one day of recall was used for each subject to allow comparison over time. Participants whose records were incomplete, unreliable, or were being breastfed were excluded. Data from NHANES 1976–1980 included only children 6 months and older, whereas the more recent NHANES included children from birth.

Beverage categories were identified as: milk [total as well as white, flavored (sugar–sweetened), whole, 2%, 1%, and fat–free separately], 100% fruit juice, fruit drinks, soft drinks, tea and soy beverages. White milk included all non–flavored fluid milk consumed as beverages as well as milk consumed with other foods such as cereal and macaroni and cheese. Flavored milk was defined as flavored fluid milk purchased as such (i.e., chocolate, strawberry) or milk prepared by the addition of flavored syrups or powder. Fruit drinks were beverages with less than 100% juice; most contained added sugars. Soft drinks included both "regular" and "diet" sodas. The percentage of the population that consumed each beverage category as well as the amounts consumed by those children was determined by age. The mean intake and percent of total intake of calories, total fat, saturated fat, protein, calcium, phosphorus, magnesium (for NHANES 1988–1994 and 2001–2006 only), and potassium were determined for each beverage category using the respective nutrient content information from each NHANES study period. NHANES 1976–1980 did not measure magnesium intake.

All analyses of percentages and/or means with standard errors were conducted using SAS version 9.2/SUDAAN version 10.0 (SAS Institute, Cary, NC/RTI, Research Triangle Park, NC) to adjust for the complex sampling design of each NHANES survey. Pair–wise comparisons between NHANES 1976–1980 and NHANES 1988–1994, 10–14% NHANES 1988–1994 and NHANES 2001–2006 and NHANES 1976–1980 and NHANES 2001–2006 were performed on mean percents and means using z scores to identify significant differences.

4.3 RESULTS

A total sample of 3,998 children <1–5 years of age was utilized from NHANES 1976–1980, 6,871 from NHANES 1988–1994 and 4,430 from NHANES 2001–2006.

4.3.1 BEVERAGE CHOICE (PERCENTAGE USE)

Across all three decades, milk was the beverage consumed by most young children (Figure 1). There was a significant decrease ($p<0.001$) in milk consumption in the most recent NHANES surveys compared to the previous surveys. During the NHANES 1976–1980 and 1988–1994 periods, approximately 84–85% of children in this age group were consuming milk, whereas only 77% were consuming milk during NHANES 2001–2006. Flavored milk intake was relatively low during NHANES 1976–1980 and NHANES 1988–1994, but increased to 14% during the last decade ($p<0.001$). Fruit juice consumption increased dramatically in this age group during NHANES 2001–2006 to more than 50% of the population compared to about 30% in the older surveys ($p<0.001$). No significant changes were observed in fruit drink intake across all three decades. On average, 35–37% of this population consumed fruit drinks. In the case of soft drinks, at least 30% of children consumed this drink on any given day over the past 30 years. During the NHANES 1988–1994 period, there

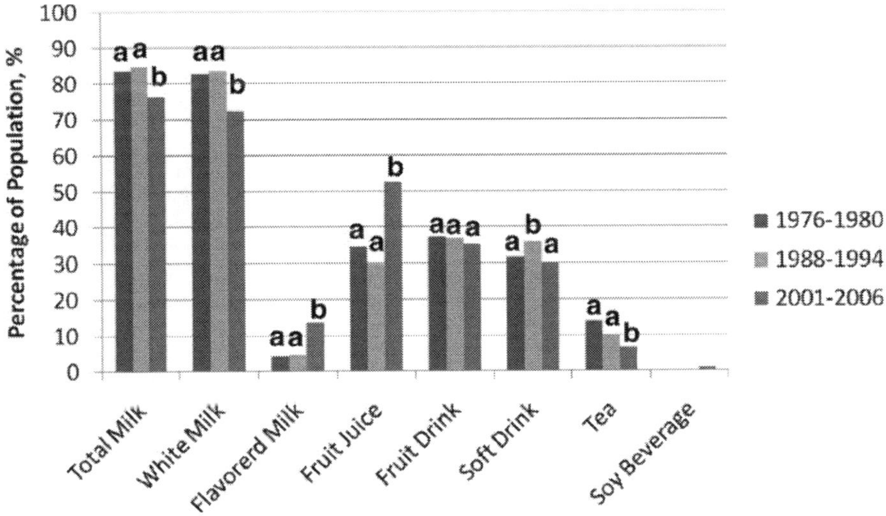

FIGURE 1: Percentage of children <1–5 years of age consuming different beverages during NHANES 1976–1980, NHANES 1988–1994, and NHANES 2001–2006. NHANES – National Health and Nutrition Examination Survey. Columns with different letters within beverage category are significantly different ($p<0.01$).

was a significant increase to 36% ($p<0.001$) which leveled back to 30% in NHANES 2001–2006. Percent of children consuming tea significantly decreased from 10–14% from previous decades to 7% during NHANES 2001–2006. Consumption of soy beverages were not recorded in the earlier surveys and intake was less than 1% in NHANES 2001–2006. In the remaining analyses, tea and soy beverages were excluded because less than 10% of young children consumed these beverages.

Age differences in beverage trends across decades were assessed in relation to total milk, flavored milk, fruit juice, fruit drink, and soft drinks consumption (Table 1). Among children less than 1 year old, there was an almost 11% increase in milk consumption in NHANES 2001–2006 compared to NHANES 1988–1994. From the age of 1–5 years, the proportion of children that consumed milk was similar for the earlier two decades, whereas during NHANES 2001–2006 there was a tendency for fewer children to consume milk after the age of 3 years. For children 1–3

TABLE 1: Percent of children<1–5 years that consumed each beverage in NHANES 1976–1980, NHANES 1988–1994, and NHANES 2001–2006, categorized by age

Beverage	NHANES 1976–1980* (A)	NHANES 1988–1994 (B)	NHANES 2001–2006 (C)	Change (A–B)	p–value (A–B)	Change (B–C)	p–value (B–C)	Change (A–C)	p–value (A–C)
Total Milk									
<1 year	N/A	3.1 ± 0.7[1]	13.8 ± 1.3	N/A	N/A	10.7	<0.0001	N/A	
1 year	86.0 ± 1.8	90.9 ± 0.9	87.0 ± 1.4	4.8	0.0171	–3.8	0.0224	1.0	0.6610
2 years	82.9 ± 1.6	89.2 ± 1.1	88.3 ± 1.4	6.3	0.0012	–0.9	0.6132	5.4	0.0111
3 years	84.1 ± 1.3	89.4 ± 1.5	83.9 ± 1.8	5.3	0.0076	–5.5	0.0189	–0.2	0.9282
4 years	86.7 ± 1.2	86.5 ± 1.8	83.0 ± 1.4	–0.2	0.9263	–3.5	0.1248	–3.7	0.0448
5 years	86.2 ± 1.6	90.9 ± 1.3	84.0 ± 2.0	4.7	0.0226	–6.9	0.0038	–2.2	0.3904
Flavored Milk									
<1 year	N/A	0.1 ± 0.1	1.2 ± 0.5	N/A	N/A	1.1	0.0310	N/A	
1 year	1.4 ± 0.3	1.8 ± 0.6	6.1 ± 0.9	0.4	0.5510	4.3	0.0001	4.7	<0.0001
2 years	3.8 ± 0.8	2.2 ± 0.6	13.9 ± 0.9	–1.6	0.1096	11.7	<0.0001	10.1	<0.0001
3 years	3.6 ± 0.6	4.6 ± 1.1	15.7 ± 2.0	1.0	0.4248	11.1	<0.0001	12.1	<0.0001
4 years	5.4 ± 1.1	3.2 ± 0.8	19.8 ± 2.2	–2.2	0.1058	16.6	<0.0001	14.4	<0.0001
5 years	8.4 ± 1.1	12.3 ± 1.9	22.2 ± 2.4	3.9	0.0757	9.9	0.0012	13.8	<0.0001
100% Fruit Juice									
<1 year	N/A	31.5 ± 2.4	38.5 ± 2.2	N/A	N/A	7.0	0.0316	N/A	
1 year	36.0 ± 2.3	41.4 ± 2.3	63.3 ± 2.2	5.4	0.0969	21.9	<0.0001	27.3	<0.0001
2 years	36.1 ± 1.9	36.3 ± 2.3	61.3 ± 2.4	0.2	0.9465	25.0	<0.0001	25.2	<0.0001
3 years	32.0 ± 2.2	24.6 ± 1.9	54.7 ± 2.7	–7.4	0.0109	30.1	<0.0001	22.7	<0.0001

TABLE 1: *cont.*

Beverage	NHANES 1976–1980* (A)	NHANES 1988–1994 (B)	NHANES 2001–2006 (C)	Change (A–B)	p–value (A–B)	Change (B–C)	p–value (B–C)	Change (A–C)	p–value (A–C)
4 years	35.1 ± 2.1	25.8 ± 3.2	50.9 ± 2.7	–9.3	0.0151	25.1	<0.0001	15.8	<0.0001
5 years	32.1 ± 2.4	19.9 ± 2.1	41.7 ± 3.6	–12.2	0.0001	21.8	<0.0001	9.6	0.0265
Fruit Drink									
<1 year	N/A	2.9 ± 0.6	7.6 = 1.0	N/A	N/A	4.7	0.0001	N/A	
1 year	33.4 ± 2.7	29.7 ± 1.7	27.2 ± 1.4	–3.7	0.2462	–2.5	0.2563	–6.2	0.0415
2 years	40.5 ± 2.5	36.5 ± 2.0	35.5 ± 2.2	–4.0	0.2115	–1.0	0.7366	–5.0	0.1348
3 years	41.2 ± 2.2	37.8 ± 2.6	38.7 ± 2.4	–3.4	0.3181	0.9	0.7992	–2.5	0.4426
4 years	40.9 ± 2.1	46.5 ± 2.7	42.3 ± 3.0	5.6	0.1016	4.2	0.2981	1.4	0.7022
5 years	41.8 ± 2.7	44.1 ± 2.8	51.7 ± 2.7	2.3	0.5543	7.6	0.0507	9.9	0.0095
Soft Drinks									
<1 year	N/A	0.6 ± 0.3	2.4 ± 0.6	N/A	N/A	1.8	0.0073	N/A	
1 year	25.9 ± 2.4	22.7 ± 1.5	18.8 ± 1.7	–3.2	0.2582	–3.9	0.0854	–7.1	0.0158
2 years	34.2 ± 2.2	37.1 ± 2.1	28.9 ± 1.9	2.9	0.3403	–8.2	0.0038	–5.3	0.0683
3 years	37.4 ± 2.2	40.3 ± 3.2	38.3 ± 2.7	2.9	0.4552	–2.0	0.6329	0.9	0.7961
4 years	37.6 ± 1.4	43.0 ± 2.7	41.2 ± 2.8	5.4	0.0758	–1.8	0.6435	3.6	0.2502
5 years	35.1 ± 1.8	46.3 ± 2.2	44.7 ± 2.5	11.2	0.0001	–1.6	0.6309	9.6	0.0018

*Data only available for children 6 months or older; N/A: not available.

1Mean ± standard error; data represent percentage of children consuming any amount of described beverages in the 24–hr recall.

and 5 years, there was a significant increase ($p<0.05$) in the percent of children who were consuming milk during NHANES 1988–1994 compared to NHANES 1976–1980 for nearly all ages. The trend was reversed for NHANES 1988–1994 versus NHANES 2001–2006, where fewer 1, 3 and 5 year olds were consuming milk. Age–related patterns for flavored milk consumption did not differ for the earlier two surveys but showed a significant increase ($p<0.05$) ranging from about 1% to 17%, depending on age, during NHANES 2001–2006 compared to NHANES 1988–1994.

There was a small, but significant decline ($p<0.05$) in the percent of 3–5 year olds who consumed fruit juice during NHANES 1988–1994 compared to NHANES 1976–1980. In contrast, there was a significant increase ($p<0.05$) in the percent of all children from 1–5 years of age who consumed fruit juice in NHANES 2001–2006. There was a 22–30 percentage unit increase in fruit juice consumption in children 1 year and older. Fruit drink intake on the other hand, was not significantly different between the three decades with the exception of an increase in those less than 1 year of age. It was particularly evident in NHANES 1988–1994 and NHANES 2001–2006 that a greater percent of children consume soft drinks with increasing age. For example, in the most recent survey, approximately 19% consume soft drinks at 1 year of age, 29% at 2 years, 38% at 3 years, 41% at 4 years, and 45% at 5 years. Between NHANES 1976–1980 and NHANES 1988–1994, no clear age differences were present with the exception that there was an 11 percentage unit increase ($p<0.001$) in soft drinks consumption among 5 year olds in NHANES 1988–1994. Similarly, no significant age differences were found when NHANES 1988–1994 and NHANES 2001–2006 were compared, except for an increase ($p<0.05$) in children less than 1 year and a significant decrease in 2 year olds ($p<0.05$).

4.3.2 BEVERAGE AMOUNTS

Overall for the group, there was a small decrease ($p<0.05$) in the amount of milk consumed between NHANES 1976–1980 and NHANES 1988–1994 and an increase ($p<0.001$) between NHANES 1988–1994 and NHANES 2001–2006. Among 1 year olds there was a small increase ($p<0.05$) in the mean amount consumed between NHANES 1976–1980 and NHANES

1988–1994, no significant change in 2 and 3 year olds, and a small to moderate decline in 4 ($p<0.05$) and 5 ($p<0.001$) year olds. In contrast there was a significant increase in the quantity of milk drank by <1–5 year olds when NHANES 2001–2006 was compared to NHANES 1988–1994. Milk consumption increased from 17.3 to 20.3 fl oz ($p<0.001$) in 1 year olds, 13.1 to 16.6 fl oz ($p<0.001$) in 2 year olds, 13.3 to 14.9 fl oz ($p<0.05$) in 3 year olds, and 13.0 to 15.0 fl oz ($p<0.05$) in 5 year olds. No significant changes were observed for 4 year olds. In the most recent survey, 2 and 3 year olds were consuming about 2 cups of milk a day and 4 and 5 year olds were consuming a little less than 2 cups a day.

Fruit juice consumption increased ($p<0.001$) in NHANES 1988–1994 compared to NHANES 1976–1980 and then remained stable through NHANES 2001–2006 for the entire sample. Children 1–3 years of age consumed significantly more fruit juice in NHANES 1988–1994 than in NHANES 1976–1980. Among 1, 2, and 3 year olds, mean fruit juice intake increased from 6.6 to 9.3 fl oz ($p<0.05$), 8.3 to 11.1 fl oz ($p<0.001$), and 7.5 to 10.6 fl oz ($p<0.05$), respectively. Changes were not significantly different for 4 and 5 year olds. The change in amounts consumed from NHANES 1988–1994 to 2001–2006 were not significantly different for any of the ages tested. During the NHANES 2001–2006 period, children less than 1 year were consuming 5.4 fl oz, children 1–2 years were consuming about 11 fl oz, children 3 years were consuming 12 fl oz, and children 4–5 years were consuming about 10 fl oz on a daily basis.

The amount of fruit drink ingested did not change significantly for the entire group during the first two decades, but significantly increased ($p<0.05$) during the most recent one. None of the age groups demonstrated significant changes in the amount of fruit drink consumed between NHANES 1976–1980 and NHANES 1988–1994. Comparison of NHANES 1988–1994 to NHANES 2001–2006 indicated significant increases from 9.0 to 12.3 fl oz ($p<0.05$) and 10.0 to 12.8 fl oz ($p<0.05$) for 1 and 2 year olds, respectively. No changes between these surveys were observed for 3–5 year olds. The NHANES 2001–2006 survey found that children less than 1 year consumed 8.8 fl oz of fruit drink, and children 1–5 years consumed about 10–13 fl oz/ day.

Analysis of the group as a whole revealed a small increase ($p<0.05$) in the amount of soft drinks consumed between NHANES 1976–1980 and NHANES 1988–1994 as there were trends towards an increase among 3

TABLE 2: Energy and macronutrient intake contributed by milk, fruit juice, fruit drink, and soft drinks in the diets of children<1–5 years participating in NHANES 1976–1980, NHANES 1988–1994, and NHANES 2001–2006

Nutrient/ Survey	Milk		100% Fruit Juice		Fruit Drink		Soft Drinks	
	Mean ± S.E.	% Daily Intake	Mean ± S.E.	% Daily Intake	Mean ± S.E.	% Daily Intake	Mean ± S.E.	% Daily Intake
Calories (kcal/d)								
NHANES 1976–1980	281 ± 5	20.8[a]	114 ± 3	8.4[a]	128 ± 4	9.0[a]	82 ± 3	5.9[a]
NHANES 1988–1994	234 ± 4	16.8[b]	143 ± 6	10.6[b]	129 ± 4	8.7[a]	86 ± 3	5.9[a]
NHANES 2001–2006	279 ± 6	18.7[c]	149 ± 5	9.8[c]	150 ± 5	9.4[b]	85 ± 4	5.3[a]
Total Fat (g/d)								
NHANES 1976–1980	13.5 ± 0.2	25.8[a]	0.2 ± 0.0	0.6[a]	0.0 ± 0.0	0.0	0.0 ± 0.0	0.0
NHANES 1988–1994	10.9 ± 0.2	21.4[b]	0.3 ± 0.0	0.8[b]	0.0 ± 0.0	0.1	0.0 ± 0.0	0.0
NHANES 2001–2006	12.0 ± 0.3	22.5[b]	0.4 ± 0.0	0.7[ab]	0.1 ± 0.1	0.2	0.0 ± 0.0	0.0
Saturated Fat (g/d)								
NHANES 1976–1980	7.5 ± 0.1	26.3[a]	0.0 ± 0.0	0	0.0 ± 0.0	0.0	0.0 ± 0.0	0.0
NHANES 1988–1994	6.8 ± 0.2	32.0[b]	0.0 ± 0.0	0.3	0.0 ± 0.0	0.1	0.0 ± 0.0	0.0
NHANES 2001–2006	7.2 ± 0. 2	32.8[b]	0.1 ± 0.0	0.4	0.0 ± 0.0	0.0	0.0 ± 0.0	0.0
Protein (g/d)								
NHANES 1976–1980	16.3 ± 0.3	31.4[a]	1.3 ± 0.1	2.9[a]	0.0 ± 0.0	0.0	0.0 ± 0.0	0.0
NHANES 1988–1994	13.8 ± 0.2	26.5[b]	0.5 ± 0.0	1.2[b]	0.1 ± 0.0	0.2	0.0 ± 0.0	0.0
NHANES 2001–2006	15.9 ± 0.4	29.1[c]	± 0.0	2.0[c]	0.1 ± 0.0	0.3	0.1 ± 0.0	0.2

For each nutrient, different subscripts down a column (a, b, c) indicate statistically significant differences (p<0.05).

and 5 years olds. No significant changes in the amounts of soft drinks consumed were noted between NHANES 1988–1994 and NHANES 2001–2006. During the most recent NHANES analysis, children less than 1 year, 1 year, 2 years, 3 years, 4 years, and 5 years consumed 1.8, 5.4, 6.2, 7.6, 8.8 and 9.8 fl oz, respectively (p for trend < 0.0001). Increasing age was associated with increased amounts of soft drinks consumed.

4.3.3 BEVERAGE CALORIE AND NUTRIENT CONTRIBUTIONS

Among the four main beverages consumed by children <1–5 years, milk was the largest daily beverage calorie contributor in all three decades surveyed, although there was a significant reduction in the last two surveys compared to NHANES 1976–1980 (Table 2). These calorie contributions include milk used on cereal or in food mixtures too. Milk was the primary contributor for all the macronutrients (Table 2) as well as calcium, phosphorus, magnesium, and potassium (Table 3). About 52%, 55%, and 62% of the daily calcium intake were provided by milk in NHANES 2001–2006, NHANES 1988–1994, and NHANES 1976–1980, respectively. The reduced calcium intake from milk in the more recent surveys is consistent with reduced milk calories. Phosphorus intake from milk ranged from 37–42% across the three decades. Magnesium intake from milk was in the range of 27–28% in the last two surveys. Potassium intake from milk ranged from 31–37%.

Percent daily caloric contribution of fruit juice and fruit drink was similar across all three surveys, in the range of 8–11%. (Table 2) Fruit juice was an important provider (16–19%) of potassium in the three surveys and magnesium (11%) in the most recent survey (Table 3). Fruit drinks provided 5% or less of the daily intake of calcium, phosphorus, magnesium, or potassium in all three surveys (Table 3). The caloric contribution of soft drinks in the range of 5–6% did not significantly differ among the three surveys (Table 2). Soft drinks provided very little of the nutrients evaluated (Table 3). The combined calorie contribution from fruit juice, fruit drinks and soft drinks over this thirty year period has increased, but given the parallel rise in total calorie intake the proportion from these beverages has remained relatively consistent at about a quarter of total daily calorie intake (1976–1980: 324 kcals, 23% of total kcal intake; 1988–1994: 358 kcals, 25% of total kcal intake; 2001–2006: 384 kcals, 25% of total kcal intake; data not shown).

Use of different types of milks (Figures 2 and 3) and eating occasion (data not shown) for milk were evaluated in NHANES 1988–94 and NHANES 2001–2006. In both surveys, whole milk was the predominant milk introduced at less than 1 year of age and consumed at 1 year of age.

TABLE 4: Calcium, phosphorus, magnesium, and potassium intake contributed by milk, fruit juice, fruit drink, and soft drinks in the diets of children<1–5 years participating in NHANES 1976–1980, NHANES 1988–1994, and NHANES 2001–2006

Nutrient/ Survey	Milk		100% Fruit Juice		Fruit Drink		Soft Drinks	
	Mean ± S.E.	% Daily Intake	Mean ± S.E.	% Daily Intake	Mean ± S.E.	% Daily Intake	Mean ± S.E.	% Daily Intake
Calories (mg/d)								
NHANES 1976–1980	553 ± 10	61.7[a]	20.6 ± 0.7	3.7[a]	29.1 ± 2.2	4.7[a]	0.0 ± 0.0	0.0[a]
NHANES 1988–1994	504 ± 8	54.5[b]	21.8 ± 1.0	3.5[a]	31.5 ± 1.5	4.9[a]	6.9 ± 0.0	1.3[b]
NHANES 2001–2006	565 ± 13	52.3[c]	55.2 ± 4.0	5.7[b]	34.7 ± 1.9	4.9[a]	5.8 ± 0.3	1.0[b]
Phosphorus (mg/d)								
NHANES 1976–1980	435 ± 8	41.8[a]	32.8 ± 1.2	4.0[a]	42.9 ± 3.0	4.7[a]	11.0 ± 1.4	1.5[a]
NHANES 1988–1994	396 ± 7	37.1[b]	22.0 ± 0.9	2.8[b]	26.6 ± 1.8	3.0[b]	17.1 ± 0.9	2.0[b]
NHANES 2001–2006	456 ± 10	39.7[c]	31.6 ± 1.1	3.2[bc]	11.5 ± 0.7	1.5[c]	11.8 ± 0.9	1.8[b]
Magnesium (mg/d)								
NHANES 1976–1980	N/A	N/A	N/A	N/A	N/A	N/A	N/A	N/A
NHANES 1988–1994	56.2 ± 0.9	27.8[a]	12.9 ± 0.6	7.1[a]	5.2 ± 0.2	3.1[a]	2.3 ± 0.1	1.5[a]
NHANES 2001–2006	53.7 ± 1.3	27.4[a]	22.0 ± 0.8	11.1[b]	7.6 ± 0.3	4.9[b]	1.7 ± 0.1	1.1[b]
Potassium (mg/d)								
NHANES 1976–1980	679 ± 13	36.9[a]	386 ± 14	18.8[a]	26.2 ± 3.4	1.7[a]	0.0 ± 0.0	0.0[a]
NHANES 1988–1994	645 ± 11	31.4[b]	332 ± 14	16.1[b]	43.8 ± 2.2	2.7[b]	2.6 ± 0.3	0.2[a]
NHANES 2001–2006	727 ± 17	33.5[c]	436 ± 14	19.2[a]	74.9 ± 3.0	4.5[c]	4.0 ± 03	0.3[a]

For each nutrient, different subscripts (a, b) for each survey indicate statistically significant differences (p<0.05).

Whole milk intake declined with increasing age in NHANES 2001–2006 compared to NHANES 1988–1994. Two percent or reduced–fat milk was consumed in almost equivalent amounts by age 5 in the most recent survey. One–percent milk and skim milk were consumed in very low amounts for all the years surveyed. Breakfast was the eating occasion where milk was most consumed in both NHANES 1988–1994 and NHANES 2001–2006. Snack time was the next meal occasion in which milk was consumed in both surveys. Slightly less milk was consumed at lunch or dinner than

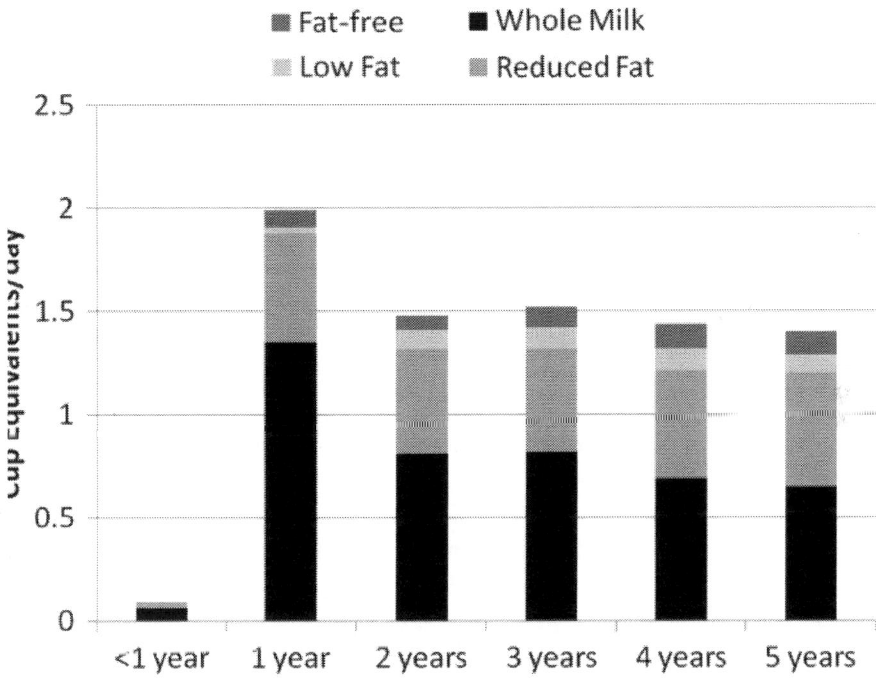

FIGURE 2: Servings of milk and type of milk consumed by children <1–5 years of age in NHANES 1988–1994. NHANES – National Health and Nutrition Examination Survey. Cup equivalents represent 8 oz. units.

at other meal times. In the context of total dairy intake, relatively small amounts of milk are consumed during meal occasions.

4.4 DISCUSSION

The 2005 Dietary Guidelines for Americans indicated that the nutrients of concern for children were vitamin E, calcium, magnesium, potassium, and dietary fiber [22]. Beverages are a potential source of calcium, magnesium, and potassium, hence these nutrients were evaluated in our study, along with phosphorus, a non–fortified nutrient provided by milk. After our analysis was completed, the 2010 Dietary Guidelines for Americans were released, which noted vitamin D, calcium, potassium, and dietary fiber were nutrients of concern for children above the age of 2 [6]. Milk is

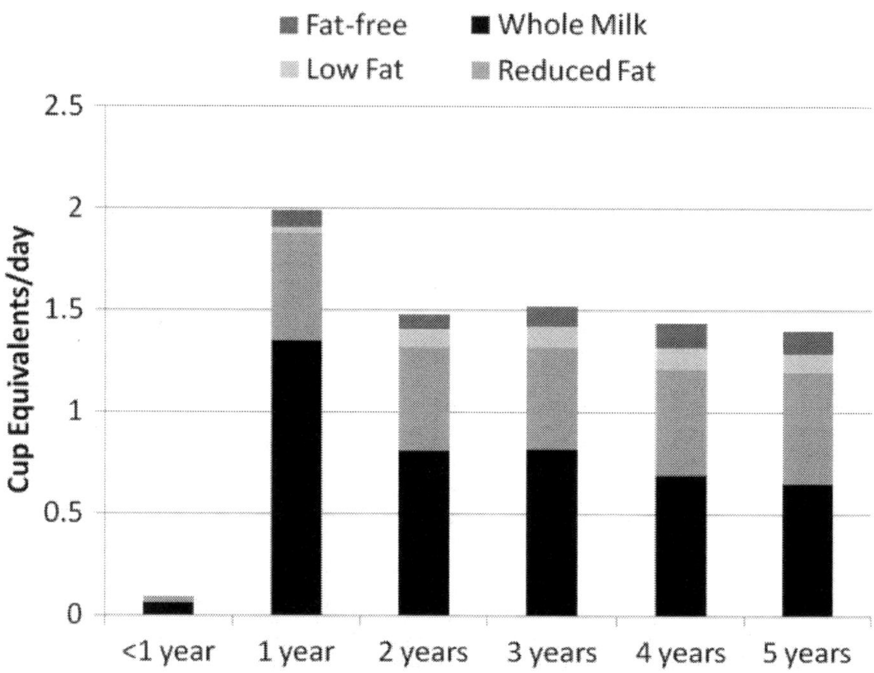

FIGURE 3: Servings of milk and type of milk consumed by children <1–5 years of age in NHANES 2001–2006. NHANES – National Health and Nutrition Examination Survey. Cup equivalents represent 8 oz. units.

the primary source of vitamin D in children's diets, but it was not possible to evaluate vitamin D trends because vitamin D intake was not assessed in NHANES prior to 2002.

Milk has been and still is the beverage consumed by most preschool children. However, there was a trend towards fewer children consuming milk in the last decade as evidence by this data and previous research [13]. Only 77% of children <1–5 years consumed milk of any type during NHANES 2001–2006 compared to 84–85% in NHANES 1988–1994 and NHANES 1976–1980. To ensure optimal growth and health, the 2010 Dietary Guidelines for Americans recommends children aged 2–3 years drink 2 cups (16 oz) and children aged 4–8 years drink 2 Â½ cups (20 fl oz) of fat–free or low–fat milk or equivalent dairy servings a day [6].

On average 2–3 year olds in NHANES 2001–2006 consumed the recommended amount of milk daily. However, children 4 and 5 years of age were consuming 14–15 fl oz/day; to meet the Dietary Guidelines, recommendation through fluid milk intake alone, 20 fl oz/day would need to be consumed for this age group. Data from Krebs–Smith, et al. indicate about one out of four young children 2–8 years old do not meet milk group recommendations [23]. Although the Dietary Guidelines recommend everyone 2 years of age or older consume low–fat (i.e., 1%) or fat–free milk; the majority of milk currently consumed by 2–5 year olds was whole or 2% milk. Several studies have shown potential health benefits of starting consumption of low–fat/fat–free milk early in life [24,25] and reduced–fat milk would be appropriate for overweight or obese children between 12 months and 2 years of age or those with a family history of obesity, dyslipidemia, or cardiovascular disease [26]. Milk consumption has been reported to increase the likelihood of achieving the recommended intake of calcium by 37% and increase intakes of vitamin A, folate, vitamin B12, and magnesium in 2–5 year olds participating in CSFII 1994–96 [14]. This study found that in addition to calcium and magnesium, milk is a major contributor to the daily intake of phosphorus and potassium in young children. Research has shown young children who drink flavored milk have comparable or higher intakes of many of these nutrients compared to those who exclusively drink plain milk, in part because flavored milk consumers have higher total milk intakes [27].

Over 50% of children 1–5 years consumed fruit juice in NHANES 2001–2006, representing an increase of 23% from the previous decade. Wang et al. [13] found a non–significant increase from 41% to 46% from NHANES 1988–1994 to NHANES 1999–2004 in their study of children 2–5 years of age. Fruit juice was an important source of potassium and magnesium in the diets of these children and has been positively linked to achieving recommended intakes of vitamin C and folate in 2–5 year old children in previous research [14]. Mean daily fruit juice consumption in 2001–2006 was 5.4 fl oz/day for children less than 1 year and 10–12 fl oz/day for children 1–5 years. For the entire group, mean intake was 10.3 fl oz/day in NHANES 2001–2006 compared to 9.5 and 7.4 fl oz/day in NHANES 1988–1994 and NHANES 1976–1980, respectively. Consistent with our findings, Wang et al. [13] observed a significant increase from 9.9 fl oz/day in NHANES 1988–1994 to 11.1 fl oz/day in NHANES 1999–2004 in their preschool sample. This is higher than the fruit juice amount recommended by the AAP, American Heart Association, and Dietary Guidelines (4–6 fl oz/day), and some children were introduced to fruit juice or drinks before the recommended 6 months of age [5,6,28].

In the last three decades, at least 35% of young children consumed fruit drinks, which is similar to the percentage reported by other studies [14,20]. The data indicate a significant increase in the amount of fruit drinks consumed in the last decade, particularly in children 1 and 2 years of age. Children less than 1 year consumed 8.8 fl oz/day and children 1–5 years consumed 10–13 fl oz/day in this analysis of NHANES 2001–2006, which was higher than estimates in CSFII 1994–1996, 1998 (4 fl oz/day) for children 6 months to 6 years [28] and 8 fl oz/day in CSFII 1994–1996 for children 2–5 years. The contribution of calcium, phosphorus, magnesium, and potassium from fruit drinks was negligible. Generally, the amount of fruit drinks consumed was greater than fruit juice for children of all ages analyzed, yet the caloric contribution of fruit drink and fruit juice to these children's diets was similar. Across all three decades, fruit drinks provided considerably more calories in the diets of preschool children than soft drinks. Although fruit drinks were not a source of the nutrients we measured, many fruit drinks are fortified with vitamin C which was positively associated with achieving recommended intakes of vitamin C in the study by Ballew et al. [14]. Most fruit drinks contain 10% or less fruit juice and a substantial amount of added sugars.

Approximately one–third of young children consumed soft drinks in all three NHANES analyses. These results are consistent with some previous from CSFII 1994–1996 [14], but lower than others including studies of NHANES 1999–2002 (39% 2–5 year olds) [20] and CSFII 1994–1995 (51%) [29]. As children age from 1 year to 5 years, a greater percentage of the population consumed soft drinks. In NHANES 2001–2006, 19% of 1 year olds consumed soft drinks, which increased to 45% by age 5 years. The mean amount of soft drinks marginally increased from 7 fl oz/day in NHANES 1976–1980 to 7.8 fl oz/day in NHANES 1988–1994 and 7.9 fl oz/day in NHANES 2001–2006. These findings are similar to data from CSFII 1994–1996 [14] but appreciably higher than 2.9 fl oz/day indicated by Rampersaud et al. [29] for CSFII 1994–1996, 1998. Carbonated beverages were negatively associated with meeting the recommended levels of vitamin A, vitamin C, and calcium [14].

The cumulative consumption of beverages with added sugar has been inversely linked to overall diet quality and meeting the adequacy of several nutrients [15]. In our analysis of the most recent NHANES, an additional 85 empty calories or 5% of total daily intake was consumed as soft drinks. The decline in milk consumption and the steady intake of nutrient poor beverages such as fruit drinks and soft drinks in our analysis is corroborated by others. Carbonated beverages and added sugar juice drinks were inversely associated with milk intakes in children 1–5 years of age in the Iowa Fluoride Study [15]. Harnack et al. [30] reported in their analysis of CSFII 1994 that high soft drink consumption appears to displace milk and fruit juice in the diets of preschool aged children. For example, those that consumed 9 fl oz or more soft drinks per day were 3.8 times more likely to consume less than 8 oz of milk per day compared to those that did not drink any soft drinks. Although the issue of displacement of milk by other beverages cannot be specifically addressed in cross–sectional data, evidence from longitudinal studies validates this phenomenon. Fiorito and colleagues [31] followed non–Hispanic white girls from age 5–15 years starting in 1996 and observed that early differences in carbonated beverage intake were predictive of later carbonated beverage and milk intake as well as selected nutrients. Girls who consumed carbonated beverages at age 5 years had higher intakes of carbonated beverages, lower milk intake, higher intake of added sugars, lower protein, fiber, vitamin D, calcium, magnesium, phosphorous, and potassium for the 10 years of follow–up. Another study of non–Hispanic white girls who were followed from age 5–9 years observed that girls who consumed adequate amounts of calcium consumed on average twice as much milk, had smaller decreases in milk intake, and consumed 18% less sweetened beverages [32].

The link between beverage consumption, energy intake and body weight have been evaluated in this young population, and is a significant concern because obesity tends to track over time [33]. High intake of sweetened beverages is linked to increased energy intake, weight gain and adiposity in children. Regular consumption of sweetened beverages between meals by children 2.5–4.5 years more than doubled the odds of being overweight at 4.5 years [12]. In a longitudinal study of girls, sweetened beverage consumption (sodas, sport drinks, fruit drinks, and sweetened coffee or tea) at age 5, but not milk or 100% juice, predicted adiposity in childhood and adolescence [19]. Higher intake of sweetened beverages at 5 years was correlated with a higher weight, percent of body fat, and waist circumference from 5–15 years of age [19]. Sweetened beverages in both of these longitudinal studies did not include flavored milk but included sugar–sweetened or artificially sweetened fruit drinks and carbonated beverages. Excessive fruit juice consumption in preschool children has been associated with obesity in some [34] but not all studies [11,35]. Wiley [36] found milk intake was positively associated with BMI among children aged 2–4 years in an analysis of NHANES 1999–2004 data while a cross–sectional analysis of NHANES 1999–2002 data of 2–5 year olds revealed that energy intakes of flavored milk drinkers were higher than those that consumed plain milk or non milk drinkers, but BMI indices did not differ among the three groups [27]. Huh et al. [37] reported in their study that milk intake at age 2 years, whether full or reduced–fat, was not linked to the risk of overweight at age 3 years.

Our analysis was limited by the use of cross–sectional data and associated analytical data across three time periods. But in the case of nationally, representative dietary data, NHANES data is the only source currently available. Given the cross–sectional approach of NHANES causality cannot be determined. Also the data was based on a single 24–hour dietary recall which may not be a fair representation of an individual's usual intake. The 24–hour dietary recall has been shown to be prone to both under– an over–reporting. Nevertheless, this method has been shown to be valid in estimating the mean intake of a population [38,39].

In summary, the number of young children consuming milk has significantly declined while those consuming fruit juice has increased dramatically in the last decade compared to the previous two decades. The proportion of children that consume fruit drinks and soft drinks has remained high and relatively stable across the three decade time period. Milk is the main source of calcium in the diets of these children, which has been declining alongside increased reduced–fat milk consumption. Milk is also a major source of magnesium and potassium, short–fall nutrients identified by the 2005 and

2010 Dietary Guidelines for Americans. The amount of milk that is consumed in this age group is less than recommended, particularly among 4 and 5 year olds, where fruit juice consumption is significantly higher than recommended. Fruit drinks are a significant source of calories. A greater number of children consume soft drinks as they age from 1–5 years. Since dietary patterns established as young children are carried throughout childhood and adolescence, and a link between non–milk sweetened beverages and obesity has been increasingly demonstrated, it is prudent that parents, educators and child caretakers replace some of the nutrient poor beverages young children are currently consuming with low–fat and fat–free milk.

REFFRENCES

1. Birch L, Savage JS, Ventura A: Influences on the development of children's eating behaviors: from infancy to adolescence. Can J Diet Pract Res 2007, 68:s1–s56.
2. Savage JS, Fisher JO, Birch LL: Parental influence on eating behavior: conception to adolescence. J Law Med Ethics 2007, 35:22–34.
3. Gartner LM, Morton J, Lawrence RA, Naylor AJ, O'Hare D, Schanler RJ, Eidelman AI: American Academy of Pediatrics Section on Breastfeeding. Breastfeeding and the use of human milk.Pediatrics 2005, 115:496–506.
4. Committee on Nutrition and the Council on Sports Medicine and Fitness: Sports drinks and energy drinks for children and adolescents: are they appropriate? Pediatrics 2011, 127:1182–1189.
5. American Academy of Pediatrics Committee on Nutrition: The use and misuse of fruit juice in pediatrics. Pediatrics 2001, 107:1210–1213.
6. U.S. Department of Health and Human Services/ U.S. Department of Agriculture: Dietary Guidelines for Americans. 2010. http://www.cnpp.usda.gov/DGAs2010–PolicyDocument.htm (accessed February 2011).
7. Skinner JD, Ziegler P, Ponza M: Transitions in infants' and toddlers' beverage patterns. J Am Diet Assoc 2004, 104(1 Suppl):s45–s50.
8. Popkin BM: Patterns of beverage use across the lifecycle. Physiol Behav 2010, 100:4–9.
9. Wiley AS: Consumption of milk, but not other dairy products, is associated with height among US preschool children in NHANES 1999–2002.Ann Hum Biol 2009, 36:125–138.
10. Lee WT, Leung SS, Lui SS, Lau J: Relationship between long–term calcium intake and bone mineral content of children aged from birth to 5 years. Br J Nutr 1993, 70:235–248.
11. Skinner JD, Carruth BR: A longitudinal study of children's juice intake and growth: the juice controversy revisited. J Am Diet Assoc 2001, 101:432–437.

12. Dubois L, Farmer A, Girard M, Peterson K: Regular sugar–sweetened beverage consumption between meals increases risk of overweight among preschool–aged children. J Am Diet Assoc 2007, 107:924–935.
13. Wang YC, Bleich SN, Gortmaker SL: Increasing caloric contribution from sugar–sweetened beverages and 100% fruit juices among US children and adolescents, 1988–2004. Pediatrics 2008, 121(6):e1604–e1614.
14. Ballew C, Kuester S, Gillespie C: Beverage choices affect adequacy of children's nutrient intakes. Arch Pediatr Adolesc Med 2000, 154:1148–1152.
15. Marshall , Marshall TA, Eichenberger Gilmore JM, Broffitt B, Stumbo PJ, Levy SM: Diet quality in young children is influenced by beverage consumption. J Am Coll Nutr 2005, 24:65–75.
16. Keller KL, Kirzner J, Pietrobelli A, St–Onge MP, Faith MS: Increased sweetened beverage intake is associated with reduced milk and calcium intake in 3– to 7–year–old children at multi–item laboratory lunches. J Am Diet Assoc 2009, 109:497–501.
17. Ogden C, Carrroll M: CDC (Centers for Disease Control): NCHS Health E–Stat. Prevalence of Obesity Among Children and Adolescents: United States, Trends 1963–1965 Through 2007–2008. [http:/ / www.cdc.gov/ nchs/ data/ hestat/ obesity_child_07_08/ obesity_child_07_08.htm#table 1 webcite] (accessed December 2010)
18. Vartanian LR, Schwartz MB, Brownell KD: Effects of soft drink consumption on nutrition and health: a systematic review and meta–analysis. Am J Public Health 2007, 97:667–675.
19. Fiorito LM, Marini M, Francis LA, Smiciklas–Wright H, Birch L: Beverage intake of girls at age 5 y predicts adiposity and weight status in childhood and adolescence. Am J Clin Nutr 2009, 90:935–942.
20. O'Connor TM, Yang SJ, Nicklas TA: Beverage intake among preschool children and its effect on weight status. Pediatrics 2006, 118:e1010–e1018.
21. (NCHS) National Center for Health Statistics: Overview: NHANES Sample Design. http://www.cdc.gov/nchs/tutorials/Nhanes/SurveyDesign/SampleDesign/intro.htm webcite (accessed December, 2010).
22. U.S. Department of Health and Human Services/ U.S. Department of Agriculture: Dietary Guidelines for Americans. 2005. [http://www.cnpp.usda.gov/DGAs-2005Guidelines.htm webcitehttp:/ / www.health.gov/ dietaryguidelines/ dga2005/ report/ HTML/ D10_Conclusions.htm webcite] (accessed December 2010).
23. Krebs–Smith SM, Guenther PM, Subar AF, Kirkpatrick SI, Dodd KW: Americans do not meet federal dietary recommendations. J Nutr 2010, 140(10):1832–1838.
24. Kaitosaari T, Ronnemaa T, Raitakari O, Talvia S, Kallio K, Volanen I, Leino A, Jokinen E, Välimäki I, Viikari J, Simell O: Effect of 7–year infancy–onset dietary intervention on serum lipoproteins and lipoprotein subclasses in healthy children in the prospective, randomized Special Turku Coronary Risk Factor Intervention Project for Children (STRIP) study. Circulation 2003, 108(6):672–677.
25. Kaitosaari T, Ronnemaa T, Viikari J, Raitakari O, Arffman M, Marniemi J, Kallio K, Pahkala K, Jokinen E, Simell O: Low–saturated fat dietary counseling starting in infancy improves insulin sensitivity in 9–year–old healthy children: the Special Turku Coronary Risk Factor Intervention Project for Children (STRIP) study. Diabetes Care 2006, 29(4):781–785.
26. Daniels SR, Greer FR, Committee on Nutrition: Lipid screening and cardiovascular health in childhood. Pediatrics 2008, 122:198–208.

27. Murphy MM, Douglass JS, Johnson RK, Spence LA: Drinking flavored or plain milk is positively associated with nutrient intake and is not associated with adverse effects on weight status in US children and adolescents. J Am Diet Assoc 2008, 108:631–639.
28. Gidding SS, Dennison BA, Birch LL, Daniels SR, Gillman MW, Lichtenstein AH, Rattay KT, Steinberger J, Stettler N, Van Horn L: American Heart Association; American Academy of Pediatrics: Dietary recommendations for children and adolescents: a guide for practitioners: consensus statement from the American Heart Association. Circulation 2005, 112:2061–2075.
29. Rampersaud GC, Bailey LB, Kauwell GP: National survey beverage consumption data for children and adolescents indicate the need to encourage a shift toward more nutritive beverages. J Am Diet Assoc 2003, 103:97–100.
30. Harnack L, Stang J, Story M: Soft drink consumption among US children and adolescents: nutritional consequences. J Am Diet Assoc 1999, 99:436–441.
31. Fiorito LM, Marini M, Mitchell DC, Smiciklas–Wright H, Birch LL: Girls' early sweetened carbonated beverage intake predicts different patterns of beverage and nutrient intake across childhood and adolescence. J Am Diet Assoc 2010, 110:543–550.
32. Fisher JO, Mitchell DC, Smiciklas–Wright H, Mannino ML, Birch LL: Meeting calcium recommendations during middle childhood reflects mother–daughter beverage choices and predicts bone mineral status. Am J Clin Nutr 2004, 79:698–706.
33. Guo SS, Chumlea WC: Tracking of body mass index in children in relation to overweight in adulthood. Am J Clin Nutr 1999, 70(1 Part 2):145S–148S.
34. Dennison BA, Rockwell HL, Baker SL: Excess fruit juice consumption by preschool–aged children is associated with short stature and obesity. Pediatrics 1997, 99:15–22.
35. Alexy U, Sichert–Hellert W, Kersting M, Manz F, Schoch G: Fruit juice consumption and the prevalence of obesity and short stature in German preschool children: results of the DONALD study. Dortmund Nutritional and Antropometrical Longitudinally Designed. J Pediatr Gastroenterol Nutr 1999, 29:343–349.
36. Wiley AS: Dairy and milk consumption and child growth: Is BMI involved? An analysis of NHANES 1999–2004. Am J Hum Biol 2010, 22:517–525.
37. Huh SY, Rifas–Shiman SL, Rich–Edwards JW, Taveras EM, Gillman MW: Prospective association between milk intake and adiposity in preschool–aged children. J Am Diet Assoc 2010, 110:563–570.
38. Woteki CE: Measuring dietary patterns in surveys. Vital Health Stat 1992, 4:101–108.
39. Beer–Borst S, Amadò R: Validation of a self–administered 24–hour recall questionnaire used in a large–scale dietary survey. Z Ernährungswiss 1995, 34:183–189.

This chapter was originally published under the Creative Commons License. Fulgoni, V. L., and Quann, E. E. National Trends in Beverage Consumption in Children from Birth to 5 Years: Analysis of NHANES Across Three Decades. Nutrition Journal 2012, 11:92 doi:10.1186/1475-2891-11-92.

CHAPTER 5

THE RELATIONSHIP BETWEEN PHYSICAL ACTIVITY, PHYSICAL FITNESS AND OVERWEIGHT IN ADOLESCENTS: A SYSTEMATIC REVIEW OF STUDIES PUBLISHED IN OR AFTER 2000

ANNETTE RAUNER, FILIP MESS, and ALEXANDER WOLL

5.1 BACKGROUND

Overweight and obesity has been called a global epidemic by the World Health Organization [1]. The prevalence of overweight and obesity is especially dramatic in economically developed countries [2] and not only in adults but also in children and adolescents. In Germany for instance, 17% of adolescents aged 14 to 17 years are overweight and nearly 9% are obese [3]. Similarly, in the United States, 18% of adolescents aged 12 to 19 years were obese in 2007/2008 [4]. In accordance with the literature [5–9], the term overweight includes obesity in this review.

Several health conditions and disorders have been attributed to being overweight in children and adolescents [10]. For instance, overweight children and adolescents are more likely to suffer from cardiovascular, metabolic, pulmonary, skeletal or psychosocial disorders [11]. Even if these conditions or disorders are not manifested during childhood, being overweight in childhood increases the risk of illness in adulthood [10]. Hence, it is critical to identify risk factors for overweight in children and adolescents and to address overweight during childhood and adolescence.

Being overweight may originate from many different factors ranging from environmental influences to genetic variations [12]. The heritability of predisposition for a high body mass index (BMI) or body fat content is between 25 and 40% [13], which suggests that other factors such as environmental factors may also play a critical role. According to Bouchard et al. [13], both the family environment and genetic predisposition influence the development of body fat content and distribution. Other important factors include lifestyle factors such as physical activity (PA), nonsmoking, high–quality diet, sedentary activities and normal weight [14]. Lifestyle factors are also important in the description of the obesogenic environment that is based on the four pillars family, sport and leisure time, eating behavior and social education [15].

Several epidemiological and intervention studies [16,17] have identified the role of physical activity and physical fitness for overweight in children and adolescents, and hence we focused on the role of sport during leisure time. Previous reviews [18–20] provided an overview of studies on the relationship either between physical activity and overweight or between fitness and overweight in children or adolescents. Despite of the influence of physical activity and fitness similarly on health outcomes including overweight, to date results of studies on the interaction between all three parameters have not been synthesized although these parameters cannot be considered independently [21]. In addition, most reviews omitted studies on adolescents and young adults or did not include longitudinal studies.

The purpose of this systematic review was to provide an overview of cross–sectional and longitudinal studies published in or after 2000 on physical activity, fitness, and overweight in adolescents, and to identify mediator and moderator effects in the interrelationship among these three parameters particularly considering gender differences because of the significant differences in these parameters between boys and girls [22].

5.2 DEFINITIONS

Physical activity comprises all modes of movement caused by muscle activity resulting in increased energy expenditure [19,23].

Physical fitness consists of the three components muscle strength, endurance and motor ability, and is a prerequisite for completing daily activities without fatigue and for participating in leisure time activities [24].

Overweight and obesity are defined as abnormally high fat content that may impair health and as high bodyweight (exceeding the standard measure) caused by an increased fat consumption [11].

5.3 METHODS

5.3.1 DATA COLLECTION

One author (AR) searched the electronic academic databases PubMed, SportDiscus, web of knowledge and Ovid for relevant studies. The following search terms were used: ["physical activity" or "fitness" or "exercise"] and ["obes*" or "overweight" or "weight gain" or "BMI"] and ["youth" or "adolescents"]. The data collection was completed in October 2011 (date of last search: 28/10/2011).

The four step search strategy is illustrated in Figure 1. In step 1, articles were screened based on title; in step 2, articles were selected based on the abstracts; in step 3 full versions of included articles were ordered; and all information was summarized in step 4. The abstracts formed an important element of the selection process and were used as decisive criterion for ordering full versions of the articles.

5.3.2 INCLUSION CRITERIA

We included only cross–sectional studies with study populations (prospective cohort studies with random sample) aged 11 to 19 years and longitudinal studies with an upper age limit of 23 years. However, two cross–sectional studies with a target group aged 7 to 12 years were also included because the age range of the study population overlapped with the target age range and the results were comparable with findings of other included studies. The search was limited to articles published in or after 2000 with physical activity and physical fitness as exercise components because

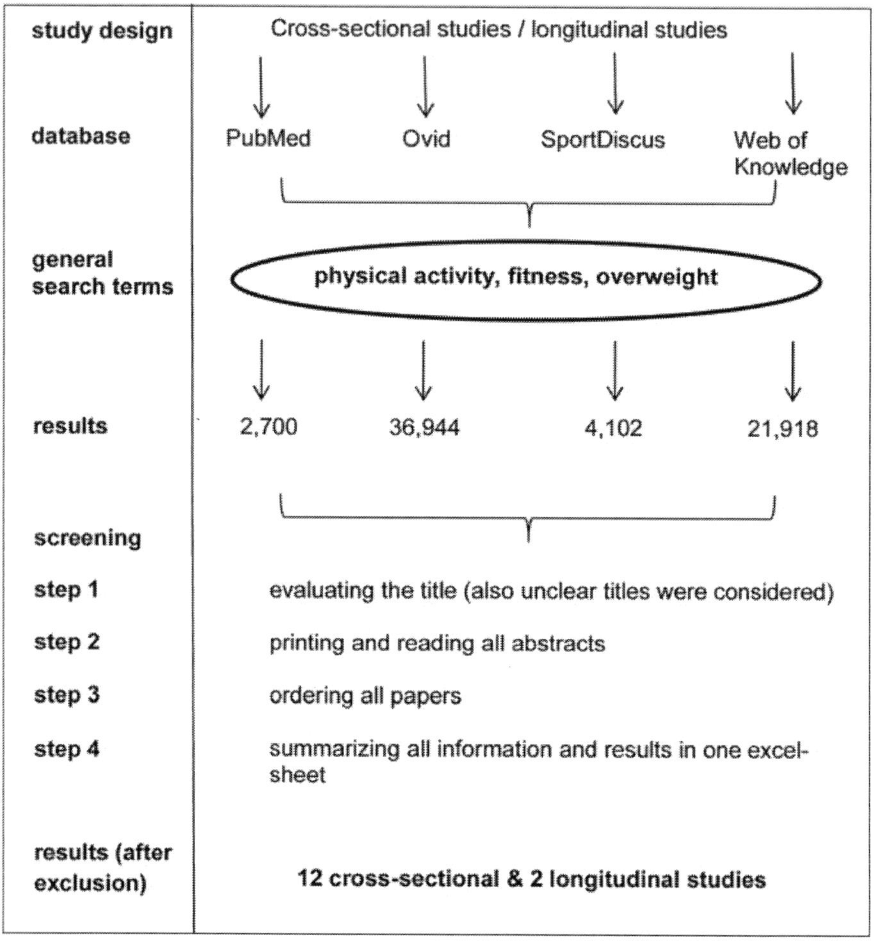

FIGURE 1: Diagram illustrating the four–step search strategy.

research on childhood and adolescence overweight and its interaction with physical activity and physical fitness has greatly increased since 2000. Only articles published in English were included.

5.3.3 EXCLUSION CRITERIA

Intervention studies, clinical trials, overviews and summarizing reviews and studies that did not analyze all three parameters physical activity, physical fitness (motor or cardiorespiratory fitness) and overweight were excluded.

5.4 RESULTS

The literature search of the four databases yielded 65,664 hits (Figure 1). Twelve cross–sectional and two longitudinal studies fulfilled all criteria and were included after the screening process.

5.4.1 MEASUREMENTS

5.4.1.1 ASSESSMENT OF OVERWEIGHT

All included cross–sectional studies used BMI as measurement of overweight or obesity [6–9,25–32]. Height and weight were quantified in ten studies [7,8,25–32] and self–reported in two studies [6,9]. In both longitudinal studies, BMI was used to determine overweight or obesity [5,33]. In two studies, waist circumference was also determined [26,30], and in five studies [4,29–32] skinfold thickness was measured. Only one study used bioelectrical impedance analysis (BIA) [8] and one study used Dual Energy X–ray Absorptiometry (DXA) [32] for determining overweight or obesity.

5.4.1.2 MEASUREMENT OF PHYSICAL FITNESS

Four cross–sectional studies included both cardiorespiratory and motor fitness [6,28,29,31]. The other eight cross–sectional studies [7–9,25–27,30,32] and the two longitudinal studies [5,33] assessed only cardiorespiratory fitness.

5.4.1.3 MEASUREMENT OF PHYSICAL ACTIVITY

The included studies measured physical activity using several different methods. Five studies used objective measurements such as accelerometry [7–9,32] and pedometry [30]. Ten studies (eight cross–sectional and two longitudinal studies) used subjective measurements derived from questionnaires with items relating to the setting (at school, outside school; divided into leisure time physical activities at sport clubs and leisure time physical activities outside of sports clubs) and intensity of physical activities [6,8,25–29,31]. Only one study collected both objective and subjective data on physical activity [8]. Most studies analyzed the relationship between overweight, physical activity and physical fitness using analyses of variance (ANOVA) and (linear and logistic) regression analysis.

5.4.1.4 ASSOCIATIONS BETWEEN PHYSICAL ACTIVITY, PHYSICAL FITNESS AND OVERWEIGHT IN CROSS-SECTIONAL STUDIES

Twelve studies met the inclusion criteria of this review. While all twelve studies assessed physical activity, physical fitness and overweight, only four studies analyzed the interaction among these three parameters. Because some studies did not report actual data but only an interpretation of their findings, statistical parameters could only be included for some studies. The results of all included studies are summarized in (see Additional file 1: Table S1). For completeness, we also report the results of those studies that assessed all three parameters but not their interaction. Throughout this article, we distinguish between genders because of the significant differences in physical activity, physical fitness and overweight between boys and girls [22]. The effects of physical activity and physical fitness on overweight and the strength of the association between physical activity, physical fitness and overweight by gender are summarized in. We defined four categories based on the results of the statistical tests ($p>0.05$; $p<0.05$; $p<0.01$; $p<0.001$) and interpreted the associations regarding overweight: for instance, a positive relationship between fat free mass and physical activity was interpreted as a negative association between overweight and physical activity and the corresponding statistical result transcribed by the corresponding statistical result. The results in this table clearly show a stronger relationship and a more pronounced gender effect on the relationship between physical fitness and overweight than between physical activity and overweight.

5.4.1.5 RELATIONSHIP BETWEEN PHYSICAL ACTIVITY, PHYSICAL FITNESS AND OVERWEIGHT

Because the statistical data evaluation in the included studies was heterogeneous, the central outcomes of all studies cannot be summarized, and hence we present the results of each study. Ortega et al. [26] reported a higher BMI in adolescents with lower cardiorespiratory fitness independent of their sedentary and leisure time activities (physical activities outside school). In boys and girls, BMI was negatively correlated with cardiorespiratory fitness independent of their leisure–time physical activity and sedentary activities (boys: $p=0.006$; girls: $p<0.001$). Similarly, cardiorespiratory fitness was inversely related to waist circumference (boys: $p=0.001$; girls: $p=0.005$) independent of physical activity. Up to 10% of variance in waist circumference in boys and 18% of variance in waist circumference in girls was explained by their sedentary activities (television viewing and video/computer time). Variability in cardiorespiratory fitness explained up to 13% of the variance in waist circumference in boys and up to 16% in girls.

In contrast, Fogelholm et al. [6] found that variance in physical activity better explained the variability in physical fitness (ß–coefficients between −0.33 and 0.49) than that in overweight (ß–coefficients between −0.27 and 0.24). The associations between overweight and physical activity and between physical activity and physical fitness were comparable for both genders. The intensity of physical activity and being overweight predicted physical fitness in adolescents. Fogelholm et al. [6] described that physically active persons who are overweight cannot achieve better physical fitness values because of the negative association between being overweight and physical fitness. Thus, overweight acts as a mediator for the relationship between physical activity and physical fitness.

In another study, Ortega [9] showed that cardiorespiratory fitness influences the association between being overweight and physical activity. Hence, cardiorespiratory fitness acts as a moderator for the relationship between overweight and physical activity. The association between physical activity, physical fitness and overweight did not differ between genders. Lohman et al. [32] reported that girls with high levels of physical activity (one standard deviation above the mean) and average body composition (fat free and fat mass) had a higher physical fitness level (+3.5%) and girls with low levels of physical activity (one standard deviation below the mean) and average body composition have a lower physical fitness level (–3.5%) compared with that of girls with average levels of physical activity and average fat free and fat mass.

5.4.1.6 RELATIONSHIP BETWEEN PHYSICAL FITNESS AND OVERWEIGHT

All studies showed an inverse relationship between physical fitness and overweight and overweight and physical fitness respectively, except for Huotari et al. [27], where no data were available. In two studies, cardiorespiratory fitness [6,31] was more strongly related to overweight than motor fitness, and two studies [28,29] showed a stronger relationship between BMI and cardiorespiratory fitness than between BMI and motor fitness. Interpreting and comparing these results is difficult because these studies used different analytic strategies and measurement instruments. The four studies [6,7,28,31] used three to seven different tests for measuring several aspects of motor capacity.

All twelve studies reported an inverse relationship between cardiorespiratory fitness and overweight. Seven studies [6,7,26,28–31] used shuttle run tests, and one study each used the cooper test [8], a submaximal treadmill test [25], the PWC 170 [32], the maximal cycle test [9] or the 2000–m (boys) and 1500–m (girls) running test for assessing cardiorespiratory fitness.

Pate et al. [25] found no difference in the relationship between physical fitness and BMI between genders, and Ortega et al. [26] observed a similar relationship between overweight and physical fitness for both genders. In addition, BMI adjusted by waist circumference was significantly negatively associated with cardiorespiratory fitness only in overweight boys ($p \leq 0.05$) but not in normal weight adolescents and overweight girls. Cardiorespiratory fitness was inversely associated with BMI in boys and in girls ($p<0.001$) and with waist circumference (boys: $p=0.001$; girls: $p=0.005$). Variance in cardiorespiratory fitness explained up to 13% of variability in waist circumference in boys and up to 16%

in girls. In addition, the comparison of cohorts collected in 1976 and 2001 by Huotari et al. [27] confirmed these findings. In girls, the influence of BMI on cardiorespiratory fitness was smaller (ß=–0.42, $p<0.001$, $R^2=0.165$) than that in boys (ß=–0.36, $p<0.001$, $R^2=0.127$) in the 2001 study. In comparison, in the 1976 study no significant relationship between BMI and cardiorespiratory fitness was found for girls or boys. The results by Gonzales–Suarez and colleagues [28] were not stratified by gender. Fogelholm et al. [6] did not find significant differences in ß–coefficients for endurance capacity between genders.

Ara et al. [29] reported that skinfold thickness was most strongly related to cardiorespiratory fitness in both boys and girls, and the ß–coefficient for this relationship was greater in boys (ß=–3.334; $p<0.001$) than in girls (ß=–2.571; $p<0.001$). The next strongest predictors of cardiorespiratory fitness were truncal subcutaneous fat (boys: ß=–1.78, $p<0.001$; girls: ß=–1.77, $p<0.001$) and BMI (boys: ß=–0.047, $p<0.001$; girls: ß=–0.059, $p<0.001$).

The results by Deforche et al. [31] are comparable to those reported by Aires et al. [7], and the endurance capacity in obese boys was higher than that in obese girls ($F=22.5$; $p<0.001$). Haerens et al. [8] analyzed the difference in overweight and cardiorespiratory fitness by gender and detected a significant ($F=6.08$; $p\leq0.05$) difference in running capacity between overweight boys and girls. Ortega [9] reported an inverse association between waist circumference and cardiorespiratory fitness without a significant gender effect. Fogelholm [6] showed marginally stronger relationships between ball skills ($ß_{boys}$=–.12, $p<0.001$; $ß_{girls}$=–0.10, $p=0.003$), jumping back and forth ($ß_{boys}$=–0.17, $p<0.001$; $ß_{girls}$=–0.14, $p<0.001$) and five–jump ($ß_{boys}$=–0.27, $p<0.001$; $ß_{girls}$=–0.26, $p<0.001$) and overweight in boys than in girls. In comparison, the influence of overweight on number of sit–ups ($ß_{boys}$=–0.20, $p<0.001$; $ß_{girls}$=–0.21, $p<0.001$) and the coordination test ($ß_{boys}$=–0.22, $p<0.001$; $ß_{girls}$=–0.24, $p<0.001$) was stronger in girls than in boys. Deforche et al. [31] found a significant interaction between gender and obesity in the sit and reach test ($F=4.3$; $p<0.05$), bent–arm–hang ($F=45.8$; $p<0.001$) and endurance shuttle run ($F=22.5$; $p<0.001$). Ng et al. [30] did not stratify weight groups by gender, Lohman et al. [32] only included females in their study, and Gonzales–Suarez et al. [28] did not perform a separate analysis of the relationship between overweight and motor fitness for genders.

5.4.1.7 RELATIONSHIP BETWEEN PHYSICAL ACTIVITY AND OVERWEIGHT

In comparison to physical fitness, the relationship between physical activity and overweight is less clear. Three studies collected physical activity data but did not analyze the relation between physical activity and overweight [6,25,27]. Six studies did not find any relationships between overweight and physical activity [7,26,29–31,34]. Two studies [28,32] analyzed relations between physical activity and overweight and between overweight and physical activity, respectively. Lohman et al. [32] found a negative significant correlation between BMI and physical activity, whereas Gonzales et al. [28] did not find differences between BMI and physical activity scores in overweight and normal youth.

Objective and subjective measurement instruments yielded comparable results. While one study detected a relationship between objectively measured physical activity and overweight [32], two studies did not find any relationship between overweight and objectively measured physical activity [7,30]. We found similar results for studies using subjective measurement instruments. One study reported a significant relationship between overweight and subjectively measured physical activity [28], and three studies did not find relationships between overweight and subjectively measured physical activity [26,29] and subjectively measured physical activity and overweight [31]. Two studies [8,9] used both objective and subjective instruments for assessing physical activity. While Ortega et al. [9] did not find any relationship between overweight and physical activity, Haerens [8] detected significant relationships between overweight and physical activity dependent of the method of data evaluation. Similarly, categorized physical activity (active versus non–active) was not related to overweight [29]. In comparison, the intensity of physical activity was related to overweight [7,8].

Five studies [8,26,28,29,31] analyzed differences in the relationship between physical activity and overweight between genders. Two studies [26,29] found a stronger relationship between physical activity and overweight for boys than for girls. In contrast, three studies [8,28,31] did not find a gender effect on the relationship between (total) physical activity and overweight. Ortega and colleagues [26] performed separate median value comparisons between BMI and waist circumference, and activity pattern and cardiorespiratory fitness by gender. The strongest relationship in boys ($p=0.006$) was that between waist circumference and sedentary activities was. In girls, the strongest association was that between waist circumference and active commuting to school (no information was provided on type of active commuting; $p=0.002$). A significant relationship between BMI and sedentary activities ($\leq$ 2 hours; ß=−0.72; $p=0.043$) was found only in boys, whereas waist circumference was negatively associated with sedentary activities ($\leq$ 2 hours) in boys (ß=−2.46; $p=0.024$) and in girls (ß=−1.47; $p=0.028$). Up to 10% of variance in waist circumference in boys and up to 18% in girls were explained by variability in sedentary activities. In contrast, Gonzales–Suarez [28] did not find an effect of gender on the relationship between being overweight and physical activity. Ara et al. [29] analyzed the differences in weight (measured using various methods) between active and non–active adolescents. BMI was higher in active boys than in non–active boys ($p=0.05$), and the sum of skin fold test scores was slightly higher in active than in non–active boys. In contrast, while fat mass was lower in active girls than in non–active girls ($p<0.05$), both groups had comparable BMI. Deforche et al. [31] reported a higher sport index in non–obese boys compared to obese boys ($F=3.7$; $p<0.05$), and a comparable sport index in obese and non–obese girls. Aires et al. [7] did not report a gender specific analysis. Haerens et al. [8] did not find significant differences between body weight groups in objectively ($F=0.08$; $p>0.05$) or subjectively ($F=0.03$; $p>0.05$) measured total physical activity analyzed by gender. However, moderate physical activity significantly differed between boys and girls ($F=4.25$; $p\leq0.001$). The results for overfat (measured via skinfold thickness) and normal fat boys and girls were comparable. Objectively ($F= 0.47$; $p>0.05$) and subjectively ($F=2.13$; $p>0.05$) measured total physical activity, light physical activity ($F= 0.18$; $p>0.05$) and moderate physical activity ($F=1.4$; $p>0.05$) did not differ significantly between overfat and normal fat boys and girls.

5.4.1.8 ASSOCIATIONS BETWEEN PHYSICAL ACTIVITY, FITNESS AND OVERWEIGHT IN LONGITUDINAL STUDIES

Two longitudinal studies captured physical activity, physical fitness and overweight.

5.4.1.9 RELATIONSHIP BETWEEN PHYSICAL ACTIVITY, PHYSICAL FITNESS AND OVERWEIGHT

Both longitudinal studies [5,33] analyzed only the relationship between physical activity and overweight and between physical fitness and overweight and not the interaction among all three parameters. However, separate analyses by Aires et al. [33] showed that while physical activity influenced cardiorespiratory fitness and cardiorespiratory fitness influenced BMI, BMI was not related to physical activity. Therefore cardiorespiratory fitness acts as a mediator in the relationship between physical activity and BMI.

5.4.1.10 RELATIONSHIP BETWEEN PHYSICAL FITNESS AND OVERWEIGHT

He et al. [5] and Aires et al. [33] reported an inverse relationship between BMI and physical fitness and between physical fitness and BMI respectively. Subjects with a low fitness level at baseline had a higher risk of becoming overweight or obese compared to those who had high initial fitness levels (data not shown) [5].

Aires et al. [33] did not report a potential gender difference. In contrast, He [5] found that boys with low fitness at baseline were more likely to be overweight 3–years later than girls (boys: OR=8.71, $p<0.001$; girls: OR=6.87, $p=0.055$).

5.4.1.11 RELATIONSHIP BETWEEN PHYSICAL ACTIVITY AND OVERWEIGHT

The questionnaire used by Aires et al. [33] provided information on sedentary activities. Adolescents with low physical activity levels did not experience a significant increase in BMI over time [33]. Similarly, He et al. [5] did not reveal significant associations between changes in BMI and physical activity. None of the studies investigated the influence of gender on the relationship between physical activity and overweight.

5.5 DISCUSSION

The purpose of this systematic review was to provide an overview of cross–sectional and longitudinal studies published in or after 2000 on physical activity, physical fitness and overweight in adolescents, and to identify mediator and moderator effects in the interrelationship among these three parameters particularly considering gender differences. Objectivity of self–reported physical activity has been questioned because of potential over– or underestimation [32] and thus should be considered with caution. However, because only few studies examined the interaction between physical activity, physical fitness and overweight, we combined results of objectively and subjectively assessed physical activity.

To the best of our knowledge, this article is the first review on the interrelationship between physical activity, physical fitness and overweight, and hence our results cannot be related to the literature or to other study populations. Synthesizing the interaction between all three parameters was difficult because only four studies specifically investigated this interaction. While the literature reported inconsistent results, all studies showed an interaction between these parameters. Several studies [6,9,26,32] confirmed that physical activity and physical fitness are equally important for health [21]. In the following the results will be discussed with reviews analyzing only the relationship between two parameters, because no comparable reviews (reviews analyzing the interaction) were found.

The different strengths of the correlations between the three parameters may be at least in part attributed to the different measurements of physical activity. For instance, two studies [6,26] assessed physical activity via questionnaire, one via accelerometer [32] and one via activity monitor and questionnaire [9], and the collection period of objectively measured physical activity ranged from three [9] to six [32] days. In addition, the two studies that measured physical activity subjectively omitted reporting details on their measurement instruments. Further, Ortega et al. [26] measured physical activity outside of school for only four days. While Fogelholm [6] measured the activity during leisure–time in and outside of sports clubs, they only reported frequency and duration and not intensity or setting of physical activity. Hence, reliable and valid questionnaires assessing frequency, duration, intensity and setting of the different physical activities are still needed [35] especially because, for instance, intensity is an important aspect in overweight prevention [36]. Interestingly, studies that used unspecific measurement instruments for physical activity reported weak or no relationships between physical activity and overweight [9,26,29]. The poor quality of physical activity measurement instruments may also explain the stronger influence of cardiorespiratory fitness than that of physical activity on overweight. The main limitation of subjective measurement instruments is potential over– and underestimation of physical activity [32]. In comparison, objective measurement instruments for physical activity can only capture specific activities and require a high effort by the participants. For instance, subjects have to regularly wear the accelerometers or pedometers for extended periods of time and on different days.

The data on the relationship between physical activity and overweight are inconsistent. Specifically, the different levels of physical activity (measured by objective measurement methods) showed different relationships to overweight. In addition, the effect of gender on the relationship between physical activity and overweight was inconsistent. While Deforche et al. [31], Haerens et al. [8] and Gonzales–Suarez et al. [28] reported no gender effect on the relationship between overweight and physical activity, other studies [26,29] revealed that gender affected the relationship between overweight and physical activity but that this association depended on the anthropometric measurement method used to measure overweight. Similar to our observations in adolescents, Must et al. [19] found inconsistent results in children with a higher tendency to an inverse relationship between physical activity level and overweight in cross–sectional studies and differences in the relationship between physical activity and overweight between boys and girls emphasizing the inconsistent state of research not only in adolescents but also in children. The previously discussed large number and poor quality of methods for measuring physical activity might explain this observation. In addition, these results show that capturing physical activity in youth is difficult. In a review of cross–sectional studies that analyzed self–reported and objectively measured physical activity in overweight children and adolescents, Winkler et al. [36] reported inconsistent results and that the intensity of physical activity played a critical role independent of age and gender. In addition, Winkler et al. [36] reported that physical activity was related to overweight in two longitudinal studies, which contradicts the results of the two longitudinal studies [5,33] included in our review that found no relationships between physical activity and overweight. In contrast, Must et al. [19] reviewed longitudinal studies and reported comparable results to our findings in longitudinal studies. Similarly to adolescents, the results in children are inconsistent and low physical activity level was not related to changes in BMI [5,19,33]. However, according to Must et al. [19], most cross–sectional and longitudinal studies showed no relationships between physical inactivity and overweight in adolescents and inconsistent gender specific results.

All studies included in our review observed inverse relationships between physical fitness and overweight. Because of the different measurement instruments for cardiorespiratory fitness used in these studies (shuttle run: [5–7,26,28–31,33], maximal treadmill test: [25], maximal cycle test: [9], cooper test: [8], PWC 170: [32], 2,000/1,500 m: [27]), final comparisons are difficult. Adolescents with lower cardiorespiratory fitness were more likely to be overweight or obese than those with high cardiorespiratory fitness [5–9,25,26,28–32]. However, gender influenced the relationship between overweight and cardiorespiratory fitness. These results are in agreement with the results of other studies including those reported by Ostojic et al. [37]. Similar results were observed for motor fitness and overweight. The measurement instruments were also inconsistent in motor fitness (Eurofit [29,31]: two studies; unknown [6,28]: two studies). While motor fitness in overweight and obese adolescents was lower than that in normal weight adolescents [6,28,29,31,33], the influence of gender on the relationship between motor fitness and overweight was heterogeneous.

Interestingly, some studies [6,25,27,31–33] included weight as independent parameter in their statistical models while other studies [5,7–9,26,28–30] used weight as dependent parameter. This observation illustrates that the causality between physical activity and overweight and between physical fitness and overweight is still unclear. For instance, Metcalf et al. [38] suggested that overweight influences level of physical activity but not vice versa. Similar data for the causal relationship between physical fitness and overweight are not available. Hence, future longitudinal studies are warranted to tease out this causal relationship. Furthermore, additional longitudinal analyses are necessary to determine the interrelationship (mediator or moderator effect) between physical activity, physical fitness and overweight which has important implications for public health policy making and developing optimal obesity prevention or treatment programs.

5.6 LIMITATIONS

Because of the small number of studies the results were not categorized based on objective or subjective physical activity measurement. In addition, studies on metabolic syndrome or cardiovascular diseases were not included (even if physical activity, physical fitness and overweight measures were used), and only studies with the primary goal of analyzing the relationship between the three parameters were included.

5.7 CONCLUSION

The small number of longitudinal studies emphasizes the lack of longitudinal research, and further prospective studies are necessary for determining cause and effect and the type (correlation, mediator and moderator effect) of the interrelationship among physical activity, physical fitness and overweight.

Overall, a concluding evaluation is difficult because several studies did not state effect or effect size and hence the reported information on significant relationships should be interpreted with caution. In addition, the studies used different methods to measure physical activity, and the objectivity of self–reported physical activity is questionable [39] and may result in over– or underestimation [32].

REFERENCES

1. World Health Organisation: Obesity: preventing and managing the global epidemic. Report of a WHO consultation. World Health Organ Tech Rep Ser 2000, 894:1–253.
2. Wang Y, Lobstein T: Worldwide trends in childhood overweight and obesity. IJPO 2006, 1(1):11–25.
3. Kurth B–M, Schaffrath Rosario A: The prevalence of overweight and obesity in children and adolesents in Germany. Results of the nation–wide children and youth health survey (KiGGS). Bundesgesundhbl–Gesundheitsforsch–Gesundheitsschutz 2007, 50:736–743.
4. Prevalence of Obesity Among Children and Adolescents: United States, Trends 1963–1965 trough 2007–2008. http:/ / www.cdc.gov/ nchs/ data/ hestat/ obesity_child_07_08/ obesity_child_07_08.htm
5. He QQ, Wong TW, Du L, Jiang ZQ, Yu TS, Qiu H, Gao Y, Liu WJ, Wu JG: Physical activity, cardiorespiratory fitness, and obesity among Chinese children. Prev Med 2011, 52(2):109–113.
6. Fogelholm M, Stigman S, Huisman T, Metsamuuronen J: Physical fitness in adolescents with normal weight and overweight. Scand J Med Sci Sports 2008, 18(2):162–170.
7. Aires L, Silva P, Silva G, Santos MP, Ribeiro JC, Mota J: Intensity of physical activity, cardiorespiratory fitness, and body mass index in youth. J Phys Act Health 2010, 7(1):54–59.

8. Haerens L, Deforche B, Maes L, Cardon G, De Bourdeaudhuij I: Physical activity and endurance in normal weight versus overweight boys and girls. J Sports Med Phys Fitness 2007, 47(3):344–350.
9. Ortega FB: Cardiovascular fitness modifies the associations between physical activity and abdominal adiposity in children and adolescents: the European Youth Heart Study. BJSM 2010, 44:256–262.
10. Daniels SR: The consequences of childhood overweight and obesity. Future Child 2006, 16(1):47–67.
11. Overweight and obesity. http://www.who.int/mediacentre/factsheets/fs311/en
12. Hebebrand J, Wermter A–K, Hinney A: Obesity, genetics and interaction between genes and the environment. Monatsschr Kinderheilkd 2004, 152(8):870–876.
13. Bouchard C, Malina RM, Pérusse L: Genetics of Fitness and Physical Performance. Champaign: Human Kinetics; 1997.
14. Pronk NP, Anderson LH, Crain AL, Martinson BC, O'Connor PJ, Sherwood NE, Whitebird RR: Meeting recommendations for multiple healthy lifestyle factors. Prevalence, clustering, and predictors among adolescent, adult, and senior health plan members. Am J Prev Med 2004, 27(2):25–33.
15. Wabitsch M: Children and adolescents with obesity in Germany. Call for action. Bundesgesundhbl – Gesundheitsforsch – Gesundheitsschutz 2004, 47(3):251–255.
16. Wareham NJ, van Sluijs EM, Ekelund U: Physical activity and obesity prevention: a review of the current evidence. Proc Nutr Soc 2005, 64(2):229–247.
17. DiPietro L: Physical activity, body weight, and adiposity: an epidemiologic perspective. Exerc Sport Sci Rev 1995, 23:275–303.
18. Jimenez–Pavon D, Kelly J, Reilly JJ: Associations between objectively measured habitual physical activity and adiposity in children and adolescents: Systematic review.IJPO 2010, 5(1):3–18.
19. Must A, Tybor DJ: Physical activity and sedentary behavior: a review of longitudinal studies of weight and adiposity in youth. Int J Obes (Lond) 2005, 29(Suppl 2):S84–96.
20. Oja P, Titze S, Bauman A, de Geus B, Krenn P, Reger–Nash B, Kohlberger T: Health benefits of cycling: a systematic review. Scand J Med Sci Sports 2011, 21(4):496–509.
21. Brandes M: Physical activity or fitness: what is more important for health? Bundesgesundhbl – Gesundheitsforsch – Gesundheitsschutz 2012, 55:96–101.
22. Trost SG, Pate RR, Sallis JF, Freedson PS, Taylor WC, Dowda M, Sirard J: Age and gender differences in objectively measured physical activity in youth. Med Sci Sports Exerc 2002, 34(2):350–355.
23. Ortega FB, Ruiz JR, Castillo MJ, Sjostrom M: Physical fitness in childhood and adolescence: a powerful marker of health. Int J Obes (Lond) 2008, 32(1):1–11.
24. Malina RM, Katzmarzyk PT: Physical activity and fitness in an international growth standard for preadolescent and adolescent children. Food Nutr Bull 2006, 27(4):S295–313.
25. Pate RR, Wang CY, Dowda M, Farrell SW, O'Neill JR: Cardiorespiratory fitness levels among US youth 12 to 19 years of age: findings from the 1999–2002 National Health and Nutrition Examination Survey. Arch Pediatr Adolesc Med 2006, 160(10):1005–1012.
26. Ortega FB, Tresaco B, Ruiz JR, Moreno LA, Martin–Matillas M, Mesa JL, Warnberg J, Bueno M, Tercedor P, Gutierrez A, et al.: Cardiorespiratory fitness and sed-

entary activities are associated with adiposity in adolescents. Obesity (Silver Spring, Md) 2007, 15(6):1589–1599.
27. Huotari PR, Nupponen H, Laakso L, Kujala UM: Secular trends in aerobic fitness performance in 13–18–year–old adolescents from 1976 to 2001. BJSM 2010, 44(13):968–972.
28. Gonzalez–Suarez CB, Grimmer–Somers K: The association of physical activity and physical fitness with pre–adolescent obesity: an observational study in metromanila, Philippines. J Phys Act Health 2011, 8(6):804–810.
29. Ara I, Moreno LA, Leiva MT, Gutin B, Casajus JA: Adiposity, physical activity, and physical fitness among children from Aragon, Spain. Obesity (Silver Spring, Md) 2007, 15(8):1918–1924.
30. Ng C, Marshall D, Willows ND: Obesity, adiposity, physical fitness and activity levels in Cree children. Int J Circumpolar Health 2006, 65(4):322–330.
31. Deforche B, Lefevre J, De Bourdeaudhuij I, Hills AP, Duquet W, Bouckaert J: Physical fitness and physical activity in obese and nonobese Flemish youth. Obes Res 2003, 11(3):434–441.
32. Lohman TG, Ring K, Pfeiffer K, Camhi S, Arredondo E, Pratt C, Pate R, Webber LS: Relationships among fitness, body composition, and physical activity. Med Sci Sports Exerc 2008, 40(6):1163–1170.
33. Aires L, Andersen LB, Mendonca D, Martins C, Silva G, Mota J: A 3–year longitudinal analysis of changes in fitness, physical activity, fatness and screen time. Acta paediatr (Oslo, Norway: 1992) 2010, 99(1):140–144.
34. Ogden C: Prevalence of Obesity Among Children and Adolescents: United States, Trends 1963–1965 trough 2007–2008. Atlanta: Centers for Disease Control and Prevention; 2010. Global physical activity surveillance. http://www.who.int/chp/steps/GPAQ/en/index.html
35. Winkler S, Hebestreit A, Ahrens W: Physical activity and obesity. Bundesgesundhbl – Gesundheitsforsch – Gesundheitsschutz 2012, 55:24–34.
36. Ostojic SM, Stojanovic MD, Stojanovic V, Maric J, Njaradi N: Correlation between fitness and fatness in 6–14–year old Serbian school children. J Health Popul Nutr 2011, 29(1):53–60.
37. Metcalf BS, Hosking J, Jeffery AN, Voss LD, Henley W, Wilkin TJ: Fatness leads to inactivity, but inactivity does not lead to fatness: a longitudinal study in children (EarlyBird 45). Arch Dis Child 2011, 96(10):942–947.
38. Melanson EL Jr, Freedson PS: Physical activity assessment: a review of methods. Crit Rev Food Sci Nutr 1996, 36(5):385–396.

CHAPTER 6

CHILDHOOD AND FAMILY INFLUENCES ON BODY MASS INDEX IN EARLY ADULTHOOD: FINDINGS FROM THE ONTARIO CHILD HEALTH STUDY

ANDREA GONZALEZ, MICHAEL H. BOYLE, KATHOLIKI GEORGIADES, LAURA DUNCAN, LESLIE R. ATKINSON, and HARRIET L. MACMILLAN

6.1 BACKGROUND

Excess body weight is associated with numerous adverse health consequences including coronary heart disease, type II diabetes and cancer, among others [1,2]. In 2008, an estimated 1.46 billion adults worldwide were classified as overweight; of these, 502 million were obese [3]. These numbers are steadily increasing with the greatest prevalence of obesity occurring in high–income countries such as the United States and Canada. Obesity and its associated health complications have a significant economic impact on healthcare with annual national costs estimated at \$4.6 to \$7.1 billion in Canada [2], and \$92.6 billion in the U.S. [4].

Obesity is a multifactorial condition influenced by diverse factors operating across the lifespan. Various family–level factors, as well as individual–level characteristics, have been identified as potential determinants of BMI in childhood and adulthood. Several prospective studies highlight the importance of childhood socioeconomic disadvantage as a major predictor of obesity [5–8]. For example, Power et al. (2003) showed that family SES in early childhood (birth to age 7) was significantly associated with obesity at age 33; this finding was not explained by parental BMI or the individual's own education. These prospective studies illustrate that childhood SES has long–lasting effects that are not easily reversed by changes in

SES occurring in adulthood [6]. Other family indicators, closely linked to SES, have been identified as risk factors for obesity in childhood. Children from single parent households have significantly higher BMIs compared to those from dual parent households [9]. In addition, parental educational attainment is inversely associated with adulthood BMI [10].

At an individual level, birth weight, as a crude estimate of in utero environment, is related to elevated BMI in childhood [11] and adulthood [12,13]. Various childhood psychosocial risk factors are also associated with elevated BMI in childhood and adulthood. In particular, diagnoses of depression, anxiety and conduct disorders in childhood and adolescence are related to increased BMI in adulthood [14–17]. Furthermore, obesity in young adulthood has been associated with behavioral problems exhibited at ages 5 to 14 years [18,19]. Individuals with a history of childhood sexual abuse [20] and physical abuse [21–23] are more likely to be overweight or obese later in life. Finally, there is some evidence to suggest that childhood learning difficulties, below average scholastic proficiency and having received special education are risk factors for obesity in young adulthood [24,25].

There is considerable evidence pointing to the importance of early life factors in the development of obesity in children and adults; however, the family as a contextual unit is rarely studied [26]. The family environment is considered key in the development of obesity [27], yet few studies have prospectively examined the impact of multiple family and childhood risks on BMI in early adulthood. We consider risk factors occurring at the child– and family levels simultaneously. In this study, we assessed a set of risk and protective variables, including prenatal risk (with low birth weight serving as marker), risk integral to the child (such as psychiatric disorder, medical complications and functional limitations), and a number of family variables including: sociodemographic factors, parental educational achievement, socio–emotional and physical functioning and family functioning. Lastly, we included retrospective self–reports of exposure to childhood physical and sexual abuse. The objective of this study was to examine the associations between individual and family–risk factors assessed in a sample of 4–16 year olds in 1983 and elevated BMI assessed in 2001 when they were young adults, at 21–35 years of age. The following issues were addressed. (1) What is the association between family contextual influences assessed in childhood/adolescence and BMI assessed in young adulthood? (2) Because of evidence that the influence of childhood risk factors may vary by gender, we also explored whether gender modifies the association between childhood risk factors on BMI in adulthood.

Typically, previous studies have used cross–sectional or retrospective reports of childhood risks to examine a limited number of factors. Few studies have examined the prospective relationship between multiple individual and family characteristics measured in childhood and BMI assessed in early adulthood [see [6,23,24,28]. Understanding childhood contextual influences operating at different levels is essential for determining which factors should be program targets for developing policies and early interventions to reduce obesity.

6.2 METHODS

6.2.1 SAMPLE

This study uses data from the initial (1983) and third (2001) waves of the Ontario Child Health Study (OCHS) – a prospective, longitudinal study of child and adolescent health in a cohort of 3,294 children ages 4–16 years, living in 1,869 households across Ontario, Canada [29]. The target population included all children born from January 1, 1966 through January 1, 1979, whose usual place of residence was a household in Ontario. A stratified, clustered, and random sample was selected from all household dwellings identified in the 1981 Census of Population. Sample weights were devised for the first wave based on the probabilities of selection and enlistment so that subject responses would be linked numerically back to the target population, improving the accuracy of statistical estimates. During a home interview, data were collected from parents (95% mothers) and adolescents aged 12–16 years by trained field staff from the Special

Surveys Division of Statistics Canada. This study was approved by the Research Ethics Board at Hamilton Health Sciences, McMaster University.

6.2.2 VARIABLES AND MEASURES

6.2.2.1 OUTCOME VARIABLES

Information was collected from OCHS follow–up participants during a structured interview, administered in the home when participants were 21–35 years of age. BMI measured in 2001 was derived using self–reports of weight and height. Several studies have shown high correlation between self–reported and measured BMI [30–33]; however, self–reported BMI yields lower rates of obesity and overweight [34–36]. We calculated a corrected BMI [37] and ran all analyses on reported and corrected BMIs in parallel. We used correction equations based on the 2005 Canadian Community Health Survey (CCHS). These equations were generated using socio–demographic variables that were significantly associated with discrepancies between self–reported and measured values of BMI by sex. Because the results were identical for all models, we highlight findings based on the self–report BMI data only.

6.2.2.2 CONFOUNDING VARIABLES

BMI has been linked to current income [6], education attainment [24] and physical and mental health [38,39]; therefore, these variables were included as potential confounders in all models. Our measures of potential confounders, measured in 2001, included: number of years of education (excluding grade repetition), household income in $1,000 s of dollars, and the SF–12® mental and physical health summary measures. The SF–12® is a valid and reliable standardized tool for assessing mental and physical functioning and overall health–related quality of life [40,41]. There are 12 questions, all selected from the SF–36®, which assess indicators of health, including: role limitation due to physical problems, general health perceptions, vitality, bodily pain, social function, role limitations due to emotional problems, and general mental health. These indicators are used to calculate two summary component scores: mental component score (MCS) and the physical component score (PCS). Lower scores indicate poorer levels of health functioning.

6.2.2.3 FAMILY VARIABLES

Maternal self–reports in 1983 provided key information on family variables including: (1) household income in 1,000 s of dollars; (2) receipt of social assistance (0=no, 1=yes); (3) both parents born outside of Canada (0=no, 1=yes); (4) average years of education for both parents in two–parent households or mother's or father's years of education in lone–parent households; 5) one or both parents with a functional limitation (0=no, 1=yes); (6) one or both parents with a chronic medical health problem (0=no, 1=yes); (7) one or both parents hospitalized for "nerves" or a nervous condition (0=no, 1=yes); and (8) one or both parents ever treated for "nerves" or a nervous condition (0=no, 1=yes). A single variable assessing family functioning was measured using the general functioning subscale of the McMaster Family Assessment Device (FAD) [42,43]. Statements described family behavior and relationships across six dimensions: problem solving, communication, roles, affective responsiveness, affective involvement, and behavioral control. Scale scores were summed and converted to z scores. The FAD has adequate one–week test–retest reliability (.66 to .76, depending on subscale), low correlations with social desirability scales (–.06 to –.19), moderate correlations with other self–report measures of family functioning (most expected correlations exceeded .50), and the FAD differentiates significantly between clinician–rated healthy and unhealthy families [43].

6.2.2.4 CHILD VARIABLES

Child level variables include: gender (0=female, 1=male), age in years in 1983, as covariates, and low birth weight (0=>2500 grams, 1=<2500 grams) as a measure of prenatal risk. Four additional child risk variables were included for estimation of child health and functioning in 1983: (1) the presence or absence of a functional limitation (0 = no, 1 = yes); (2) the presence/absence of a medical condition (0 = no, 1 = yes); (3) the presence/absence of a psychiatric disorder (0 = no, 1 = yes); and (4) school performance. Mothers were the principal informants for measures of functional limitations and medical conditions. Measures of limitation of normal function and chronic illness or medical condition were adapted from various sources including the Rand Corporation's Measure of Children's Health and the Canada Health Survey [44]. Child functional limitation consisted of one or more limitations of normal functioning in physical activity (i.e. vigorous activity, bending, climbing), mobility (use of transportation and getting around the neighborhood), and self–care (daily activities—i.e. eating, dressing, bathing) due to illness, injury or medical condition, and/or limitation in role performance (kind or amount of ordinary play or schoolwork) due to physical, emotional or learning problems. Chronic illness or medical condition consisted of one or more illnesses/conditions present for at least 6 months' duration derived from a list of 22 separate conditions [44]. Assessments of child psychiatric disorders via problem checklists were provided by mothers and teachers for children aged 4–11 years, and mothers and youths, for adolescents aged 12–16 years. Classification of child psychiatric disorder consisted of the presence of one or more conditions, including conduct disorder, emotional disorder, and attention–deficit disorder. The checklists were originally developed to screen for psychiatric disorder among children in the general population [29]. Our measure of school performance included both teacher and maternal assessment of school performance. The teacher assessment was based on a 4–item rating scale to the questions: "How would you describe the child's current performance in the following categories: reading and English, spelling, arithmetic or math, and overall?" with response options 1 = far below grade to 5 = far above grade. Responses were summed and converted to z scores. Maternal assessments were used when teacher reports were missing and included the question: "Which of the following statements best describes how well ___ has done in school during the past 6 months": ranging from 1 = very well, excellent to 5 = not well at all, very poor student. Responses were reverse coded and converted to z scores.

6.2.2.5 RETROSPECTIVE REPORTS OF CHILDHOOD MALTREATMENT

Most risk variables measured at the family and child levels were assessed in 1983; however, exposure to childhood physical and sexual abuse prior to the age of 16 years was assessed in 2001 using the Childhood Experiences of Violence Questionnaire Short–Form (CEVQ–SF) [45]. The CEVQ–SF is a brief, reliable, valid, retrospective self–report measure assessing exposure to victimization and maltreatment [45,46]. Assessment of child physical abuse (PA) consists of 3 items assessing the frequency with which the individual was exposed to: (i) an adult slapping their head or face, or spanking with an object, (ii) having something thrown at them or being shoved, and (iii) being physically attacked, burned, choked or punched before age 16. Assessment for exposure to child sexual abuse (SA) consisted of a single question, "before age 16 when you were growing up, how many times did an adult ever do any of the following things when you didn't want them to: touch the private parts of your body or make you touch their private parts, threaten or try to have sex with you or sexually force themselves on you?" Each item consisted of a 5–point response option ranging from 1 (never) to 5 (10+ times). The cut–off score for severe physical abuse (PA) exposure was >10 times for the first two PA items (score of 5), and 1–2 times (score of 2 or above) for the third PA question. The cut–off for sexual abuse exposure was a response of 1–2 times (score of 2 or above) on the SA question. If a respondent met the cut–off criteria for severe PA or SA, they were coded as 1. The two–week test reliabilities of the CEVQ–SF for measuring PA, severe PA, and SA in an earlier study were: $\kappa=0.61$, $\kappa=0.72$ and $\kappa=0.91$ respectively [46].

6.2.2.6 MULTIPLE IMPUTATION AND ATTRITION WEIGHTS

Overall, 1,928 (58.5%) of the original 3,294 children were complete respondents in 2001 and an additional 427 (18.1%) completed abbreviated interviews; 910 were non–respondents, and 29 were excluded due to death (n=26) or institutional placement (n=3). The final sample for analysis included 1,928 participants. There were 549/1,928 (28.5%) participants with missing values on individual variables (i.e. 256 on only one variable and 293 on two or more variables). To estimate values for missed responses, we used multiple imputation in SPSS 19.0 to create five complete data sets. Missing family variables were imputed at the family level, whereas child variables were imputed at the child level. Models from each of these data sets were individually run in MLwiN 2.24 [47] and parameter estimates and the estimated standard errors (SEs) were combined using Rubin's rules [48].

Attrition weights were developed and applied to the original 1983 sample weights to recapture the original sample characteristics [28,49] using weighted complete–case analysis [50]. Fourteen variables measured in 1983 were selected to model non–response in 2001. The 1983 variables included were: child health status, functioning, and health service use; measures of parental health, family structure and functioning; and numerous indicators of family socioeconomic disadvantage. Several selected variables were associated with attrition, including child use of mental–health social services: 3.1% respondents, 7.5% no respondents (4.4% in 1983); and family in rental housing; 17.6% respondents, and 30.2% no respondents (21.3% in 1983).

To test the accuracy of the attrition weights, baseline characteristics of the OCHS sample in 1983 were compared with estimates derived using attrition weights applied to respondents at follow–up in 2001. This comparison yielded very similar estimates; for example, on the two variables identified above, this comparison yielded 4.4% versus 4.0% for child utilization of mental health–social services, and 21.3% versus 21.4% for family in rental housing. Health and functioning of OCHS respondents as young adults in 2001 were also compared with an independent probability sample of age–matched peers (N=5,718) living in Ontario and participating in the CCHS (Statistics Canada, 2004). Weighted estimates on identical variables derived from OCHS versus the CCHS were very close, for example, male sex (51.7, 50.1%), at work last week (80.6, 80.3%), personal income<$15,000 (21.6, 22.9%), excellent health (32.6, 32.3%), has asthma (11.4, 11.5%), smokes daily or occasionally (34.3, 35.2%) [49].

6.2.3 STATISTICAL ANALYSIS

The information collected on children in this study form a hierarchical structure consisting of individual children (Level 1, or child level) nested in families (Level 2, or family level). In this study, we use multilevel linear regression and the statistical software MLwiN 2.24 [47] to estimate the extent to which BMI assessed in 2001 is associated with child and family level risk factors assessed in 1983. In multilevel modeling, residual error is partitioned across levels, thereby capturing the extent to which variation in response is associated with each level. Fixed effect estimates in regression using linear multilevel modeling are interpreted in the same fashion as fixed effects in ordinary least squares regression: the intercept is the estimated mean response and the beta coefficients denote an increase or decrease in the dependent variable associated with one unit of change in each of the independent variables. Our modeling strategy consisted of introducing all confounding variables, including the 2001 variables (current education, income and mental and physical health scores), and child gender and age in 1983, as confounding factors, followed by family risk indicators (Model 1). This was then followed by the introduction of the child variables (Model 2). To examine whether there was a differential effect based on the age of the child in 1983, we initially stratified by childhood age, childhood (4–11 years) and adolescence (12–16 years). Because the results were consistently similar across age categories, we only report findings on the entire sample.

6.3 RESULTS

Table1 presents the sample characteristics, including the percent of families (n=1,270) and children (n=1,928) classified by each contextual variable, along with BMI in 2001. In young adulthood, the mean level of BMI was 25.38 (Table 1). Using cut–offs categorized by the World Health Organization, 2.5% of the sample were underweight (BMI<18.5 kg/m^2), 51.2% were normal weight (BMI>18.5 and<25 kg/m^2), 31.1% of the sample were overweight (BMI≥25 kg/m^2) and 15.2% were obese (BMI ≥30 kg/m^2). These findings are comparable to prevalence rates in self–reported BMI data in other Canadian studies [51,52].

TABLE 1: Sample Characteristics

Characteristic	
Families, $N = 1{,}270$	
Family income in $1,000 s (*M, SD*)	32.37, 15.50
Receipt of Social Assistance (*n*)	5.5% (106)
Both parents born > Canada (*n*)	19.5% (375)
Parent education in years (*M, SD*)	11.93, 3.38
Parent medical health problem (*n*)	21.2% (409)
Parent functional limitation (*n*)	8.2% (159)
Parent treated for "nerves" (*n*)	16.2% (311)
Parent hospitalized for "nerves" (*n*)	5.9% (119)
Family Functioning (*M, SD*)	36.14, 5.20
Children, $N = 1{,}928$	
Child Health in 1983	
Male child (*n*)	48.8% (941)
Age (*M, SD*)	10.08, 3.68
Medical condition (*n*)	16.2% (309)
Functional limitation (*n*)	4.5% (85)
Psychiatric disorder (*n*)	10.2% (197)
School performance—teacher/parent (*M, SD*)	3.36, 0.95
Low birth weight (*n*)	2.7% (53)
Retrospective Assessment in 2001 of Childhood Abuse	
Severe physical abuse (*n*)	18.2% (351)
Sexual abuse (*n*)	5.3% (107)
Outcomes and Covariates in 2001	
Body Mass Index (*M, SD*)	25.38, 4.78
Education in 2001	15.15, 2.70
Income in 2001 in $1,000 s (*M, SD*)	33.79, 21.97
SF–12® Mental Component Score (*M, SD*)	18.23, 2.11
SF–12® Physical Component Score (*M, SD*)	22.32, 3.32

6.3.1 *INFLUENCE OF FAMILY AND CHILD VARIABLES ON BMI*

Variability in BMI attributable to family–level differences is estimated by the intraclass correlation coefficient (ICC). The ICC is derived from the random effects variances reported in the multilevel null model and represents the total unexplained variance in BMI associated with between–family differences: 39.05% from 8.92/(8.92+13.92) (not shown).

Multilevel regressions for BMI are presented in Table 2. Controlling for current education, income and mental and physical health status, and for child age and sex, we found that in Model 1 (Family Variables), individuals from families who received social assistance during their childhood had higher BMI in early adulthood (2.02, $p<0.001$), while being the child of immigrant parents (born outside of Canada) was associated with lower BMI (−1.20, $p<0.001$). Family income was not significantly associated with BMI (.004). In addition, parental education (positive), parent hospitalized for "nerves" and family functioning (negative) exhibited significant associations with BMI. Every one year increase in parental education was associated with 0.18 decrease in BMI ($p<0.001$), whereas having a parent hospitalized for nerves was associated with increases in BMI of 1.20 ($p<0.05$).

In Model 2 (Child Variables), individuals classified with a childhood psychiatric disorder and poorer school performance exhibited higher BMI ($p<0.01$ and 0.05, respectively). Low birth weight status was not significantly associated with BMI. The total proportional reduction in error (explained variance) associated with the predictor variables in Model 2 is 1–(7.60+12.87)/(8.92+13.92) or 10.4%.

To explore if associations between family and child variables and adult BMI were modified by gender, we tested statistical interactions between gender and all of the variables in Model 2 in Table2. Each interaction was tested on its own (i.e., added separately to Model 2). There were two statistically significant interactions: one involving gender by receipt of social assistance ($\beta=-2.44$, $SE=1.05$) and the other involving gender by childhood history of medical condition (β=−1.36, SE=.58). As shown in Figure1, there was a positive association between receipt of social assistance and BMI for females but not for males. Similarly, as illustrated in Figure2, females with a childhood history of a medical condition had higher BMIs than those with no history of a medical condition. This effect was not seen in males. As mentioned above, we replicated these findings using corrected BMI (data not shown).

TABLE 2: Multilevel Models Neighborhood, Family, and Childhood Influences on BMI (b and (95% Confidence Interval))

	Model 1 Family Variables	**Model 2 Child Variables**
Fixed effects		
Intercept	24.76 (24.35 to 25.16)	25.62 (24.64 to 26.59)
Family Variables		
Family income in $1,000 s	0.004 (–0.01 to 0.02)	0.0038 (–0.01 to 0.02)
Social assistance	2.02 (0.92 to 3.11)***	1.89 (0.26 to 3.52)*
Immigrant parents	–1.20 (–1.78 to –0.61)***	–1.20 (–1.73 to –0.66)***
Education in years	–0.18 (–0.25 to –0.09)***	–0.17 (–0.26 to –0.09)***
Parent medical problem	0.47 (–0.12 to 1.06)	0.46 (–0.13 to 1.06)
Parent functional limitation	–0.50 (–1.35 to 0.36)	–0.50 (–1.40 to 0.37)
Parent treated for "nerves"	–0.10 (–0.74 to 0.55)	–0.16 (–0.85 to 0.53)
Parent hospitalized for "nerves"	1.20 (0.15 to 2.24)*	1.19 (–0.01 to 2.39)
Family functioning	–0.05 (–0.10 to –0.007)*	–0.06 (–0.11 to –0.01)*
Child variables		
Age in years	0.11 (0.05 to –0.22)	0.10 (0.04 to 0.17)**
Male	1.28 (0.87 to 1.70)***	1.27 (0.66 to 1.59)***
Psychiatric disorder		1.12 (0.31 to 1.94)**
Functional limitation		–0.11 (–1.43 to 1.21)
Medical condition		–0.03 (–0.71 to 0.66)
School performance		–0.26 (–0.50 to –0.01)*
Low birth weight		–0.77 (–2.19 to 0.64)
Physical abuse		0.12 (–0.47 to 0.72)
Social abuse		0.50 (–1.33 to 0.64)
Random effects (*SE*)		
Level 2, Family	7.65 (0.79)	7.60 (1.27)
Level 1, Child	13.03 (0.64)	12.87 (0,99)
–2*log likelihood	11,519	11,498

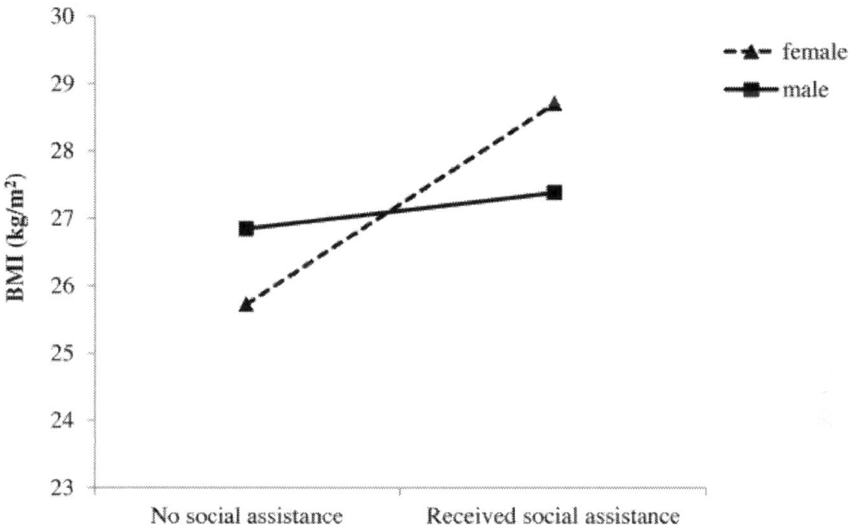

FIGURE 1: Gender moderates the association between receipt of social assistance in childhood/adolescence and BMI in early adulthood.

$* p<.05$, $** p<.01$, $*** p<.001$.

6.4 DISCUSSION

Using a comprehensive model, simultaneously incorporating child and family variables in a prospective design, we examined associations between a number of risk factors experienced in childhood/adolescence and BMI in early adulthood, adjusting for respondents' age, sex, education, income and health. This study provides several important findings at the family level: 1) socioeconomic adversity (measured by receipt of social assistance) was related to increased BMI, whereas parental education was associated with lower BMI in early adulthood; 2) parental immigrant status was associated with lower BMI; 3) family functioning was negatively related to BMI (higher family function was associated with lower BMIs) and 4) parental mental health problems were associated with increased BMI. At the child level, presence of child mental disorders and poor school performance were both related to higher BMI, even after controlling for cur-

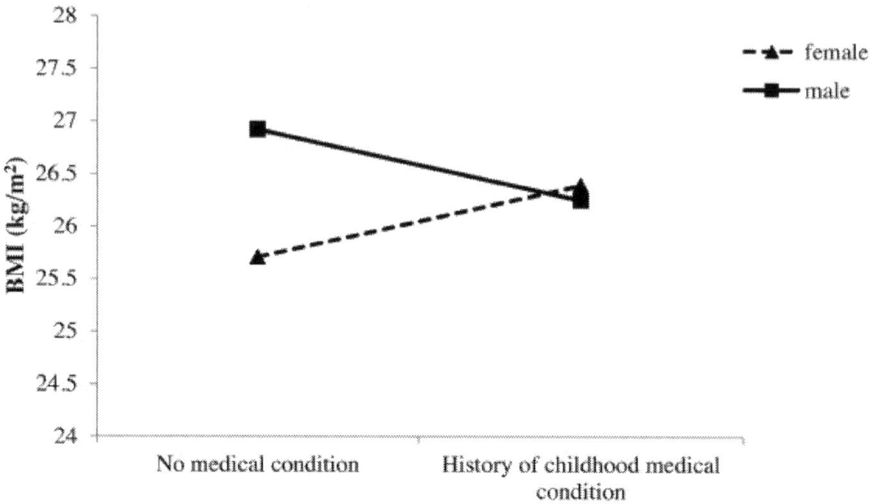

FIGURE 2: Gender moderates the association between presence of a medical condition in childhood/adolescence and BMI in early adulthood.

rent education and mental health status. Altogether, the predictor variables explained about 10% of respondent variability in BMI at young adulthood. In the fully adjusted model, the effects of variables such as immigrant status and social assistance converted to standard deviation units exhibited what would be considered small to medium effects (d=0.25 and 0.40) based on Cohen's criteria [53].

In this study, we found that 39.05% of the variation in BMI was associated with between–family differences. The familial aggregation of BMI reflects a multifaceted interplay between genetic susceptibility to weight gain and shared environmental influences within families. Some of these shared influences are linked to measured family risk factors, such as socioeconomic disadvantage, parental education and family functioning. Recent studies have found that genetics play an increasingly important role in explaining variation in BMI over time, with its greatest influence in late adolescence; whereas environmental influences decrease over time, exerting their strongest effects early in childhood and adolescence [54–56].

Our finding that family status factors (parental education and receipt of social assistance) are associated with BMI is consistent with previous research [5–8,10]. In children and adolescents, lower SES, regardless of how it is measured (parental education, occupation, or income), is associated with increased BMI and obesity [57–59]. Economic disadvantage in families may be directly related to a number of factors that pose a risk for higher BMI, such as engaging in less physical activity, poorer nutrition and eating habits, and lack of participation in organized sports [60,61]. In addition, early in childhood, healthy lifestyle trajectories may be set via modeling by caregivers [61]. We also found that parental immigrant status was negatively related to BMI in adulthood. This finding is in agreement with a recent report indicating that first generation immigrants have lower BMI compared to second generation or Canadian born children [62]. Interestingly, children from that study lived in a multi–ethnic, disadvantaged inner city community, where many immigrants initially settle. These areas are typically characterized by lower levels of education and income. Despite these socioeconomic risks, immigrant status still conferred a protective influence in this sample; however, this protective health advantage may be lost over time with exposure to unhealthy lifestyle habits in host communities.

Family functioning was associated with lower BMI in early adulthood. Our measure of family functioning was comprised of questions on problem solving, affective responsiveness and involvement and behavioral control within the family. These questions tap into parenting practices. Research investigating the impact of parenting practices on BMI in children and adolescents has found that authoritative homes, characterized by a family context of warmth, high emotional support, encouragement, monitoring and bidirectional communication, is related to healthier eating habits, increased physical activity and lower BMI [63–67]. Interestingly, our distal family status factors (parental education and receipt of social assistance) remained significant even after adding family functioning to the model. This implies that family functioning does not act as a mediator between family status and BMI but exerts its own independent effects. Pathways relating family functioning to BMI are likely complex and may impact weight through its direct influence on diet and physical activity [66] or through more indirect mechanisms such as child self–regulation capabilities [68].

Parental mental health problems predicted elevated BMI, although this association became non–significant when childhood variables were added to the model. Consistent with previous findings [15,19,27], we found that childhood psychiatric disorder and school difficulties were associated with greater BMI in early adulthood, even after controlling for current mental health status and years of education. Although there is some debate regarding the magnitude of the effect, starting at an early age [15] mental health problems are established predictors of elevated BMI through the lifespan [16–18]. Given the established nature of this association, there is likely an underlying mechanism shared by both mental health and obesity. Potential candidates include genetic, behavioral, and/or psychological factors which are common to both phenotypes [27,69]. Contrary to previous research [21,22], we did not find an association between history of childhood physical or sexual abuse and adult BMI. This may be due to the fact that we have such a comprehensive array of risk factors measured. Also contrary to previous research [12,13], we did not find an association between low birth weight and adult BMI. This may be due to reporting inaccuracies: birth weights were based on self–reports by mothers several years after the birth of their child. The prevalence of low birth weight in this sample (2.5%) is lower than prevalence rates reported for Ontario (4.8%), suggesting that low birth weight may be under represented in our sample.

We found that gender moderated the effect of two risk factors on BMI: receipt of social assistance and presence of a medical condition in childhood. In females, but not in males, the presence of these risk factors was associated with higher BMI in early adulthood. Our results support other findings that childhood socioeconomic disadvantage is associated with later obesity in women [70–72]. The association between BMI and later chronic disease in adulthood is fairly well characterized; however research linking medical problems in childhood to later BMI is relatively scarce. One recent study found that childhood leukemia was associated with increased BMI in adulthood but only in females [73]. Presence of medical conditions or chronic illnesses in childhood may place greater limitations on physical activity, leading to a more sedentary lifestyle. In addition, it is noteworthy that within the OCHS, the prevalence of a chronic illness or medical condition in 1983 was not evenly distributed across socioeconomic groups. Children from low income households, defined as below the Statistics Canada poverty line, had higher rates of medical conditions compared to children of families above the poverty line [50]. It is possible that these gender differences are linked to variations in underlying biological mechanisms, such as HPA axis function given that there are sex differences in stress system structure and function [74] and that the HPA is linked to BMI [75,76], various medical conditions [77] and disparities in SES [78]. Theoretically, it is argued that females have a heightened predisposition to stress–related disease with exaggerated sensitivity to stress which pushes females over the "disease threshold" [79]. If this is true, medical conditions or socioeconomic disadvantage experienced in childhood may lead to biological sensitivities to stress, especially in females, putting them at greater risk of elevated BMI later in life.

6.4.1 LIMITATIONS

Despite strengths of the OCHS in assessing childhood risks associated with BMI in early adulthood, this study has limitations. First, approximately 30% of 1983 participants were lost over the 18–year follow–up. Because this loss was selective to socioeconomic disadvantage, Boyle and colleagues [49] devised attrition weights that integrated original sample selection probabilities from 1983. We believe that any potential systematic bias is likely to be small and would be more related to underestimation of the influence of risk factors on outcomes. Second, there are a few measurement limitations. All risk factors were measured at one point in time, and we are not able disentangle their temporal associations or to assess intervening variables between 1983 and 2001. Our assessment of risk factors was collected prospectively over twenty years ago. Despite this lapse of time, we believe our findings are currently applicable to young adults who are likely to experience the same risk factors. We are unaware of any changes that would mitigate the association between childhood risk factors and BMI, as found in this study. Two, we do not have measures of child or parent BMI in 1983. It is well established in studies tracking weight status in children that those with higher BMIs early on tend to maintain these trajectories throughout adolescence and into adulthood, indicating some stability for most individuals [11,80,81]. Our measure of BMI was also based on self–report. Self–reported BMI yields lower rates of obesity and overweight [35–37]. Nevertheless, self–reported BMI remains an important tool for health surveillance [51]; commonly used because it is a simple, economical, and non–invasive method of collecting data from large samples [82]. Moreover, self–reported BMIs are related to morbidity and mortality [83–85]. Applying a BMI correction factor did not alter our findings. Three, our abuse risk indicators were measured retrospectively. In general population studies, such assessments are very difficult to obtain from children prospectively and there is no reason to believe that current BMI would influence the recall of these experiences. Fourth, in exploring interactions between gender and childhood risk factors, we are vulnerable to obtaining a significant effect because of multiple testing. We chose to explore these interactions because of the lack of consistent research evidence on this question and the recognition that differential weight and body fat composition are integral to gender differences.

6.5 CONCLUSIONS

In conclusion, this study provides support for the notion that a significant proportion of the variation in BMI is attributable to family factors. More studies are needed to disentangle the influence of specific family factors on BMI, including genetic by environment interactions. To date, most preventive programs for childhood obesity are school–based and focus, with limited effectiveness, on healthy–eating, active living, and mental well–being initiatives [86–88]. Our findings suggest that preventive interventions and policy practices need to target family environments early in childhood, especially environments where multiple risk indicators are present in the family. Family based programs targeting parenting styles and skills, as well as dietary behaviors and physical activity have demonstrated positive effects on children's weight loss [89,90]. More research and development of family–based prevention and treatment programs is needed.

REFERENCES

1. WHO: Obesity and overweight. http://www.who.int/mediacentre/factsheets/fs311/en/

2. Public Health Agency of Canada: Obesity in Canada. A joint report from the Public Health Agency of Canada and the Canadian Institute for Health Information, Ottawa; 2011.
3. Fiucane MM, Stevens GA, Cowan MJ, Danaei G, Lin JK, Paciorek CJ, et al.: on behalf of the Global Burden of Metabolic Risk Factors of Chronic Diseases and Collaborating Group (Body Mass Index): National, regional, and global trends in body–mass index since 1980: systematic analysis of health examination surveys and epidemiological studies with 960 country–years and 9.1 million participants. Lancet 2011, 377(9765):557–567.
4. Finkelstein EA, Fiebelkorn IC, Wang G: National Medical Spending Attributable to Overweight and Obesity: How Much, and Who's Paying? Health Affairs 2003, 3:219–223.
5. Poulton R, Caspi A, Milne BJ, Thompson A, Sears MR, Moffitt TE: Association between children's experience of socioeconomic disadvantage and adult health: a life–course study. Lancet 2002, 360(9346):1640–1645.
6. Power C, Manor O, Matthews S: Child to adult socioeconomic conditions and obesity in a national cohort. Int J Obes Relat Metab Disord 2003, 27(9):1081–1086.
7. Tamayo T, Herder C, Rathmann W: Impact of early psychosocial factors (childhood socioeconomic factors and adversities) on future risk of type 2 diabetes, metabolic disturbances and obesity: a systematic review. BMC Public Health 2010, 2010(10):525–540.
8. Wells NM, Evans GW, Beavis A, Ong AD: Early childhood poverty, cumulative risk exposure, and body mass index trajectories through young adulthood. Am J Public Health 2010, 100(12):2507–2512.
9. Huffman FG, Kanikireddy S, Patel MJ: Parenthood – A contributing factor to childhood obesity. Int J Environ Res Public Health 2010, 7:2800–2810.
10. Kestila L, Rahkonen O, Martelin T, Lahti–Koski M, Koskinen S: Do childhood social circumstances affect overweight and obesity in early adulthood? Scand J Public Health 2009, 37:206–219.
11. Pryor LE, Tremblay RE, Boivin M, Touchette E, Dubois L, Genolini C, Liu X, et al.: Developmental trajectories of body mass index in early childhood and their risk factors: An 8–year longitudinal study. Arch Pediatr Adolesc Med 2011, 165(10):906–912.
12. Druet C, Ong KK: Early childhood predictors of adult body composition. Best Pract Res Clin Endocrinol Metab 2008, 2008(22):489–502.
13. Parsons TJ, Power C, Logan S, Summerbell CD: Childhood predictors of adult obesity: a systematic review. Int J Obes Relat Metab Disord 1999, 23:S1–S107.
14. Brook JS, Zhang C, Saar NS, Brook DW: Psychosocial predictors, higher body mass index, and aspects of neurocognitive dysfunction. Percept Mot Skills 2009, 108(1):181–195.
15. Durante CS, Sourander A, Niklakaros G, Pihlajamaki H, Helenius H, Piha J, et al.: Child mental health problems and obesity in early adulthood. J Pediatr 2010, 156:93–97.
16. Gaysina D, Hotopf M, Richards M, Colman I, Kuh D, Hardy RJ: Symptoms of depression and anxiety, and change in body mass index from adolescence to adulthood: results from a British birth cohort. Psychol Med 2011, 41:175–184.

17. Pine DS, Cohen P, Brook J, Coplan JD: Psychiatric symptoms in adolescence as predictors of obesity in early adulthood: a longitudinal study. Amer J Public Health 1997, 87:1303–1310.
18. Anderson SE, He X, Schoppe–Sullivan S, Must A: Externalizing behavior in early childhood and body mass index from age 2 to 12 years: longitudinal analyses of a prospective cohort study. BMC Pediatr 2010, 10:49.
19. Mamun AA, O'Callaghan MJ, Cramb SM, Najman JM, Williams GM, Bor W: Childhood behavioral problems predict young adults' BMI and obesity: evidence from a birth cohort study. Obesity 2009, 17:761–766.
20. Irish L, Kobayashi I, Delahanty DL: Long–term physical health consequences of childhood sexual abuse: A meta–analytic review. J Pediatr Psychol 2010, 35(5):450–461.
21. Bentley T, Widom CS: A 30–year follow–up of the effects of child abuse and neglect on obesity in adulthood. Obesity 2009, 17(10):1900–1905.
22. Greenfield EA, Marks NF: Violence from parents in childhood and obesity in adulthood: Using food in response to stress as a mediator of risk. Soc Sci Med 2009, 68(5):791–798.
23. Thomas C, Hyppönen E, Power C: Obesity and Type 2 diabetes risk in midadult life: The role of childhood adversity. Pediatrics 2008, 121(5):e1240–e1249.
24. Lawlor DA, Clark H: Smith Davey G, Leon DA: Childhood intelligence, educational attainment and adult body mass index: findings from a prospective cohort and within sibling–pairs analysis. Int J Obesity 2006, 30:1758–1765.
25. Lissau I, Sorensen TI: School difficulties in childhood and risk of overweight and obesity in young adulthood: a ten year prospective study. Int J Obes Relat Metabol Disord 1993, 1993(17):169–175.
26. Sund ER, Jones A, Midthjell K: Individual, family, and area predictors of BMI and BMI change in an adult Norwegian population: Findings from the HUNT study. Soc Sci Med 2010, 70:1194–1202.
27. Vamosi M, Heitmann BL, Kyvik KO: The relation between an adverse psychological and social environment in childhood and the development of adult obesity: a systematic literature review. Obes Rev 2010, 11(3):177–184.
28. Boyle MH, Georgiades K, Racine Y, Mustard C: Neighborhood and family influences on educational attainment: results from the Ontario child health study follow–up 2001. Child Dev 2007, 78:168–189.
29. Boyle MH, Offord D, Hofmann HF, Catlin GP, Byles JA, Cadman DT, et al.: Ontario child health study: I. Methodology: Arch Gen Psychiatry 1987, 44:826–831.
30. Brene ND, McManus T, Galuska DA, Lowry R, Wechsler H: Reliability and validity of self–reported height and weight among high school students. J Adolesc Health 2003, 32:281–287.
31. Bolton–Smith C, Woodward M, Tunstall–Pedoe H, Morrison C: Accuracy of the estimated prevalence of obesity from self–reported height and weight in an adult Scottish population. J Epidemiol Commun Health 2000, 54:143–148.
32. Niedhammer I, Bugel I, Bonenfant S, Goldberg M, Leclere A: Validity of self–reported weight and height in the French GAZEL cohort. Int J Obes Relat Metabol Disord 2000, 24:1111–1118.

33. Rimm EB, Stampfer MJ, Colditz GA, Chute CG, Litin LB, Willett WC: Validity of self–reported waist and hip circumferences in men and women. Epidemiology 1990, 1:466–473.
34. Connor Gorber S, Tremblay M, Moher D, Gorber B: A comparison of direct versus self–report measures for assessing height, weight and body mass index: a systematic review. Obes Rev 2007, 8(4):373–374.
35. Elgar FJ, Roberts C, Tudor–Smith C, Moore L: Validity of self–reported height and weight and predictors of bias in adolescents. J Adolesc Health 2005, 37:371–375.
36. Rowland ML: Self–reported weight and height. Am J Clin Nutr 1990, 52(6):1125–1133.
37. Shields M, Gorber Connor S, Janssen I: Tremblay MS: Bias in self–reported estimates of obesity in Canadian health surveys. 3rd edition. An update on correction equations for adults, Catalogue no.82–003–XPE. Statistics Canada; 2011:22.
38. Guh DP, Zhang W, Bansback N, Amarsi Z: Birmingham Laird C. Anis AH: The incidence of co–morbidities related to obesity and overweight: A systematic review and meta analysis. BMC Public Health 2009, 9:88.
39. Tan MLS, Wee H–L, Lee J, Ma S, Heng D, Tai E–S, et al.: Association of anthropometric measures with SF–36v2 PCS and MCS in a multi ethnic Asian population. Qual Life Res; 2012.
40. Resnick B, Parker R: Simplified scoring and psychometrics of the revised 12–item Short–Form Health Survey. Outcomes Manag Nurs Pract 2001, 5:161–166.
41. Ware JE, Kosinski M, Keller SD: A 12 item Short Form Health Survey. Construction of scales and preliminary tests of reliability and validity. Med Care 1996, 34:220–233.
42. Byles JA, Byrne C, Boyle MH, Offord DR: Ontario child health study: Reliability and validity of the general functioning subscale of the McMaster Family Assessment Device. Fam Process 1988, 27:97–104.
43. Epstein NB, Baldwin LM, Bishop DS: The McMaster Family Assessment Device. J Marital Fam Ther 1983, 9:171–180.
44. Cadman DT, Boyle MH, Offord DR, Szatmari P, Rae–Grant NI, Crawford JW, et al.: Chronic illness and functional limitations in Ontario children: Findings of the Ontario Child Health Study. CMAJ 1986, 135:761–767.
45. Walsh CA, MacMillan HL, Trocmé N, Jamieson E, Boyle MH: Measurement of victimization in adolescence: development and validation of the Childhood Experience of Violence Questionnaire. Child Abuse Negl 2008, 32:1037–1057.
46. Tanaka M, Wekerle C, Leung E, Waechter R, Gonzalez A, Jamieson E, MacMillan HL: Preliminary Evaluation of the Childhood Experiences of Violence Questionnaire Short Form. J Interpers Violence 2011, 27(2):396–407.
47. Rabash J, Steele F, Browne W, Goldstein H: A user's guide to MLwiN, Version 2.14. Institute of Education, London; 2009.
48. Rubin DB: Multiple imputation for nonresponse in surveys. John Wiley & Sons, New York; 1987.
49. Boyle MH, Hong S, Georgiades K, Duku E, Racine YA, Mustard C: Ontario child health study follow–up 2001. Evaluation of sample loss. Hamilton, Ontario, Offord Centre for Child Studies, McMaster University; 2006.
50. Little RJA, Rubin DB: Statistical analysis with missing data. John Wiley & Sons, New York; 2002.

51. Elgar F, Stewart J: Validity of self–report screening for overweight and obesity: Evidence from the Canadian Community Health Survey. Can J Public Health 2008, 99(5):423–427.
52. Statistics Canada. http://www.statcan.gc.ca/pub/82–625–x/2010002/article/11255–eng.htm
53. Cohen J: A coefficient of agreement for nominal scales. Educational and Psychological Measurement 1960, 20(1):37–46.
54. Dubois L, Kyvik KO, Girard M, Tatone–Tokuda F, Perusse D, Hjelmborg J, et al.: Genetic and environmental contributions to weight, height, and BMI from birth to 19 years of age: An international study of over 12,000 twin pairs. PLOS one 2012, 7(2):e30153.
55. Silventoinen K, Rokhom B, Kaprio J, Sorensen TIA: The genetic and environmental influences on childhood obesity: a systematic review of twin and adoption studies. Int J Obesity 2010, 34:29–40.
56. Lajunen H–R, Kaprio J, Rose RJ, Pulkkinen L, Silventoinen K: Genetic and environmental influences on BMI from late childhood to adolescence are modified by parental education. Obesity 2012, 20:583–589.
57. Bauman LJ, Silver EJ, Stein RE: Cumulative social disadvantage and child health. Pediatrics 2006, 117(4):1321–1328.
58. Chen E, Matthews KA, Boyce WT: Socioeconomic differences in children's health: How and why do these relationships change with age? Psychol Bull 2002, 128:295–329.
59. Larson K, Russ SA, Crall JJ, Halfon N: Influence of multiple social risks on children's health. Pediatrics 2008, 121:337–44.
60. Bradley RH, Corwyn RF: Socioeconomic status and child development. Ann Rev Psychol 2002, 53:371–399.
61. Hanson MD, Chen E: Socioeconomic status and health behaviors in adolescence: a review of the literature. J Behav Med 2007, 30(3):263–285.
62. Maximova K, O'Loughlin J, Gray–Donald K: Healthy weight advantage lost in one generation among immigrant elementary schoolchildren in multi–ethnic, disadvantaged, inner–city neighborhoods in Montreal, Canada. Ann Epidemiol 2011, 21:238–244.
63. Berge JM, Wall M, Loth K, Neumark–Sztainer D: Parenting style as a predictor of adolescent weight and weight–related behaviors. J Adolesc Health 2010, 46(4):331–338.
64. Bradley RH, McRitchie S, Houts RM, Nader P, O'Brien M: NICHD Early Child Care Research Network: Parenting and the decline of physical activity from age 9 to 15. Int J Behav Nutr Phys Act 2011, 15(8):33.
65. Rhee KE, Lumeng JC, Appugliese DP, Kaciroti N, Bradley RH: Parenting styles and overweight status in first grade. Pediatrics 2006, 117(6):2047–2054.
66. Sleddens EF, Gerards SM, Thijs C, de Vries NK, Kremers SP: General parenting, childhood overweight and obesity–inducing behaviors: a review. Int J Pediatr Obes 2011, 6(2–2):e12–e27.

67. Rodenburg G, Kremers SP, Oenema A, van de Mheen D: Psychological control by parents is associated with a higher child weight. Int J Pediatr Obes 2011, 6(5–6):442–449.
68. Evans GW, Fuller–Rowell TE, Doan SN: Childhood cumulative risk and obesity: the mediating role of self–regulator ability. Pediatrics 2012, 129(1):e68–e73.
69. Repetti RL, Robles TF, Reynolds B: Allostatic processes in the family. Dev Psychopathol 2011, 23:921–938.
70. Maty SC, Lynch JW, Raghunathan TE, Kaplan GA: Childhood socioeconomic position, gender, adult body mass index, and incidence of Type 2 diabetes mellitus over 34 years in the Alameda County Study. Am J Public Health 2008, 98:1486–1494.
71. Muennig P, Lubetkin E, Jia H, Franks P: Gender and the burden of disease attributable to obesity. Am J Public Health 2006, 96(9):1662–1668.
72. Senese LC, Almeida ND, Fath KA, Smith BT, Loucks EB: Associations between childhood socioeconomic position and adulthood obesity. Epidemiol Rev 2009, 31:21–51.
73. Meacham LR, Gurney JG, Mertens AC, Ness KK, Sklar CA, Robison LL, et al.: Body mass index in long–term adult survivors of childhood cancer: a report of the Childhood Cancer Survivor Study. Cancer 2005, 103(8):1730–1739.
74. Cahill L: Why sex matters for neuroscience. Nat Rev Neurosci 2006, 7:477–484.
75. Hillman JB, Dorn LD, Loucks TL, Berga SL: Obesity and the hypothalamic–pituitary–adrenal axis in adolescent girls. Metabolism 2012, 61(3):341–348.
76. Pasquali R, Vicennati V, Gambineri A, Pagotto U: Sex–dependent role of glucocorticoids and androgens in the pathophysiology of human obesity. Int J Obes 2008, 32:1764–1779.
77. McEwen BS, Gianaros PJ: Stress– and allostasis–induced brain plasticity. Ann Rev Med 2011, 62:431–445.
78. Seeman T, Epel E, Gruenewald T, Karlamangla A, McEwen BS: Socio–economic differentials in peripheral biology: cumulative allostatic load. Ann NY Acad Sci 2010, 1186:223–239.
79. Becker JB, Monteggia LM, Perrot–Sinal TS, Romeo RD, Taylor JR, Yehuda R, Bale TL: Stress and disease: Is being female a predisposing factor? J Neurosci 2007, 27(44):11851–11855.
80. Hesketh K, Wake M, Waters E, Carlin J, Crawford D: Stability of body mass index in Australian children: a prospective cohort study across the middle childhood years. Public Health Nutr 2003, 7(2):303–309.
81. Whitaker RC, Wright JA, Pepe MS, Seidel KD, Dietz WH: Predicting obesity in young adulthood from childhood and parental obesity. New Engl J Med 1997, 337(13):869–873.
82. Engstrom JL, Paterson SA, Doherty A, Trabulsi M, Speer K: Accuracy of self–reported height and weight in women: An integrative review of the literature. J Midwifery Womens Health 2003, 48(5):338–345.
83. Af Sillen U, Nilsson J, Mansson N, Nilsson PM: Self–rated health in relation to age and gender: influence on mortality risk in the Malmo Preventive Project. Scand J Public Health 2005, 33:183–189.

84. Larsson D, Hemmingsson T, Allebeck P, Lundberg I: Self–rated health and mortality among young men: What is the relation and how may it be explained? Scand J Public Health 2002, 30:259–266.
85. Stommel M, Schoenborn CA: Accuracy and usefulness of BMI measures based on self–reported weight and height: findings from the NHANES & NHIS 2001–2006. BMC Public Health 2009, 9:421–431.
86. Birch LL, Ventura AK: Preventing childhood obesity: what works? Int J Obes 2009, 33:S74–S81.
87. Flynn MAT, McNeil DA, Maloff B, et al.: Reducing obesity and related chronic disease risk in children and youth: a synthesis of evidence with 'best practice' recommendations. Obes Rev 2006, 7(1):7–66.
88. Whitlock EP, O'Connor EA, Williams SB, Beil TL, Lutz KW: Effectiveness of weight management interventions in children: A targeted systematic review for the USPSTF. Pediatrics 2010, 125:e396.
89. Kitzman–Ulrich H, Wilson DK, St George SM, Lawman H, Segal M, Fairchild A: The integration of a family systems approach for understanding youth obesity, physical activity, and dietary programs. Clin Child Fam Psych Rev 2010, 13(3):231–253.
90. Gerards S, Dagnelie PC, Jansen MWJ, de Vries NK, Sanders MR, Kremers SPJ: Lifestyle Triple P: a parenting intervention for childhood obesity BMC Public Health. 2012, 12:267.

PART II

ADVERSE HEALTH CONSEQUENCES

CHAPTER 7

CHILDHOOD OBESITY AND OBSTRUCTIVE SLEEP APNEA

INDRA NARANG and JOSEPH L. MATHEW

7.1 INTRODUCTION

The epidemic of pediatric obesity has caused serious concern all over the world as the prevalence has increased alarmingly over time, not only in developed countries but also in developing countries [1, 2]. Furthermore, there is increasing recognition that childhood obesity is occurring at progressively younger ages [3]. Recent publications have highlighted the challenge of defining childhood obesity in a manner that is both evidence based as well as uniformly applicable across different settings [4]. In general, a statistical definition using BMI for age is used wherein >85th percentile is defined as overweight and >95th percentile as obesity [5]. In contrast, the WHO defines obesity as BMI for age *Z*–score >3 and overweight as *Z*–score >2. A large–scale multicentric study calculated BMI in children and adolescents and extrapolated the cut–off values for adult obesity (BMI > 30) and overweight (BMI > 25), to the corresponding values in childhood and adolescence. Based on this, they were able to tabulate age– and gender–specific cut–off values for children and adolescents [6]. At the present time, waist circumference is not used routinely to define obesity in children and adolescents.

The high prevalence of obesity is believed to be a complex interplay of genetic, environmental (life–style), socioeconomic, cultural, and psychological factors which are beyond the scope of this paper. However, interestingly the pattern of in utero growth may program the pattern of subsequent body fat deposition and neuroendocrine interactions that promote eating behavior. Specifically, there is an observed increase in childhood obesity with increasing birth weight [7]. Counterintuitively, infants with low birth weight and an early adiposity rebound are also predisposed to higher rates of obesity in later childhood [8].

There are several well–documented adverse consequences of childhood obesity. Specific morbidities associated with obesity include hypertension, left ventricular abnormalities, insulin resistance, type 2 diabetes, dyslipidemia, nonalcoholic fatty liver disease, and obstructive sleep apnea [9]. Further, in obese individuals, the clustering of dyslipidemia, hypertension, and impaired glucose tolerance/insulin resistance is referred to as the Metabolic Syndrome (MetS), which further increases the risk of atherosclerotic heart disease [10]. In a followup of over 200,000 Norwegian adolescents, the relative risk of death due to ischaemic heart disease for those with a BMI > 85 percentile was 2.9 for males and 3.7 for females when compared to those with lower BMI percentiles [11]. These may underrepresent the true cardiovascular burden in adulthood given the knowledge that 75% of obese children will become obese adults [12, 13]. Additional complications of obesity include menstrual problems and polycystic ovarian disease, gallstones, orthopedic issues, and psychological stress compounding poor quality of life [14, 15] in these children.

7.2 OBSTRUCTIVE SLEEP APNEA (OSA)

During sleep in normal individuals, there is reduction in the tone of airway musculature; however pharyngeal dilator activity keeps the airway patent. Therefore, although normal children can have occasional pauses in breathing for up to 10–15 seconds, there is no significant airflow limitation. Therefore paO_2 may fall only by 2–4 mm Hg, and end–tidal CO_2 may increase marginally by 3–4 mm Hg. More importantly, there is no arousal from sleep [16, 17].

Obstructive sleep apnea (OSA) is part of the spectrum of clinical conditions comprising sleep–disordered breathing (SDB). The spectrum of SDB ranges from partial to complete upper airway obstruction. In children, obstructive apnea is defined by the absence of nasal airflow despite the presence of chest wall and abdominal wall movements, for a duration of at least two breaths. In contrast, the term "obstructive hypopnea" refers to decrease in nasal airflow by 50% from the baseline accompanied by fall in oxygen saturation of 3% and/or arousal. The number of apneic and hypopneic events per hour of sleep is expressed as apnea/hypopnea index (AHI) on polysomnography [18]. In adults, AHI < 5/hour is considered normal. However, in children, AHI > 1event/hour is regarded abnormal. In general, the same criteria can be used for adolescents in the age group 12–15 years. For those beyond 18 years, it is recommended that adult criteria be used. In children, AHI is also used to categorize the severity of OSA; AHI up to 1.5 events/hr is classified as mild, 1.5–5.0 events/hr as moderate, and >5.0 events/hr as severe. This is in contrast to adults, where the corresponding values are 5–15 events/hr, 15–30 events/hr, and >30 events/hr.

OSA is characterised by snoring, recurrent partial (hypopneas), or complete (apneas) obstruction of the upper airway. OSA is associated with intermittent oxyhemoglobin desaturation, sleep disruption, and fragmentation [19]. About 3–12% of the "healthy" pediatric population has habitual snoring, whereas only 1–3% has OSA [20, 21]. However it should be noted that snoring is not synonymous with OSA. Habitual snorers typically do not have obstructive apnea, hypopnea, respiratory effort–related arousals, or abnormal gas exchange. This is because neuromuscular compensation in these children prevents significant airway obstruction.

In OSA, the episodes of airway obstruction can be related to increased airway collapsibility on account of mechanical and neuronal factors. The most common mechanical factor in children is hypertrophy of adenoids and/or tonsils narrowing the airway lumen [22]. Approximately 2% of otherwise healthy children have large tonsils and adenoids that mechanically obstruct airways [20, 23]. However, OSA is a balance of mechanical obstruction and decreased activity of pharyngeal dilator muscle activity. During sleep, children with OSA have reduced airway muscle tone which critically narrows and obstructs the airway, resulting in upper airway obstruction. The consequent hypoxemia results in arousal with restoration of airway tone and relief of the obstruction. The frequency of episodic apnea determines the diagnosis and severity of OSA.

7.3 DIAGNOSIS OF OSA

OSA can be suspected by the presence of both nocturnal as well as day–time symptoms. The most common nighttime symptoms are snoring during sleep; sometimes parents are able to describe characteristic episodic pauses in breathing despite movement of the chest or abdomen. Other descriptions include gasping, restlessness during sleep, nighttime sweating, sleeping in unusual positions, parasomnias (sleep terrors, sleep walking), and secondary nocturnal enuresis.

The daytime symptoms are based on functional consequences of disturbed sleep and/or hypoxemia/hypercarbia. These include early morning headache and sometimes nausea or vomiting, excessive daytime sleepiness, and fatigue. Recent reports have also highlighted neurocognitive consequences of OSA including decreased concentration, diminished memory, difficulty in making decisions, learning difficulties, and also behavioural manifestations such as hyperactivity mimicking ADHD, unusual aggressiveness, and even social withdrawal.

Children with OSA are often mouth breathers and sometimes have hyponasal speech. Children with severe OSA can also have growth stunting. Clinical examination usually reveals a crowded oropharynx, enlarged tonsils, and reduced peritonsillar space. Sometimes, a large tongue may also contribute to airway obstruction. Endoscopic examination identifies hypertrophied adenoids.

Currently, polysomnography (PSG) or a sleep study is the gold standard for a specific diagnosis of OSA [24]. Given the complexity of this investigation in terms of skill, resources, and time, some investigators have tried to use alternate approaches such as clinical questionnaires, objective and/or subjective measures of daytime sleepiness, overnight oximetry, audio recording, video recording, and nap polysomnography (during the daytime). While some of these methods can identify children with OSA, they have poor negative predictive value [25].

7.4 OBESITY AND OBSTRUCTIVE SLEEP APNEA

There is now ample data confirming that OSA associated with obesity is highly prevalent in children and adolescents. The association between obesity and OSA emerges from two sets of observations; the first is the observed high prevalence of OSA among obese children and adolescents, and the second is the higher proportion of children with OSA who are obese. Thus it appears that both conditions can coexist and yet potentiate the adverse impact of each. It is believed that the prevalence of OSA among obese children and adolescents can be as high as 60% [26]. In one study of obese children undergoing polysomnography, it was observed that 46% had OSA [27]. Likewise, in another study OSA was observed in 59% of obese children undergoing evaluation for symptoms of sleep–disordered breathing [28]. Yet another study reported that 55% children scheduled for bariatric surgery for morbid obesity had OSA [29]. In fact, in many children and adolescents with OSA, the severity of OSA parallels the severity of obesity [30]. A recent population–based study involving 400 children between 2 and 8 years of age found that obesity was the most significant risk factor for OSA with an odds ratio of 4.69 (95% CI 1.58–13.33). Further analysis suggested that for each unit increase in BMI, there was a 12% higher risk of OSA [23].

7.5 MECHANISMS FOR INCREASED RISK OF OSA IN OBESE CHILDREN AND ADOLESCENTS

There are multiple factors that interact to significantly increase the risk of OSA among obese children and adolescents.

Similar to nonobese children, airway obstruction by adenotonsillar hypertrophy is a fairly common cause of OSA among obese children [31–33] affecting approximately 45% of all obese children with OSA [34]. However, alarmingly, following adenotonsillectomy, OSA persists in about 50% of obese children [35] which is significantly higher than the observed persistence rate of 10–20% amongst nonobese children [36, 37]. Another additional interesting observation is that the prevalence of adenotonsillar hypertrophy among obese children is higher than among nonobese children, which indirectly suggests that adenotonsillar hypertrophy in obese children could be a consequence of another distinct mechanism. Possible explanations include endocrine mediated somatic growth that results in larger and/or heavier fat pads, soft palate, and tongue among adults with obesity [38]. It is possible that similar mechanisms operate in children and adolescents as well.

Functional factors that operate to promote upper airway obstruction OSA in obese individuals during sleep include altered neuromuscular tone resulting in greater upper airway collapsibility during sleep. Indeed, measurements of airway flow and mechanics have shown that in obese children, there is a positive critical closing pressure of the pharynx causing the airway to collapse during sleep with even mild negative inspiratory pressure [39].

Additional mechanical factors that predispose to functional abnormalities include central adiposity and an excess mechanical load on the chest wall. These factors interact and result in decreased chest wall excursion and decreased diaphragmatic excursion causing a reduction in, chest wall compliance with reduced functional residual capacity and tidal volumes. As a result, hypoventilation, atelectasis, and ventilation/perfusion mismatch may ensue resulting in increased work of breathing resulting in fatigue, all of which may be exacerbated during sleep and could further predispose to sleep–disordered breathing. Although these interacting physiologies are not well understood, they could in part explain why adenotonsillectomy is not curative in all obese children with hypertrophied adenoids and tonsils.

7.6 OBESITY, OSA, AND ASSOCIATED COMORBIDITIES IN OBESE CHILDREN AND ADOLESCENTS

7.6.1 EXCESSIVE DAYTIME SLEEPINESS (EDS)

EDS is prevalent among obese children with and without OSA [40]; specifically EDS increased progressively and significantly with increasing BMI. Prepubertal obese subjects with OSA have more EDS than non–obese subjects with OSA of similar severity [41].

7.6.2 QUALITY OF LIFE (QOL)

Multiple published studies demonstrate reported poor QOL among overweight and obese children and adolescents [42] and those with OSA [43].

In one study with 151 children, with a mean age of 12 years, the presence of OSA was a predictor of poor QOL in overweight children [44].

7.6.3 NEUROCOGNITIVE FUNCTION

OSA is associated with cognitive, behavioral, and functional deficits in young children [45]. Although total sleep duration may modulate behavioral function, it is believed that sleep fragmentation associated with OSA is a key determinant of behavioral alterations in pediatric OSA subjects. A recent study with 52 children reported improvement in both neurobehavioral function and daytime sleepiness in children who used an average of 3–hour positive airway pressure (PAP) at night [46]. In another small study with 6 obese adolescents, even modest level of PAP adherence displayed improved attention and school performance whereas a similar group of 7 nonadherent adolescents showed academic decline [47]. Resolution of OSA is associated with improvement in neurocognitive status.

7.6.4 PHYSICAL ACTIVITY

Physical activity levels are reduced both in obese children and those with OSA [48]. Increased physical activity may not only promote weight loss but also, secondary to weight loss, may improve the severity of OSA [49].

7.6.5 CARDIOVASCULAR BURDEN

Multiple adult studies indicate that OSA contributes to or exacerbates cardiovascular disease in the context of obesity [50]. A similar evaluation of childhood obesity–related OSA on cardiovascular structure and function is currently not available. However, indirect measurements that reflect blood pressure regulation, cardiac function, autonomic dysfunction, and endothelial properties suggest a similar pattern in obese children and adolescents [48, 51, 52]. The precise mechanisms linking cardiovascular disease both to OSA and obesity are not completely understood. However, a common mechanism is activation of the sympathetic nervous system. Specifically, repetitive arousals, episodic hypoxaemia, hypercapnia, and changes in intrathoracic pressures lead to sympathetic activation via chemoreceptor activation, impaired baroreflex sensitivity, and increased activity of the renin–angiotensin system. In obesity, increased adiposity elevates levels of free fatty acid (FFA) which with increased levels of leptin promote sympathetic activation. Chronic sympathoactivation instigates dyslipidaemia, left ventricular modelling, endothelial dysfunction and arterial stiffness, inflammation with high levels of hs–CRP, and insulin resistance with resultant glucose intolerance. All of these factors are inextricably linked and together induce significant cardiovascular morbidity. In addition, episodic hypoxemia in children with OSA causes pulmonary vasoconstriction and ultimately pulmonary artery hypertension [49, 53].

7.6.5.1 HYPERTENSION

In children, obesity is a risk factor for high BP, and OSA is independently associated with increased BP [51, 54]. In a recent study, among prepubertal, non–obese children, the presence of OSA was associated with an elevation in BP by 10–15 mmHg independently of BMI during both wakefulness and sleep when compared to nonsnoring controls (ZA). In a separate study of 140 children, children with severe OSA when compared with controls with no OSA showed significantly increased mean arterial BP during awakefulness and sleep, increased diastolic BP during wakefulness and sleep, and increased systolic BP during sleep. Almost one–third of the patients with severe OSA showed a mean 24–hour systolic BP > 95th percentile. Similarly, obese children with moderate–to–severe OSA had a significantly increased risk for hypertension than obese children with mild OSA, suggesting that OSA may be a trigger for hypertension in obese children [55]. These findings are of significance as a recent longitudinal study has shown that elevated BP in childhood tracks into adult life and is associated with an increased risk of hypertension and metabolic syndrome later in life [56].

7.6.5.2 CHANGES IN VENTRICULAR STRUCTURE AND FUNCTION

In adulthood, there is a significant association between left ventricular mass and cardiovascular mortality. Pilot data shows a significantly higher left ventricular mass index (LVMI) with reduced diastolic and systolic function among obese children without documented OSA compared with lean controls [57]. In non–obese children with OSA, abnormalities in LVMI correlate with both the presence and severity of OSA [58]. One study reported that subjects with severe OSA had an odds ratio of 11.2 for LVMI > 95th percentile [59], while another showed that relief of OSA following an adeno–tonsillectomy resulted in measured cardiac variables in the same range as controls [58]. Furthermore, in non–obese children with OSA, improvements in L diastolic function [59] and the right ventricular myocardial performance index have been observed after resolution of OSA [60]. Thus OSA in the context of obesity is likely to exacerbate abnormalities of L structure and function.

7.6.5.3 ENDOTHELIAL FUNCTION

OSA is also involved in causing endothelial dysfunction, mediated by reduced levels of nitric oxide and increased levels of mediators like endothelin–1 and plasma aldosterone.

7.6.5.4 CARDIAC AUTONOMIC ACTIVITY

Cardiac autonomic activity is usually measured using indices of heart rate variability (HRV). Low HR signifies sympathetic overdrive and has been consistently associated with the risk of incident cardiovascular disease in adults. In obese children, HR was lower than non–obese children with body weight as the strongest predictor for lower HR [61]. However, in non–obese children with moderate–to–severe OSA, HR variability was lower compared to those without OSA [62].

7.6.6 OSA AND THE METABOLIC SYNDROME

There is emerging data that OSA itself mediates insulin resistance, dyslipidemia, hypertension, and inflammation. These occur through sympathetic hyperactivity, intermittent hypoxemia, and sleep fragmentation or insufficient sleep. In other words, OSA may be a cause of obesity and not a consequence alone. This conclusion stems from independent pieces of evidence that point towards the contribution of OSA to various components of the metabolic syndrome and perhaps, more importantly, the reversibility with treatment of OSA.

Recent data showed increased levels of insulin (indicating insulin resistance) in adolescents with OSA [63]. In younger children, adeno–tonsillectomy is associated with improvement in lipid profile, insulin sensitivity, and inflammatory markers in some studies [64, 65].

7.6.7 CONTRIBUTION OF OSA TO OBESITY

OSA is associated with inadequate sleep quantity and quality in children as well as adults. A recent systematic review [66] examining the relationship between sleep duration and the development of obesity reported that in children and adults, shorter duration of sleep was associated with increased risk of obesity (odds ratio 1.89, 95% CI 1.46–2.43 in children). Thus OSA can have a direct impact by worsening obesity.

7.7 MANAGEMENT

It is clear that childhood obesity and OSA can present to a wide range of professional disciplines on account of the multisystem manifestations. Therefore successful management depends on concerted effort by a multidisciplinary team of professionals including sleep physician, ENT surgeon, respirologist, child nutritionist, child psychologist, cardiologist, and social worker, working together with the obese child and his/her family. The goals of management are enhanced quality of life and prevention of short– and long–term complications.

7.8 MANAGEMENT OF OBESITY

A detailed discussion on the various modalities for weight reduction and management of obesity is outside the scope of this paper; however a brief review of the current recommendations is presented. Although the body of evidence in children and adolescents is still being generated, it is generally recommended they should perform at least 60 minutes of moderately intense physical activity daily, in order to prevent obesity or maintain weight. This should be encouraged even if it does not result in weight loss, on account of the general health benefits of exercise. They should also be advised to eat meals at regular times and preferably free from distractions.

Pharmacological interventions are generally not recommended for children below 12 years, barring exceptional circumstances such as severe OSA or raised intracranial tension. In addition, the rare decision to use pharmacotherapy does not exclude the need for physical activity and dietary control. Some experts maintain that medication is better utilized to maintain weight loss, rather than induce it. Currently, Orlistat is the only pharmacological agent that can be considered, since Sibutramine has been withdrawn.

Bariatric surgery is also rarely recommended in children, unless they are morbidly obese (BMI $\geq$ 40 kg/m^2) or 35–40 kg/m^2 with coexisting diseases that could be improved by loss of weight. Even then, surgery is considered only after nonsurgical measures have been tried without success.

7.9 MANAGEMENT OF OSA

Based on the observation that almost half of all obese children with OSA have adeno–tonsillar hypertrophy, the American Board of Pediatrics [24] recommends adeno–tonsillectomy as the first step in management.

Although adeno–tonsillectomy results in improvement in 80% cases and improves obstructive symptoms in 80% of cases of otherwise normal children with OSA, children with morbid obesity are more likely to fail treatment than normal children. In some series, almost 50% continue to have OSA [55].

Therefore Positive Airway Pressure (PAP) therapy has become the standard of care, usually in conjunction with weight loss strategies. In adult patients, PAP therapy results in dramatic improvement in OSA symptoms. In addition, there are encouraging reports of improvement in cardiovascular status including reduction in systolic and BP, L function [67], markers of endothelial function [68], and cardiac autonomic activity [69]. In addition, withdrawal of PAP for two weeks was associated with systolic and diastolic BP increase of 4–6 mmHg [70]. PAP therapy can be administered either as continuous PAP (CPAP) or the more physiological bilevel PAP (BiPAP). In children also, the symptoms of OSA improve with PAP therapy; however there is limited data evaluating its efficacy in improving clinical outcomes and QOL.

In one study, non–obese children who had resolution of OSA, 6 months following adeno–tonsillectomy, had a reduction in diastolic BP of the order of 5 mmHg [71]. The importance of this study cannot be overemphasised in the context of recent findings of a large meta–analysis that lowering systolic SBP by 10 mmHg or diastolic BP by 5 mmHg in adults (regardless of the baseline BP) reduced fatal and nonfatal cardiac events by approximately 25% [72].

Other treatment options that are sometimes useful in adult patients with OSA include oral appliances and devices that expand the upper airway space. However these require skilled construction and are generally efficacious in mild OSA only. However it is a viable option for those who cannot or will not use CPAP. These appliances have limited value in children on account of less developed dentition [73]. Some adults use simple devices to prevent sleeping in the supine position. These devices work by promoting sleep in the lateral or prone position.

Surgical management options include uvulo–palatopharngoplasty wherein bulky soft tissues that obstruct the airway can be trimmed or excised to create a larger airway space. It has also been used to strengthen and support hypotonic pharyngeal muscles in those children where reduced neuromuscular tone is responsible for airway floppiness and obstruction. Some centres use the procedure for obese children with severe OSA, to reduce redundant oropharyngeal tissue bulk.

Presently, there is no randomized trial comparing the various modalities in children adolescents, to estimate the superiority of one over the other.

7.10 CONCLUSION

Childhood and adolescent obesity have reached epidemic proportions worldwide.

OSA significantly complicates obesity and is an independent risk factor for cardiovascular, metabolic, neuro–cognitive burden as well as negative impact on the quality of life in obese children.

All disciplines involved in the well–being of obese children must be involved in sleep surveillance strategies to highlight obese children at an increased risk for OSA. Early recognition and treatment of OSA, in addition to weight loss strategies, could provide an opportunity for cardiovascular and metabolic risk reduction in childhood which would positively impact the health of these children not only in childhood but also in adulthood.

REFERENCES

1. C. L. Ogden, M. D. Carroll, and K. M. Flegal, "High body mass index for age among US children and adolescents, 2003–2006," The Journal of the American Medical Association, vol. 299, no. 20, pp. 2401–2405, 2008.
2. L. Wang, L. Kong, F. Wu, Y. Bai, and R. Burton, "Preventing chronic diseases in China," The Lancet, vol. 366, no. 9499, pp. 1821–1824, 2005.
3. R. C. Whitaker, M. S. Pepe, J. A. Wright, K. D. Seidel, and W. H. Dietz, "Early adiposity rebound and the risk of adult obesity," Pediatrics, vol. 101, no. 3, article E5, 1998.
4. C. L. Ogden and K. M. Flegal, "Childhood obesity: are we all speaking the same language?" Advances in Nutrition, vol. 2, pp. 159S–166S, 2011.
5. The Expert Committee on Clinical Guidelines for Overweight in Adolescent Preventive Services, "Guidelines for overweight in adolescent preventive services: recommendations from an expert committee," American Journal of Clinical Nutrition, vol. 59, no. 2, pp. 307–316, 1994.

6. T. J. Cole, M. C. Bellizzi, K. M. Flegal, and W. H. Dietz, "Establishing a standard definition for child overweight and obesity worldwide: international survey," British Medical Journal, vol. 320, no. 7244, pp. 1240–1243, 2000.
7. J. J. Reilly, J. Armstrong, A. R. Dorosty et al., "Early life risk factors for obesity in childhood: cohort study," British Medical Journal, vol. 330, no. 7504, pp. 1357–1359, 2005.
8. T. D. Brisbois, A. P. Farmer, and L. J. McCargar, "Early markers of adult obesity: a review," Obesity Reviews, vol. 13, no. 4, pp. 347–367, 2012.
9. J. Chaicharn, Z. Lin, M. L. Chen, S. L. D. Ward, T. Keens, and M. C. K. Khoo, "Model–based assessment of cardiovascular autonomic control in children with obstructive sleep apnea," Sleep, vol. 32, no. 7, pp. 927–938, 2009.
10. R. H. Eckel, S. M. Grundy, and P. Z. Zimmet, "The metabolic syndrome," The Lancet, vol. 365, no. 9468, pp. 1415–1428, 2005.
11. T. Bjørge, A. Engeland, A. Tverdal, and G. D. Smith, "Body mass index in adolescence in relation to cause–specific mortality: a follow–up of 230,000 Norwegian adolescents," American Journal of Epidemiology, vol. 168, no. 1, pp. 30–37, 2008.
12. R. C. Whitaker, J. A. Wright, M. S. Pepe, K. D. Seidel, and W. H. Dietz, "Predicting obesity in young adulthood from childhood and parental obesity," The New England Journal of Medicine, vol. 337, no. 13, pp. 869–873, 1997.
13. S. R. Daniels, D. K. Arnett, R. H. Eckel et al., "Overweight in children and adolescents: pathophysiology, consequences, prevention, and treatment," Circulation, vol. 111, no. 15, pp. 1999–2012, 2005.
14. M. Raj and R. K. Kumar, "Obesity in children & adolescents," Indian Journal of Medical Research, vol. 132, no. 11, pp. 598–607, 2010.
15. D. S. Ludwig, "Childhood obesity—the shape of things to come," The New England Journal of Medicine, vol. 357, no. 23, pp. 2325–2327, 2007.
16. S. L. Verhulst, N. Schrauwen, D. Haentjens, L. van Gaal, W. A. de Backer, and K. N. Desager, "Reference values for sleep–related respiratory variables in asymptomatic European children and adolescents," Pediatric Pulmonology, vol. 42, no. 2, pp. 159–167, 2007.
17. S. Uliel, R. Tauman, M. Greenfeld, and Y. Sivan, "Normal polysomnographic respiratory values in children and adolescents," Chest, vol. 125, no. 3, pp. 872–878, 2004.
18. H. E. Montgomery–Downs, L. M. O'Brien, T. E. Gulliver, and D. Gozal, "Polysomnographic characteristics in normal preschool and early school–aged children," Pediatrics, vol. 117, no. 3, pp. 741–753, 2006.
19. American Thoracic Society, "Cardiorespiratory sleep studies in children. Establishment of normative data and polysomonographic predictors of morbidity," American Journal of Respiratory and Critical Care Medicine, vol. 160, no. 4, pp. 1381–1387, 1999.
20. N. J. Ali, D. J. Pitson, and J. R. Stradling, "Snoring, sleep disturbance, and behaviour in 4–5 year olds," Archives of Disease in Childhood, vol. 68, no. 3, pp. 360–366, 1993.
21. T. Gislason and B. Benediktsdottir, "Snoring, apneic episodes, and nocturnal hypoxemia among children 6 months to 6 years old: an epidemiologic study of lower limit of prevalence," Chest, vol. 107, no. 4, pp. 963–966, 1995.

22. R. Arens, J. M. McDonough, A. T. Costarino et al., "Magnetic resonance imaging of the upper airway structure of children with obstructive sleep apnea syndrome," American Journal of Respiratory and Critical Care Medicine, vol. 164, no. 4, pp. 698–703, 2001.
23. S. Redline, P. V. Tishler, M. Schluchter, J. Aylor, K. Clark, and G. Graham, "Risk factors for sleep–disordered breathing in children: associations with obesity, race, and respiratory problems," American Journal of Respiratory and Critical Care Medicine, vol. 159, no. 5, pp. 1527–1532, 1999.
24. C. L. Marcus, D. Chapman, S. D. Ward et al., "Clinical practice guideline: diagnosis and management of childhood obstructive sleep apnea syndrome," Pediatrics, vol. 109, no. 4, pp. 704–712, 2002.
25. M. S. Wise, C. D. Nichols, M. M. Grigg–Damberger et al., "Executive summary of respiratory indications for polysomnography in children: an evidence–based review," Sleep, vol. 34, no. 3, pp. 389–398, 2011.
26. S. L. Verhulst, L. van Gaal, W. de Backer, and K. Desager, "The prevalence, anatomical correlates and treatment of sleep–disordered breathing in obese children and adolescents," Sleep Medicine Reviews, vol. 12, no. 5, pp. 339–346, 2008.
27. C. L. Marcus, S. Curtis, C. B. Koerner, A. Joffe, J. R. Serwint, and G. M. Loughlin, "Evaluation of pulmonary function and polysomnography in obese children and adolescents," Pediatric Pulmonology, vol. 21, no. 3, pp. 176–183, 1996.
28. J. M. Silvestri, D. E. Weese–Mayer, M. T. Bass, A. S. Kenny, S. A. Hauptman, and S. M. Pearsall, "Polysomnography in obese children with a history of sleep–associated breathing disorders," Pediatric Pulmonology, vol. 16, no. 2, pp. 124–129, 1993.
29. M. Kalra, T. Inge, V. Garcia et al., "Obstructive sleep apnea in extremely overweight adolescents undergoing bariatric surgery," Obesity Research, vol. 13, no. 7, pp. 1175–1179, 2005.
30. G. B. Mallory Jr., D. H. Fiser, and R. Jackson, "Sleep–associated breathing disorders in morbidly obese children and adolescents," Journal of Pediatrics, vol. 115, no. 6, pp. 892–897, 1989.
31. E. S. Katz and C. M. D'Ambrosio, "Pathophysiology of pediatric obstructive sleep apnea," Proceedings of the American Thoracic Society, vol. 5, no. 2, pp. 253–262, 2008.
32. Y. K. Wing, S. H. Hui, W. M. Pak et al., "A controlled study of sleep related disordered breathing in obese children," Archives of Disease in Childhood, vol. 88, no. 12, pp. 1043–1047, 2003.
33. S. L. Verhulst, N. Schrauwen, D. Haentjens et al., "Sleep–disordered breathing in overweight and obese children and adolescents: prevalence, characteristics and the role of fat distribution," Archives of Disease in Childhood, vol. 92, no. 3, pp. 205–208, 2007.
34. J. E. Gordon, M. S. Hughes, K. Shepherd et al., "Obstructive sleep apnoea syndrome in morbidly obese children with tibia vara," Journal of Bone and Joint Surgery—Series B, vol. 88, no. 1, pp. 100–103, 2006.
35. R. B. Mitchell and J. Kelly, "Adenotonsillectomy for obstructive sleep apnea in obese children," Otolaryngology—Head and Neck Surgery, vol. 131, no. 1, pp. 104–108, 2004.

36. A. Tal, A. Bar, A. Leiberman, and A. Tarasiuk, "Sleep characteristics following adenotonsillectomy in children with obstructive sleep apnea syndrome," Chest, vol. 124, no. 3, pp. 948–953, 2003.
37. J. S. Suen, J. E. Arnold, and L. J. Brooks, "Adenotonsillectomy for treatment of obstructive sleep apnea in children," Archives of Otolaryngology—Head and Neck Surgery, vol. 121, no. 5, pp. 525–530, 1995.
38. R. J. Schwab, M. Pasirstein, R. Pierson et al., "Identification of upper airway anatomic risk factors for obstructive sleep apnea with volumetric magnetic resonance imaging," American Journal of Respiratory and Critical Care Medicine, vol. 168, no. 5, pp. 522–530, 2003.
39. I. C. Gleadhill, A. R. Schwartz, N. Schubert, R. A. Wise, S. Permutt, and P. L. Smith, "Upper airway collapsibility in snorers and in patients with obstructive hypopnea and apnea," American Review of Respiratory Disease, vol. 143, no. 6, pp. 1300–1303, 1991.
40. M. Tsaoussoglou, E. O. Bixler, S. Calhoun, G. P. Chrousos, K. Sauder, and A. N. Vgontzas, "Sleep–disordered breathing in obese children is associated with prevalent excessive daytime sleepiness, inflammation, and metabolic abnormalities," The Journal of Clinical Endocrinology and Metabolism, vol. 95, no. 1, pp. 143–150, 2010.
41. D. Gozal and L. Kheirandish–Gozal, "Obesity and excessive daytime sleepiness in prepubertal children with obstructive sleep apnea," Pediatrics, vol. 123, no. 1, pp. 13–18, 2009.
42. M. de Beer, G. H. Hofsteenge, H. M. Koot, R. A. Hirasing, H. A. Delemarre–van de Waal, and R. J. B. J. Gemke, "Health–related–quality–of–life in obese adolescents is decreased and inversely related to BMI," Acta Paediatrica, vol. 96, no. 5, pp. 710–714, 2007.
43. V. M. Crabtree, J. W. Varni, and D. Gozal, "Health–related quality of life and depressive symptoms in children with suspected sleep–disordered breathing," Sleep, vol. 27, no. 6, pp. 1131–1138, 2004.
44. M. A. Carno, E. Ellis, E. Anson et al., "Symptoms of sleep apnea and polysomnography as predictors of poor quality of life in overweight children and adolescents," Journal of Pediatric Psychology, vol. 33, no. 3, pp. 269–278, 2008.
45. D. W. Beebe, "Neurobehavioral morbidity associated with disordered breathing during sleep in children: a comprehensive review," Sleep, vol. 29, no. 9, pp. 1115–1134, 2006.
46. C. L. Marcus, J. Radcliffe, S. Konstantinopoulou, S. E. Beck, M. A. Cornaglia, J. Traylor, et al., "Effects of positive airway pressure therapy on neurobehavioral outcomes in children with obstructive sleep apnea," American Journal of Respiratory and Critical Care Medicine, vol. 185, no. 9, pp. 998–1003, 2012.
47. D. W. Beebe and K. C. Byars, "Adolescents with obstructive sleep apnea adhere poorly to positive airway pressure (PAP), but PAP users show improved attention and school performance," PLoS ONE, vol. 6, no. 3, Article ID e16924, 2011.

48. A. Tal, A. Leiberman, G. Margulis, and S. Sofer, "Ventricular dysfunction in children with obstructive sleep apnea: radionuclide assessment," Pediatric Pulmonology, vol. 4, no. 3, pp. 139–143, 1988.
49. R. Tauman and D. Gozal, "Obesity and obstructive sleep apnea in children," Paediatric Respiratory Reviews, vol. 7, no. 4, pp. 247–259, 2006.
50. A. I. Pack and T. Gislason, "Obstructive sleep apnea and cardiovascular disease. A perspective and future directions," Progress in Cardiovascular Diseases, vol. 51, no. 5, pp. 434–451, 2009.
51. R. S. Amin, J. L. Carroll, J. L. Jeffries et al., "Twenty–four–hour ambulatory blood pressure in children with sleep–disordered breathing," American Journal of Respiratory and Critical Care Medicine, vol. 169, no. 8, pp. 950–956, 2004.
52. P. L. Enright, J. L. Goodwin, D. L. Sherrill, J. R. Quan, and S. F. Quan, "Blood pressure elevation associated with sleep–related breathing disorder in a community sample of white and hispanic children: the Tucson children's assessment of sleep apnea study," Archives of Pediatrics and Adolescent Medicine, vol. 157, no. 9, pp. 901–904, 2003.
53. T. Shiomi, C. Guilleminault, R. Stoohs, and I. Schnittger, "Obstructed breathing in children during sleep monitored by echocardiography," Acta Paediatrica, vol. 82, no. 10, pp. 863–871, 1993.
54. R. S. C. Horne, J. S. C. Yang, L. M. Walter et al., "Elevated blood pressure during sleep and wake in children with sleep–disordered breathing," Pediatrics, vol. 128, no. 1, pp. e85–e92, 2011.
55. L. C. K. Leung, D. K. Ng, M. W. Lau et al., "Twenty–four–hour ambulatory BP in snoring children with obstructive sleep apnea syndrome," Chest, vol. 130, no. 4, pp. 1009–1017, 2006.
56. S. S. Sun, G. D. Grave, R. M. Siervogel, A. A. Pickoff, S. S. Arslanian, and S. R. Daniels, "Systolic blood pressure in childhood predicts hypertension and metabolic syndrome later in life," Pediatrics, vol. 119, no. 2, pp. 237–246, 2007.
57. L. P. Koopman, B. W. McCrindle, C. Slorach, N. Chahal, W. Hui, T. Sarkola, et al., "Interaction between myocardial and vascular changes in obese children: a pilot study," Journal of the American Society of Echocardiography, vol. 25, pp. 401–410. e1, 2012.
58. G. Attia, M. A. Ahmad, A. B. Saleh, and A. Elsharkawy, "Impact of obstructive sleep apnea on global myocardial performance in children assessed by tissue doppler imaging," Pediatric Cardiology, vol. 31, no. 7, pp. 1025–1036, 2010.
59. R. S. Amin, T. R. Kimball, J. A. Bean et al., "Left ventricular hypertrophy and abnormal ventricular geometry in children and adolescents with obstructive sleep apnea," American Journal of Respiratory and Critical Care Medicine, vol. 165, no. 10, pp. 1395–1399, 2002.
60. D. Duman, B. Naiboglu, H. S. Esen, S. Z. Toros, and R. Demirtunc, "Impaired right ventricular function in adenotonsillar hypertrophy," International Journal of Cardiovascular Imaging, vol. 24, no. 3, pp. 261–267, 2008.

61. S. M. Rodriguez–Colon, E. O. Bixler, X. Li, A. N. Vgontzas, and D. Liao, "Obesity is associated with impaired cardiac autonomic modulation in children," International Journal of Pediatric Obesity, vol. 6, no. 2, pp. 128–134, 2011.
62. D. Liao, X. Li, S. M. Rodriguez–Colon et al., "Sleep–disordered breathing and cardiac autonomic modulation in children," Sleep Medicine, vol. 11, no. 5, pp. 484–488, 2010.
63. S. Redline, A. Storfer–Isser, C. L. Rosen et al., "Association between metabolic syndrome and sleep–disordered breathing in adolescents," American Journal of Respiratory and Critical Care Medicine, vol. 176, no. 4, pp. 401–408, 2007.
64. K. A. Waters, S. Sitha, L. M. O'Brien et al., "Follow–up on metabolic markers in children treated for obstructive sleep apnea," American Journal of Respiratory and Critical Care Medicine, vol. 174, no. 4, pp. 455–460, 2006.
65. D. Gozal, O. S. Capdevila, and L. Kheirandish–Gozal, "Metabolic alterations and systemic inflammation in obstructive sleep apnea among nonobese and obese prepubertal children," American Journal of Respiratory and Critical Care Medicine, vol. 177, no. 10, pp. 1142–1149, 2008.
66. F. P. Cappuccio, F. M. Taggart, N. B. Kandala et al., "Meta–analysis of short sleep duration and obesity in children and adults," Sleep, vol. 31, no. 5, pp. 619–626, 2008.
67. N. A. Bayram, B. Ciftci, T. Durmaz et al., "Effects of continuous positive airway pressure therapy on left ventricular function assessed by tissue Doppler imaging in patients with obstructive sleep apnoea syndrome," European Journal of Echocardiography, vol. 10, no. 3, pp. 376–382, 2009.
68. L. F. Drager, L. A. Bortolotto, A. C. Figueiredo, E. M. Krieger, and G. Lorenzi–Filho, "Effects of continuous positive airway pressure on early signs of atherosclerosis in obstructive sleep apnea," American Journal of Respiratory and Critical Care Medicine, vol. 176, no. 7, pp. 706–712, 2007.
69. E. Kufoy, J. A. Palma, J. Lopez, M. Alegre, E. Urrestarazu, J. Artieda, et al., "Changes in the heart rate variability in patients with obstructive sleep apnea and its response to acute CPAP treatment," PLoS ONE, vol. 7, no. 3, Article ID e33769, 2012.
70. M. Kohler, A. C. Stoewhas, L. Ayers, O. Senn, K. E. Bloch, E. W. Russi, et al., "Effects of continuous positive airway pressure therapy withdrawal in patients with obstructive sleep apnea: a randomized controlled trial," American Journal of Respiratory and Critical Care Medicine, vol. 184, no. 10, pp. 1192–1199, 2011.
71. R. Amin, L. Anthony, V. Somers et al., "Growth velocity predicts recurrence of sleep–disordered breathing 1 year after adenotonsillectomy," American Journal of Respiratory and Critical Care Medicine, vol. 177, no. 6, pp. 654–659, 2008.
72. M. R. Law, J. K. Morris, and N. J. Wald, "Use of blood pressure lowering drugs in the prevention of cardiovascular disease: meta–analysis of 147 randomised trials in the context of expectations from prospective epidemiological studies," British Medical Journal, vol. 338, Article ID b1665, 2009.

73. F. R. Carvalho, D. Lentini–Oliveira, M. A. Machado, G. F. Prado, L. B. Prado, and H. Saconato, "Oral appliances and functional orthopaedic appliances for obstructive sleep apnoea in children," Cochrane Database of Systematic Reviews, no. 2, Article ID CD005520, 2007.

CHAPTER 8

ASSESSMENT OF ENDOTHELIAL DYSFUNCTION IN CHILDHOOD OBESITY AND CLINICAL USE

LUC BRUYNDONCKX, VICKY Y. HOYMANS,
AMARYLLIS H. VAN CRAENENBROECK, DIRK K. VISSERS,
CHRISTIAAN J. VRINTS, JOSÉ RAMFT,
and VIVIANE M. CONRAADS

8.1 INTRODUCTION

Society is faced with a new pandemic. Obesity and its associated non–communicable diseases, such as cardiovascular (CV) complications, diabetes [1], sleep apnea [2], and asthma, are a major threat to the management of health care worldwide. It may seem paradoxical but both childhood malnutrition in developing countries and the rapidly increasing prevalence of overweight and obesity in Western youth have a common denominator, low income [3].

Being obese as a child comes at a price. Childhood overweight and obesity increase the risk of obesity at adult age [4] and are associated with CV risk factors. A high BMI during childhood and adolescence has been associated with, respectively, premature death from disease and increased risk of coronary heart disease at adult age [5, 6]. Interestingly, however, in a recent meta–analysis of four studies, overweight and obese children who became nonobese in adult life were not different in terms of several risk parameters of cardiovascular disease to those patients who had never been obese [7]. This could be explained by the fact that other indicators of a healthy lifestyle, such a regular physical exercise, were not taken into account. During the past two decades, the concept of early vascular changes, which act as the primum movens for future CV complications, has been tested [8]. The development and refinement of technical tools that allow the in vivo evaluation of endothelial dysfunction, which is considered the earliest demonstrable feature of atherosclerosis, rapidly advanced pathophysiological insights. Recently, more fundamental research into disrupted bone–marrow–related endothelial repair mechanisms (i.e., endothelial progenitor cells), as well as the identification of novel markers of endothelial damage (i.e., endothelial microparticles), has created an entirely new line of research [9, 10]. These biomarkers hold promise in terms of designing effective strategies and to evaluate their effect in the combat against the devastating consequences of childhood obesity.

A long–lasting change of unhealthy lifestyles is fundamental in helping obese children and their parents to fight this disease. Therefore, a multidisciplinary approach to adopt physical activity and balanced calorie restriction into daily life is inevitable to counterbalance the attraction of highly processed food, motorized transportation, and TV–related sedentarism.

This paper will provide a critical overview on both in vivo as well as in vitro markers for endothelial integrity. As an introduction we will first describe the obesity–induced imbalance of endothelial repair and damage, resulting in early endothelial dysfunction. Currently available data on the effect of lifestyle changes as well as possible pharmacological interventions to counteract obesity–induced endothelial malfunction will be recapitulated. Although the focus of this review is on childhood obesity, it is important to stress that data from obese adults are preponderant and therefore will be incorporated in the text. Comparison of the available literature will expose differences and stresses the fact that the scientific community should invest more in childhood obesity research since, indeed, children are not small adults. In addition, the value of longitudinal studies, starting early in childhood, cannot be underestimated when it comes to exactly decipher the timing of pathophysiological events.

8.2 ENDOTHELIAL DYSFUNCTION

The endothelial cell–layer is located at the border between circulating blood and vascular smooth muscle cells (VSMC) [11]. In response to stimuli that indicate increased blood flow demand (e.g., increased shear stress), activation of the PhosphoInositol 3–Kinase (PI3K)/Akt pathway

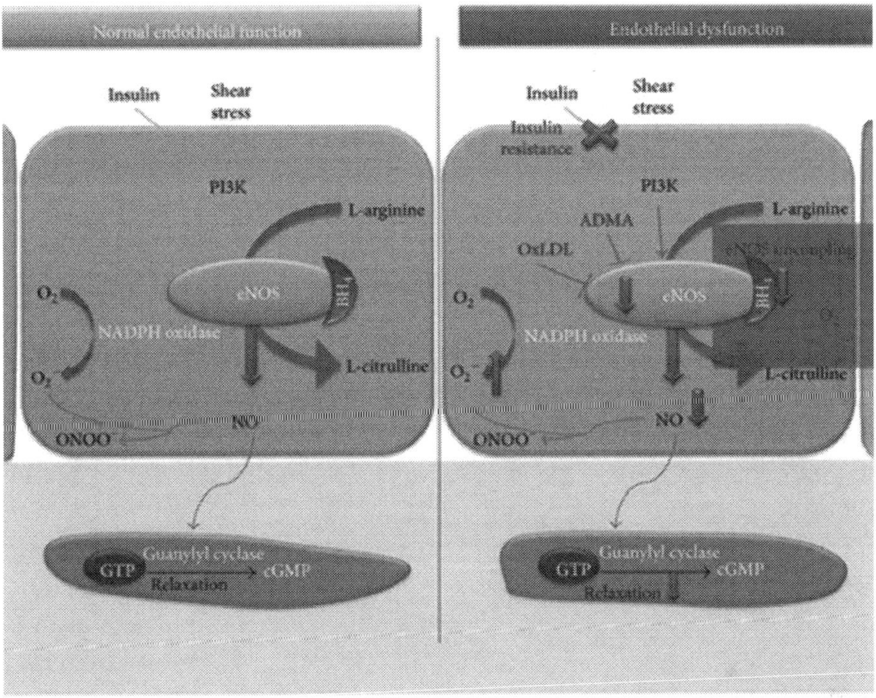

FIGURE 1: Normal endothelial function versus endothelial dysfunction. Schematic overview of nitric oxide (NO) production and relaxation of Vascular Smooth Muscle Cells (VSMC). In response to increased shear stress or as a result of insulin signaling, the phosphoinositol 3 kinase (PI3K)/akt pathway is activated leading to phosphorylation of endothelial Nitric Oxide Synthase (eNOS). eNOS, together with the necessary cofactor tetrahydrobiopterin (BH_4), converts L–arginine to L–citrulline and NO. NO activates guanylyl cyclase, which induces smooth muscle relaxation, through increased production of cyclic Guanosine MonoPhosphate (cGMP). Superoxide reduced NO bioavailability by reacting with NO to form peroxynitrite ($ONOO^-$), which has strong oxidant properties. Endothelial dysfunction in obese children is characterized by insulin resistance impairing insulin–mediated NO production and subsequent vasodilation. Furthermore, oxidized LDL and ADMA are inhibitors of eNOS activation. In the situation of diminished availability of BH4, eNOS becomes "uncoupled" and paradoxically leads to increased reactive oxygen species (ROS) generation, which also contributes to reduced bioavailability of NO and vasoconstriction. ADMA, Asymmetric DiMethylArginine; PI3K, PhosphatidylInositol 3–Kinase; BH_4, tetrahydrobiopterin; eNOS, endothelial Nitric Oxide Synthase; , superoxide; $ONOO^-$, peroxynitrite; GTP, Guanosine TriPhosphate; cGMP, cyclic Guanylyl MonoPhosphate; NADPH oxidase, Nicotinamide Adenine Dinucleotide Phosphate oxidase; OxLDL, Oxidized Low–Density Lipoprotein Cholesterol.

will lead to phosphorylation of endothelial Nitric Oxide Synthase (eNOS), which is necessary for its activation and the generation of Nitric Oxide (NO) (Figure 1) [12]. By fine–tuning vascular smooth muscle relaxation, NO is one of the main regulators of vascular tone [13, 14]. In addition, healthy endothelium is responsible for the maintenance of an athero-protective environment: it prevents platelet aggregation, proliferation of VSMC, and adhesion and subsequent diapedesis of leukocytes through the vascular wall [15].

The common and straightforward definition of endothelial dysfunction is "an imbalance between vasodilating and vasoconstricting substances produced by (or acting on) endothelial cells" [16]. The close interaction between endothelial cells and their environment, however, highlights the process complexity. Indeed, due to its location within the vascular wall, the endothelium is directly exposed to damaging factors such as high blood pressure and elevated lipid levels, which are common in obese subjects. The notion that endothelial dysfunction is a necessary and first step towards atherosclerosis has led to a tremendous interest and research investment in an attempt to unravel underlying pathophysiological mechanisms [17].

Fortunately, endothelial dysfunction is not an irreversible process. Upon endothelial damage and ischemia, repair mechanisms are activated. The concept of endothelial homeostasis, which is the net result of endothelial damage and repair, has provided the basis for developing novel therapeutic options, as well as a fascinating new research domain of endothelial "rejuvenation."

8.2.1 FACTORS CONTRIBUTING TO ENDOTHELIAL DYSFUNCTION IN OBESITY

Similar to adults, CV risk factors tend to cluster in obese children and include hypertension, high cholesterol and triglycerides, insulin resistance, a proinflammatory status, disturbances in adipocytokines, and physical inactivity [18]. As a result, assessing the specific contribution of one sole risk factor is extremely difficult. In the following, we will briefly cover recent insights into factors causing endothelial dysfunction in childhood obesity.

8.2.1.1 HYPERTENSION

Hypertension is prevalent in obese children. Based on data gathered in the Avon Longitudinal Study of Parents and Children (ALSPAC; 7589 children aged 8.8–11.7 years), the odds ratio for hypertension was 10.7 (95% CI 7.2–15.9) for obese boys and 13.5 (95% CI 9.4–19.5) for obese girls [18], compared to children with normal weight. The interaction between endothelial dysfunction and hypertension is complex but recent data from longitudinal population studies do not support a bidirectional relationship.

In a cross–sectional and longitudinal study of 3500 adults, Shimbo et al. showed that hypertension is more prevalent in patients with low flow–mediated dilation (FMD) [19]. However, reverse findings could not be confirmed since impaired endothelial function was not predictive of incident hypertension. Similar findings were seen in The Cardiovascular Risk in Young Finns Study, where elevated systolic blood pressure in adolescent boys predicted impaired brachial endothelial function 21 years later in adulthood [20]. A recent analysis of data derived from the ALSPAC study [21] sheds light on the vascular consequences of obesity with time. In their study of 6,576 prepubertal children (aged 10 to 11 years; 80% normal weight, 16% overweight and 4% obese), obesity was associated with higher blood pressure, higher heart rate, and higher resting and hyperemic blood flow, all of which are compatible with a higher cardiac output state. Counterintuitively, these obese children had larger baseline brachial artery diameters, higher FMD, and lower arterial stiffness, despite an unfavorable metabolic profile. Although speculative, the authors propose an initial adaptive response, which ultimately, with longer exposure to obesity, will fail and culminate in vascular damage. Data derived from the Cardiovascular Risk in Young Finns Study support this view and demonstrate an adverse impact of CV risk factors including adiposity, high LDL cholesterol and high insulin level on the progression of intima–media thickness (IMT) in young adults aged 30 years [22]. Previous results from this longitudinal study provide further evidence for this hypothesis, since a correlation was found between CV risk factors and IMT only for participants with demonstrated endothelial dysfunction [23].

Obesity increases blood pressure through multiple mechanisms, which also affect endothelial function. These factors include increased activity of the renin–angiotensin aldosterone system (RAAS), as well as sympatho–activation. Angiotensin II, one of the main hormones in the RAAS system, also directly impairs NO production by affecting eNOS activity [30]. Increased sympathetic nervous activity leads to peripheral vasoconstriction and impairment of endothelial function [31]. Importantly, hypertension reduces bioavailability of NO and increases oxidative stress, due to increased Reactive Oxygen Species (ROS) generation and lower antioxidant capacity. In addition, Asymmetric DiMethyl Arginine (ADMA), a natural inhibitor of eNOS, has been found in higher concentrations in obese individuals [32].

8.2.1.2 LIPIDS

High levels of low–density lipoprotein (LDL) cholesterol and low levels of High–Density Lipoprotein (HDL)–cholesterol are well–known independent CV risk factors. Native LDL can impair endothelial function by decreasing NO bioavailability and eNOS activity. However, the effect is more pronounced when LDL is taken up by macrophages and oxidized to oxidized LDL [33]. HDL cholesterol reduces vascular tone by increasing NO bioavailability; it reduces the expression of adhesion molecules for leukocytes and increases endothelial integrity by upregulating endothelial cell migration and proliferation [34, 35].

Results from the PEP Family Heart Study in 3038 adolescents (12 to 18 years) have demonstrated that central obesity, defined as elevated waist circumference and/or waist–to–hip ratio, is an independent predictor of hypertension, fasting glucose, elevated triglycerides, LDL–cholesterol, non–HDL–cholesterol, triglyceride/HDL–cholesterol ratio, low HDL–cholesterol, and risk factor clustering [36]. Yet, despite having central obesity, both prevalence of an unfavorable lipid profile as well as absolute concentrations of lipids were less pronounced than those found in obese adults. For instance, the prevalence of increased triglyceride concentration was 5,8% and 11,2% in obese male and female children, respectively, but reached approximately 25% in a large cohort of Italian obese adults [37].

Recent research supports the hypothesis that the well–known beneficial effects of HDL, such as macrophage cholesterol efflux and endothelial NO stimulation, are highly heterogeneous and may be altered in patients with coronary artery disease as well as diabetes. Therefore, HDL dysfunction has to be taken into account when novel HDL–targeted therapeutic interventions, such as Cholesteryl Ester Transfer Protein (CETP) inhibitors, are introduced [38].

8.2.1.3 PHYSICAL (IN)ACTIVITY

Physical activity is associated with a significant reduction in CV mortality in adult men and women [39–41], whereas physical inactivity predicts the development of overweight and obesity [42].

Using a questionnaire, physical activity in more than 6000 children between 11 and 19 years was investigated by De Bourdeaudhuij et al. Overweight children spent on average 5.27 ± 4.65 hours per week on physical activity versus 6.21 ± 5.07 hours per week for normal weight children. Significant differences were noted for vigorous and moderate physical activity as well, yet no differences in sedentary behavior were noted [43].

Similar to adults, endothelial function relates to fitness and physical activity in children, yet correlations are affected by age. In young children (5–10 years: mean age 8 years), physical activity was the strongest predictor of FMD in multivariate analysis [44]. Slightly older children (n = 485 (59 male, 86 female) with a mean age of 10.3 ± 0.03 years) demonstrated a significant correlation between FMD and the time spent at moderate and high intensity physical activity [45]. Correlation in that study with physical activity was strongest in the lowest tertile of endothelial function, suggesting that these children could benefit most from physical training. Endothelial function was also clearly related to fitness, objectively expressed as peak oxygen consumption during cardiopulmonary exercise testing. Based on a large study, involving 483 13–year–old adolescents, multivariate analysis pointed out that endothelial function correlated significantly to leisure time physical activity. This finding was only confirmed in boys, leading the authors to hypothesize that the overall lower physical activity seen in girls may explain these results [46]. A longitudinal study in children further demonstrated that an extensive increase in physical activity led to a significant improvement of FMD as well as less progression of intima–media thickness [47].

8.2.1.4 ADIPOKINES

Normal adipose tissue consists of adipocytes, and furthermore macrophages, fibroblasts, endothelial cells, and preadipocytes are present in the so–called vascular–stromal fraction [48]. Although it has long been thought that storage and release of free fatty acids were its only functions, adipose tissue is currently seen as an important endocrine organ, regulating glucose and lipid metabolism and producing a vast amount of cytokines (adipocyte–derived cytokines or adipokines) and hormones [49, 50]. Obesity is characterized by hypertrophic adipose tissue. As an effect of hypercaloric intake, adipocytes enlarge and release more free fatty acids, which activate macrophages to produce Tumor Necrosis Factor–alpha (TNF–α). As a consequence, adipocyte expression of Intercellular Adhesion Molecule–1 (ICAM–1), InterLeukin (IL)–6, and Macrophage Chemo attractant Protein–1 (MCP–1) is enhanced, promoting diapedesis of monocytes from the circulation to adipose tissue. With the exception of adiponectin, adipocytokine levels rise with obesity. Growing adipose tissue poses higher demands in terms of oxygen and nutrient delivery. Despite vasculogenesis [51], local hypoxia induces several angiogenic factors, which further suppress adiponectin expression even more and upregulate leptin production. This "hypersecretion of proatherogenic, pro–inflammatory, and prodiabetic adipocytokines which is accompanied by a decreased production of adiponectin" is called adipose tissue dysfunction [52]. Of the adipokines, leptin and adiponectin have a proven direct influence on angiogenesis and the endothelium, and are therefore further discussed [53].

Leptin

Leptin is a cytokine, which is mainly synthesized by white adipose tissue and released into the circulation. Its physiological role is to suppress appetite and to increase energy expenditure via the hypothalamus. However, circulating levels of leptin contradictorily rise with increasing body fat percentage, while obese patients have no diminished appetite [54].

Leptin has several proangiogenic effects, including enhanced Akt–mediated eNOS phosphorylation [55] as well as stimulation of endothelial cell proliferation [56], which are expected to be beneficial in the setting of endothelial dysfunction in obese patients. However, leptin also exerts proatherogenic effects, including the induction of ROS, a pro–inflammatory vascular effect, and stimulation of the proliferative capacity of VSMC [57].

These insights have led to the concept of a selective leptin resistance for both central (appetite) and peripheral pro–angiogenic effects, without any changes in terms of pro–atherogenic stimuli. The inverse correlation between leptin and endothelial function, independent of the metabolic and inflammatory disturbances associated with obesity adds, further supports to this hypothesis [58].

Adiponectin

Adiponectin is a 30 kDa protein produced by adipocytes with anti–inflammatory, antiatherogenic, and insulin sensitizing effects. This adipokine plays a central role in lipid and energy metabolism. Contrary to all other adipokines, adiponectin concentrations are lower in obese. In their elegant paper, Torigoe et al. demonstrated that even in healthy young men, circulating concentrations of adiponectin determine endothelial function [59], through eNOS mRNA stabilization and eNOS phosphorylation [60]. Moreover, high–glucose level–induced generation of ROS by endothelial cells is suppressed by adiponectin [61]. Recent human data show that adiponectin stimulates the migratory capacity of circulating angiogenic cells [62], thereby supporting animal experiments that suggested angiogenesis–stimulating effects [63] and the promotion of neovascularization [64].

8.2.1.5 INFLAMMATION

Obesity, and visceral obesity in particular, is associated with a low–grade pro–inflammatory status [65]. Upon activation, endothelial cells express adhesion molecules, which allow leukocytes to adhere and initiate a cascade of inflammatory reactions. Inflammatory cytokines produced by macrophages in adipose tissue (IL–18, TNF–α) further add to the expression of adhesion molecules. Not surprisingly inflammatory cytokines (e.g., C–Reactive Protein or CRP) are also associated with CV risk [66], and their concentration inversely correlates with endothelial function [67]. Another important mediator is IL–6 [68], which not only is a potent stimulator for the production of CRP, but also stimulates angiotensin II in VSMC and the associated production of ROS [69]. Besides reacting with NO, and thereby neutralizing its vasodilatory effect, ROSs also reduce eNOS activity and consequently NO production. Moreover, ROSs are able to directly inhibit eNOS and react with tetrahydrobiopterin (BH4), a necessary cofactor for eNOS activity [70]. In this situation, eNOS becomes "uncoupled" and paradoxically leads to increased ROS generation [71].

8.2.1.6 INSULIN RESISTANCE AND TYPE 2 DIABETES MELLITUS

Under physiological conditions, insulin is a potent vasodilator, via the stimulation of the PI3K/Akt pathway to augment NO production. However, in the setting of insulin resistance this reaction is absent [72]. Yet insulin is still able to activate the Extracellular signaling–Regulated Kinase (ERK1/2) pathway leading to production of EndoThelin 1 (ET–1) and thus vasoconstriction [73]. The same pathway also leads to expression of adhesion molecules like Vascular Cell Adhesion Molecule (VCAM)–1 on endothelial cells [74]. Endothelial dysfunction per se can even contribute to insulin resistance, since impaired microvascular vasodilation in skeletal muscle reduces delivery of insulin and glucose to skeletal muscle and thereby causes insulin resistance [75].

In type 2 diabetes mellitus, NO bioavailability is significantly diminished: a reduction in eNOS activity [76] as well as eNOS uncoupling [77] and increased generation of ROS have been demonstrated [78] and further deteriorate endothelial function. Hyperglycemia in diabetic patients leads to production of Advanced Glycation End products (AGE) after reacting with proteins, lipids, and nucleic acids. AGE can induce the expression of ROS via NF–kappa B activation and TNF–α [79]. Type 2 diabetes is additionally characterized by increased levels of ET–1 [80]. Interestingly, a recent double–blind–placebo controlled trial in 46 type II diabetes patients with microalbuminuria demonstrated that 4 weeks of treatment with an oral ET–1 receptor antagonist (bosentan) can repair endothelial function [81].

8.3 CLINICAL ASSESSMENT OF ENDOTHELIAL DYSFUNCTION

The landmark study by Ludmer et al. [82] led to the concept of endothelial dysfunction as the primum movens of the atherosclerotic process. The consequences of an abnormal vasodilator response (i.e., impaired vasodilation and even paradoxical vasoconstriction of coronary arteries upon the administration of acetylcholine) have thereafter been extensively studied. Epicardial and microvascular coronary endothelial dysfunction predicts CV events in patients with and without coronary artery disease [83]. Obviously, acetylcholine infusion can be applied during coronary angiography, but the invasive character prohibits its use in healthy individuals and children. Since then, novel, less invasive techniques have been developed, which will be discussed in detail in the following paragraphs.

8.3.1 FLOW–MEDIATED DILATION

In 1992, Celermajer et al. described a noninvasive technique that allows assessment of peripheral endothelial function [84]. Measurement of flow–mediated dilation (FMD) at the level of a large conduit artery, usually the brachial artery, has since then become the most applied technique [85]. Briefly, high–resolution ultrasound is used to measure the internal diameter of the brachial artery, from lumen–intima interface on the near and far vascular wall [86]. After assessing baseline diameter, the brachial artery is occluded during 5 minutes using a sphygmomanometer inflated to suprasystolic pressure. When the cuff is deflated; the increased flow causes endothelial–dependent dilation through raised shear stress [84]. It is this proportional response (i.e., related to baseline diameter) that is significantly impaired in the case of endothelial dysfunction.

Pathophysiological research confirms that this effect is mainly mediated by the release of NO by endothelial cells during the phase of hyperemia since it can be prevented by using N(G)–MonoMethyl–L–Arginine (L–NMMA), which is a selective inhibitor of eNOS. Importantly, NO dependence is influenced by location of the inflating cuff. Dilation upon hyperemia is completely blocked by infusion of L–NMMA when the cuff is placed distal to the measuring site, but only partially when the cuff is placed proximal to the echo probe [87]. The latter position of the probe creates an area of ischemia within the region of the probe, and therefore other vasoactive substances may play a role. Although in the past the predictive power of FMD was attributed to its NO dependency, a recent meta–analysis by Green et al. demonstrated that FMD measured after proximal occlusion is at least as predictive as measured after distal occlusion, although less NO dependent [88].

The assessment of FMD is still the most widely used method but it requires intensive training in order to achieve acceptable reproducibility. Besides technical factors, which are reviewed elsewhere, [86, 89], patient–related characteristics require a high level of standardization (Table 1). Measurements should be performed on the same time of day since FMD has a known diurnal variation [90]. A high fat meal [91] deteriorates endothelial function, and dietary components such as tea [92], Vitamin C [93], and chocolate [94] have short–term beneficial effects on endothelial function. Measurements should therefore always be performed in a fasting state before medication intake [95, 96]. Specifically for the pediatric setting, children frequently present with mild infections, which may impair endothelial function for up to 2 weeks [97]. Both mental stress [98] and room and skin temperature [99] influence endothelial function. It is therefore recommended to perform measurements in a dim–lighted and temperature–controlled (21–24°C) room and to carefully explain the procedure to studied subjects. Tobacco use should be avoided for 4 to 6 hours prior to measurements [100]. No guidelines are given regarding passive smoking, although this may be relevant for children, since childhood exposure to tobacco smoke leads to endothelial dysfunction in both children [101] and young adults [102]. Although we did not find evidence for a synergistic effect of smoking and childhood obesity on endothelial dysfunction, maternal smoking has been shown to correlate with several obesity–related risk factors in young children [103]. Gender matters in childhood since it influences endothelial function, with FMD being lower in boys [104]. Moreover, endothelial function differs during the menstrual cycle [105].

TABLE 1: Patient–related factors influencing clinical assessment of endothelial function in children.

Influencing Factor	Solution
Time of day	Perform measurements between 8 and 12 am to rule out an effect of diurnal variation of endothelial function
Food	Perform measurement after an overnight fast
Medication	Ask patients to take their medication after the test
(Mild) infection	Postpone the test for >2 weeks
Active and passive smoking	Exclude smokers and/or record (parental) smoking habits
Gender	Check whether all groups are matched for sex
Age	Check whether all groups are matched for age and pubertal stage
Menstrual Cycle	Girls in same phase of cycle or note the menstrual phase
Skin temperature	Allow sufficient patient acclimatization time; cover patients using a blanket
Mental stress	Avoid anxiety by providing patient information; perform the test in a quiet room

Contrary to adults, children present more variability in time to peak dilation in response to hyperemia. While in adults peak dilation usually occurs 60 seconds after cuff release, in children maximal dilation has been described between 40 and 120 seconds after–occlusion [106]. This effect is age dependent, and time to peak tends to drop with increasing age [104], advocating against a fixed moment in time to record maximal endothelium–dependent dilation. Effects of age on maximal dilation are conflicting with Donald et al. demonstrating no effect of age on FMD in a very large population of children (n = 7,557) [99] and Sarkola et al. mentioning a decrease in FMD, which they explained by an increasing baseline internal diameter of the brachial artery [104].

Although we found 2 articles describing the absence of endothelial dysfunction in obese children [21, 107], numerous research groups have reported on impaired endothelial–dependent vasodilation, assessed with FMD, in this population [108–122].

8.3.2 PERIPHERAL ARTERIAL TONOMETRY

Peripheral arterial tonometry was developed as a novel technique to overcome the disadvantages of user dependence of FMD. The only commercially available and validated apparatus (Endo–PAT 2000) involves finger probes to measure arterial pulse wave amplitudes at the fingertip.

Guidelines for FMD measurements are also implemented for Endo–PAT measurements, although the interference of several factors was investigated separately. Research confirmed the effect of diurnal variation [123] and the effect of dietary components such as omega–3 fatty acids [27], polyphenol rich olive oil [124], chocolate [125], and tea [126]. Furthermore, measurements were influenced by drugs [81] and mental stress [127].

The Endo–PAT probes are placed on one fingertip of both hands and are inflated to produce a subdiastolic counter pressure, in order to provide fixation and prevent venous pooling. Pressure differences secondary to dilating arterioles in the fingers are measured. The procedure is initiated with a 5–minute baseline assessment; then a manometer cuff is inflated to supra–systolic pressures and the brachial artery of the nondominant arm is occluded. After 5 minutes, the cuff is released and reactive hyperemia is observed during 5 minutes. The software provided calculates a Reactive Hyperemia Index (RHI) and is defined as the ratio of the mean Pulse Wave Amplitude (PWA) between 90 and 150 seconds after deflation divided by a preocclusion period during 210 seconds before occlusion. This ratio is then divided by the same ratio for the control arm and multiplied by a baseline correction factor. A "Framingham reactive hyperemia" is also calculated. For this ratio, the period used to assess the hyperemia is between 90 and 120 seconds after occlusion since this period most strongly correlated with CV risk factors [128]. The ratio is log transformed as data indicate that values are not normally distributed.

Since the probes can never be placed proximal to the occlusion site, provided by the cuff, there is an effect of local ischemia. In addition, the hyperemic response that is measured is not entirely caused by endothelial NO production since it is not fully abolished by L–NAME [129]. Other mediators that contribute to vasodilation include prostacyclin (PGI2), a derivative of arachidonic acid that is secreted by endothelial cells. In nor-

TABLE 2: Overview of studies using Endo–PAT to measure endothelial function in obese children.

Reference	Age (years)	Definition of obesity and overweight	Comorbidities	Parameter	Outcome
Mahmud et al. [24]	Obese: 13.4 ± 1.7; Lean: 14.0 ± 1.4	Obesity = BMI > 95th percentile	All insulin resistant, based on the HOMA score	RHI	Mean RHI was significantly lower in obese adolescents compared with controls (1.51 ± 0.4 vs. 2.06 ± 0.4
Metzig et al. [25]	12.4	Obesity = BMI > 95th percentile	15% with hypertension, 15% with dyslipidemia, 9% with OSAS, and 15% with impaired glucose tolerance	RHI	No significant effect of glucose ingestion on RHI
Kelly et al. [26]	12.7	Obesity = BMI > 95th percentile	?	RHI	No significant effect of exenatide therapy on RHI
Dangardt et al. [27]	15.7	?	?	Maximum dilation, area under the curve	Significant improvement of endothelial function after 3 months of omega–3 fatty acide supplementation
Tryggestad et al. [28]	Obese: 13.9 ± 2.5; Lean: 13.3 ± 3.0	Obesity = BMI > 95th percentile	?	RHI	No significant difference between obese and normal weight children
Landgraf et al. [29]	Obese: 11.8 ± 2.9; Lean: 12.9 ± 2.9	Obesity = BMI > 97th percentile; overweight = BMI> 90th percentile	?	RHI	Mean RHI was significantly lower in obese and overweight children compared with controls (1.28 ± 0.24 vs. 1.96 ±0.79)

mal arteries, NO has inhibitory effects on PGI2 release [130], and therefore the effect of PGI2 on endothelial dilation can vary in diseases associated with reduced NO bioavailability. A third mediator responsible for endothelial dependent dilation is Endothelium–Derived Hyperpolarizing Factor (EDHF), which has a larger influence on arterial tone in smaller vessels [131]. Fourthly, sympathetic tone can also modulate endothelial function, and an inverse correlation between sympathetic nerve activity and RHI has been demonstrated in healthy subjects [132].

Feasibility and reproducibility of Endo–PAT in adults are excellent [133]. The technique also appears highly reproducible in adolescents and causes hardly any discomfort [134]. To our knowledge, only 6 studies have been published in which endothelial function measured with Endo–PAT in obese children is described, and only 3 of them compared obese to lean children (Table 2). Results have been conflicting with Landgraf et al. [29] and Mahmud et al. [24] describing a lower RHI in obese children versus lean controls, whereas Tryggestad et al. [28] reported no difference.

Importantly, the Endo–PAT measures microvascular endothelial function, whereas FMD assesses endothelial function at the level of larger conduit arteries. Therefore, discrepant results are conceivable. Although Endo–PAT and FMD correlate well in healthy subjects [135], in patients with chest pain [136] and with coronary artery disease [137] such comparisons between the two techniques in both healthy and obese children are still missing.

Pubertal development and its associated hormonal changes complicate the study of endothelial function in children. RHI increases during puberty in both genders [138] and correlates to changes in estradiol and dehydroepiandrosterone sulfate in peripheral blood [139]. Both hormones upregulate eNOS concentration and activity [140].

Chen et al. demonstrated that age influences timing of the peak response of endothelial–dependent dilation measured with PAT [127] and noted shorter time to peak after 3–year followup, similar to FMD. These authors therefore propose to analyze the entire hyperemic response curve, instead of focusing on a very specific postocclusion time interval. By calculating the area under the curve, the problem of variability regarding time to peak dilation may be circumvented.

8.4 CELL–BASED METHODS FOR EVALUATING ENDOTHELIAL DYSFUNCTION

The cell–derived methods mentioned hereafter are considered as biomarkers for endothelial status justifying their discussion in this review. Traditional CV risk factors influence their numbers and function. Fundamental and translational research have demonstrated that these biomarkers are more than plain bystanders whose numbers drop or rise as a reflection of endothelial homeostasis, but rather appear to be active players in the process of endothelial damage and repair.

8.4.1 ENDOTHELIAL PROGENITOR CELLS

In 1997, Asahara et al, in a murine model of hind limb ischemia, described for the first time that CD34 (a stem cell marker) positive mononuclear cells, which were isolated from human peripheral blood, could selectively incorporate into new capillaries in areas of ischemic injury resulting into neovascularization of the affected limb [141]. These cells were termed Endothelial Progenitor Cells (EPC). Since then, numerous experiments have investigated the biological characteristics of EPC. It has become clear that the original term EPC actually covers 3 different cell types, denoted as Endothelial Cell Colony Forming Unit (CFU–EC), Circulating Angiogenic Cells (CAC), and "true" EPC. In this paper, only the latter 2 will be given further consideration, because great controversies still exist on the origin, the proliferative potential, and the differentiation capacity of CFU–EC [142].

True EPCs, alternatively called Endothelial Colony Forming Cells (ECFC), are released upon stimulation from bone marrow into the peripheral circulation and can incorporate into damaged endothelium (Figure 2) [143]. They are believed to be responsible for vasculogenesis and, as such, most closely fulfill the criteria for an Endothelial Progenitor Cell. EPCs appear in cultures of mononuclear cells after 7–21 days, have a cobblestone morphology, high proliferative capacity [144], and are able to form vessels in vivo [142]. Further analysis demonstrated that these cells express CD34

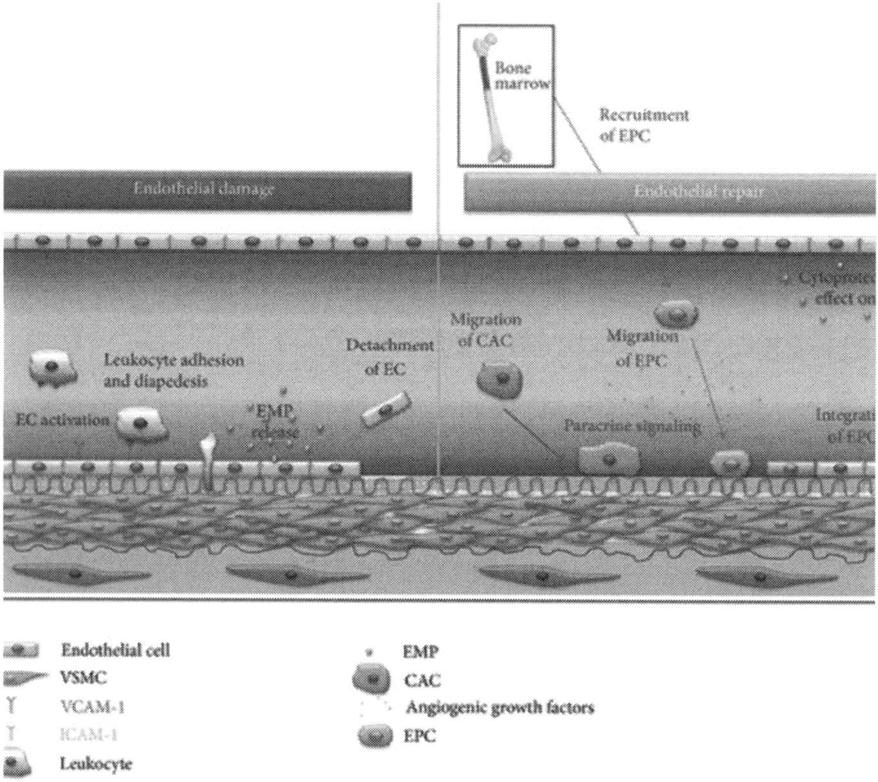

FIGURE 2: Mechanisms involved in endothelial damage and repair. Upon activation endothelial cells express adhesion molecules (i.e.; VCAM–1 and ICAM–1), which allow leukocytes to adhere, transmigrate, and initiate a cascade of inflammatory reactions and the release of EMP into the circulation. With significant endothelial damage, cells become senescent and are detached. This ultimately leads to the recruitment of CAC, monocyte–macrophage–derived cells that contribute to vascular repair by adhering to loci of endothelial damage, and producing angiogenic cytokines that induce the mobilization of EPC from the bone marrow. The angiogenic cytokines produced by CAC also serve as homing molecules with a chemotactic effect on EPC. As a consequence, EPC migrate to damaged endothelium and eventually integrate into the endothelial cell layer. Besides being released after endothelial activation, EMPs also contributes to endothelial homeostasis by cytoprotective effects on endothelial cells, including reduced apoptosis. EC, Endothelial Cell; VSMC, Vascular Smooth Muscle Cell; VCAM–1, Vascular Cell Adhesion Molecule 1; ICAM–1, InterCellular Adhesion Molecule 1; EMP, Endothelial MicroParticles; CAC, Circulating Angiogenic Cells; EPC, Endothelial Progenitor Cells.

and Vascular Endothelial Growth Factor Receptor–2 (VEGFR–2) but not CD133 or CD45 (a pan–leukocytic marker) [145], even not at the mRNA level as was demonstrated by Case et al. [146].

The mechanism by which EPCs are released from the bone marrow is not fully understood but is the result of Matrix MetalloProteinase (MMP)–9 activation [147], in a NO–dependent mechanism [148]. As a consequence, reduced NO bioavailability not only leads to endothelial dysfunction, but also to the impaired recruitment of EPC to the loci of damaged endothelium as well, starting a vicious circle as described by Van Craenenbroeck and Conraads [149].

The technique most commonly used to quantify EPC is flow cytometry. Although this method is technically challenging due to very low numbers of circulating EPC, it is minimally invasive (blood sample via venipuncture) and therefore very convenient in children. Since there is no single, specific marker to identify EPC [150], a combination of markers is applied. Many different protocols have been developed between laboratories, with, however, poor intermethod agreement [151], which may have contributed to conflicting results in the literature. Schmidt–Lucke et al. introduced a modified International Society of Hematotherapy and Graft Engineering (ISHAGE) protocol for CD34+/KDR+ EPC enumeration gated on the basis of low SSC and low–to–bright CD45 fluorescence [152]. The authors concluded that it is in fact the CD45dim positive CD34+/KDR+ PC that correlate best with clinical characteristics of the studied patients with coronary artery disease. This protocol was presented to facilitate interlaboratory comparison and speed up EPC enumeration, yet technical recommendations for rare event analysis were not taken into account [153].

Since mature endothelial cells have limited regenerative capacity, EPC are necessary for endothelial repair. This notion is supported by a mathematical model of endothelial maintenance, which predicted a critical phase of endothelial cell defects that causes serious vascular damage. Such devastating consequences can be significantly delayed by incorporation of EPC [154]. Further in vivo evidence is provided by the fact that lower numbers of circulating EPC predict CV events and death in adults

[155, 156] and by data showing that human EPCs are able to form fused vessels when implanted in immunodeficient mice [142].

Using flow cytometry it was demonstrated that EPC numbers correlate with endothelial function in CAD patients [157], in patients with type 1 diabetes [158], in young smokers [159], and in healthy subjects [160].

Müller–Ehmsen J et al. noted lower numbers of EPC in obese volunteers compared to healthy participants, while they observed a significant increase in EPC levels after weight loss [161]. Besides the direct effect of obesity, several other studies demonstrated reduced numbers of EPC in patients with obesity–related comorbidities such as hypertension [162, 163], hypercholesterolemia [164], and type II diabetes mellitus [165, 166].

Research in obese children has shown conflicting results. To our knowledge, there are only two papers comparing numbers of endothelial progenitor cells in obese children and lean controls. Unfortunately, each group used different protocols and different markers to detect these cells. Jung et al. were the first to investigate numbers of EPC in obese adolescents. They did not find significant differences in CD34+/KDR+/CD133+–cells, but did mention higher numbers of CD34–/KDR+CD133+–cells in overweight adolescents [167]. However, it has been shown that neither of these cells give rise to EPC in culture [146].

Arnold C et al. investigated whether numbers of circulating EPC related to physical fitness in obese children. Maximal oxygen uptake significantly correlated with CD34+ and CD133+/CD34+ cells in this population, yet the endothelial marker (KDR) was not assessed in this study [168].

In conclusion, studies comparing numbers of EPC between healthy and obese children that include CD45 as a marker and correlating their numbers to endothelial function in this specific population are not available and are eagerly awaited.

8.4.2 CIRCULATING ANGIOGENIC CELLS

Circulating Angiogenic Cells (CAC) are grown from peripheral blood mononuclear cells after 4 to 7 days of culture in endothelial promoting media and fibronectin–coated plates. CACs have a low proliferative capacity and do not incorporate into endothelium, yet these cells restore damaged endothelium in a paracrine fashion by aiding in the recruitment and proliferation of, respectively, EPC and endothelial cells. CACs do not demonstrate an endothelial phenotype [169], but are hematopoietic cells, which closely resemble activated M2 monocytes [170]. Cell culture is the most widely used technique to determine their numbers and allows researchers to further explore their physiological activity. The migratory capacity towards Vascular Endothelial Growth Factor (VEGF) and Stromal cell–Derived Factor (SDF)–1α is commonly quantified using 7–day–old CAC cultures [171]. Both VEGF and SDF–1α recruit CAC to hypoxic tissues [172]. After an additional 3 days of culture, the supernatant can be collected and analyzed for protein secretion (such as VEGF) [173], and thus paracrine activity can be evaluated.

Numbers of CAC and their migratory capacity correlate with endothelial function [174]. Obese adults have lower numbers of CAC compared to healthy controls [175] and these cells display lower migratory capacity and reduced secretion of angiogenic growth factors, which can be reversed by weight loss [176]. However, results for obese children are still lacking.

NO is necessary for the migration of CAC towards VEGF [177], and therefore reduction in eNOS activity and reduced bioavailability of NO in obese patients could explain impaired migratory capacity of CAC in obese patients [178].

Leptin stimulates the migratory capacity of CAC, yet CACs of obese patients display a resistance against the leptin associated pro–angiogenic effect, which again is reversed by weight loss. Leptin resistance has been attributed to higher concentrations of Protein Tyrosine Phosphatase 1B (PTP–1B), a known inhibitor of the leptin signaling cascade. Higher levels of PTP–1B were seen in obese individuals, which returned to levels comparable to lean subjects after weight loss. Furthermore, blocking PTP–1B activity pharmacologically restored responsiveness to leptin in CAC of obese patients [176].

Adiponectin similarly enhances the migratory capacity of CAC towards SDF–1α [62]. This finding has been attributed to an upregulation of the receptor for SDF–1α, C–X–C Chemokine Receptor type 4 (CXCR–4) on CAC. Lower concentrations of adiponectin, seen in obese, could therefore further contribute to the functional deficit of CAC.

8.4.3 ENDOTHELIAL MICROPARTICLES

Endothelial MicroParticles (EMP) are small particles (100 nm to 1 μm) shed from the plasma membrane by endothelial cells upon damage, activation, or apoptosis [179]. These microparticles are covered with surface antigens from the parental endothelial cell, making quantification as well as identification of the underlying process for their generation possible using specific flow cytometry markers.

Initially, EMP were considered as indicators of endothelial disruption [180], but their role has been redefined; instead of being mere markers, they also appear to elicit physiological effects. On the one hand, EMP themselves can further compromise endothelial function [181], whereas, on the other hand, they exert beneficial effects on endothelial cell survival and even promote endothelial regeneration [182].

Interestingly, EMP also act as vectors carrying RNA (including microRNA), DNA, and proteins to target cells [183]. Several studies have shown that the content of EMP depends on the trigger by which their parental cells were stimulated [184, 185].

The exact mechanisms that regulate the release of EMP in vivo are not completely understood, but involve an end in membrane phospholipid asymmetry and the expression of phosphatidylserine on the outside of the cell membrane, membrane budding, and eventually microparticle release.

EMPs are widely identified by their constitutive expression of CD31 (platelet endothelial cell adhesion molecule), but not CD42b. The absence of the CD42b (platelet–specific glycoprotein Ib marker) marker is used to avoid potential contamination with platelet microparticles that have the same size range as EMP. EMPs are to be detected in platelet poor plasma that was subjected to sequential centrifugation. In fact, careful attention has to be paid to all preanalytical and analytical steps, as variables in both phases may affect accurate microparticle enumeration and could be an important source of variability, interlaboratory discrepancies, and artifacts [186, 187].

Numbers of EMP are elevated in obese women and inversely correlate with endothelial function [188]. Van Ierssel et al. noted about 100 to

400 EMP per μL blood in healthy volunteers [189], yet numbers may vary depending on methods used for sample preparation and analysis.

Data on EMP in children are scarce. Siklar et al. [190] provided indirect evidence that EMP numbers are higher in obese versus healthy children. By using a commercially available test (STA–PROCOAG–PPL Kit (Diagnostica Stago SAS, France)), they measured the procoagulant activity of phospholipids through assembly of the prothrombinase complex. Healthy children had a significantly longer microparticle release time, which was presumably due to a lower number of EMPs. Furthermore, Gündüz et al. showed that circulating vascular endothelial cadherin (CD144)+EMP are higher in obese and overweight children than in lean controls [191], but unfortunately, endothelial function was not assessed in this study. Further research is necessary to investigate the potential role of EMP as a biomarker for endothelial damage in obese children.

8.5 PHARMACOLOGICAL AND NONPHARMACOLOGICAL INTERVENTIONS TO RESTORE ENDOTHELIAL DAMAGE

8.5.1 WEIGHT LOSS

The ultimate goal of therapy remains improvement of long–term physical health through permanent healthy lifestyle habits. Although weight maintenance is advised for younger children and a decline in BMI is achieved through growth, weight loss remains the cornerstone of therapy for obesity in adolescents [192].

Weight loss leads to an improvement of CV risk factors associated with endothelial dysfunction in childhood obesity [193, 194]. Because many studies have focused on the effect of multidisciplinary treatment approaches, little is known about the effect of weight loss alone on endothelial function in obese children. Kaufman et al. [195] studied the effect of a 5–month dietary intervention with a goal of a 5 to 8% decrease in total body mass in 15 overweight children. The study group consisted of six boys and nine girls (aged years). After the intervention, a significant decrease in weight, body fat percentage, and BMI was observed, with a trend towards improvement of endothelial function.

Woo et al. [196] compared the effects of weight loss alone to a combined treatment arm of diet and 6 weeks of exercise training in 82 overweight children, 9 to 12 years of age. In this study, an improvement of endothelial function was seen in both groups, albeit more pronounced in the diet plus exercise group.

8.5.2 EXERCISE

Exercise training appears more effective than weight loss alone to improve endothelial function in obese children. As little as eight weeks of exercise training consisting of three 1–hour sessions of circuit training each week led to a significant improvement in endothelial function, even without weight loss in a randomized cross–over study involving 19 obese adolescents [197]. Two other studies, one involving endurance training and another applying aerobic interval training, confirmed the effect of exercise alone on endothelial function [198, 199].

In addition to the effect on factors leading to endothelial dysfunction [200], exercise training can more directly affect endothelial function through its effect on endothelial shear stress. Translational research using arterial biopsies of adults demonstrated that regular exercise leads to an upregulated expression of eNOS mRNA and higher eNOS protein and increased eNOS –phosphorylation levels in endothelial cells [201].

On top of increasing NO synthesis, exercise decreases production of ROS by reducing Nicotinamide Adenine Dinucleotide Phosphate (NAD(P)H) oxidase activity [202] and by enhancement of antioxidant capacity [203]. In addition, elevated shear stress, as seen with regular exercise training, can increase levels of BH4 [204] and thereby reduces eNOS uncoupling.

Exercise training has a significant effect on endothelial repair mechanisms, including both EPC and CAC. In adults with metabolic syndrome, 8 weeks of exercise training led to a significant increase in repair capacity of CAC [205]. These findings were confirmed by transplantation of human EPC into nude mice with defined carotid endothelial injury. In adults, a multitreatment approach including physical activity restored functional impairment of CAC [176]. In children, a program consisting of 12 weeks of exercise training increased the percentages of CD34+, CD133+ and CD34+/CD133+ cells and reduced carotid IMT [206]. In addition, 1 year of exercise training led to a significant rise in CD45low/CD34+/KDR+ EPC [207].

8.5.3 PHARMACOLOGICAL INTERVENTIONS

Studies on the effect of pharmacological therapy in reversing endothelial dysfunction in children are very scarce. Therefore, potential candidates investigated in obese adults will be briefly discussed.

Since the withdrawal of both rimonabant and sibutramine, due the increased risk of psychiatric adverse events [208] and increased CV risk [209], respectively, orlistat a reversible blocker of lipase, is the only drug still available to aid weight loss. A recent meta–analysis of available data in children points out that the drug leads to 5 kg weight loss and 5 cm reduction in weight circumference after at least 6 months of therapy compared with placebo, but no improvement in the lipid profile nor insulin level was observed [210]. An open–label 3–month trial in adults consisting of a calorie–restricted diet and 120 mg of orlistat could not demonstrate a significant improvement in FMD [211].

Metformin is approved in many countries to treat insulin resistance in obese children. Although 3 months of metformin ingestion significantly improved both endothelial function and insulin resistance in adults with metabolic syndrome [212], adding metformin to a structured lifestyle intervention did not reverse insulin resistance in children [213].

8.5.4 DIETARY COMPONENTS

Dietary components have recently gained significant interest because of their potential benefit on endothelial function and presumed prevention of CV disease.

Adding vitamins C (500 mg/d) and E (400 IU/d) for 6 weeks to a program of diet and exercise further improved endothelial function in hyperlipidemic children [93]. Dangardt et al. were able to show that 3 months of omega–3 fatty acid supplementation improved vascular function in obese adolescents [27].

Although results from several randomized controlled trials with dietary components are quite promising, effects need to be confirmed in larger population–based studies.

8.5.5 FUTURE RESEARCH NEEDS

As mentioned earlier in the paper, children are not just small adults and the speed by which physiological changes occur during puberty is tremendous. Puberty modulates endothelial function, and this improvement seems to be mediated through hormonal changes. The underlying mechanisms are not yet completely understood and deserve further investigation, as it may reveal key factors capable of ameliorating endothelial function later in life.

Further observation and perfectioning of flow cytometry and culture techniques have made EPC enumeration even more reliable. Yet studies on childhood and adolescent obesity that have adequately implemented these optimized strategies are lacking. In addition, the relationship of EPC and EMP with endothelial function also needs to be fully addressed in obese and overweight children.

And last but not least, whereas pharmacological interventions in obese children have been largely disappointing, inclusion of specific dietary components has shown to be highly effective in improving endothelial function in the short term. Further research, however, is needed to confirm whether these beneficial effects still remain at long–term followup.

8.6 CONCLUSIONS

Recent research has demonstrated the relevance of obesity–induced endothelial dysfunction, both in adults and in children. Fundamental and translational studies have led to significant understanding of cellular and molecular alterations held responsible for endothelial disruption. Unfortunately, research has until now mainly focused on the progression of atherosclerosis in obese adults, who have been exposed to the consequences of endothelial dysfunction during many years. Despite technical difficulties and ethical concerns, the investigation of endothelial function in obese children is a necessary step to further examine the initiation and progression of endothelial dysfunction, which is crucial to the development of treatment strategies.

Use of novel circulating markers can further unravel the delicate balance between endothelial damage and repair that causes endothelial dysfunction in obese subjects.

Health care management should aim to avoid the catastrophic rise in CV morbidity and mortality, which will accompany obese children into adulthood. Therefore, multi–disciplinary prevention programs need to be set up and tested for their clinical effect. In order to speed up advances achieved in this domain, clinical research will largely depend on the effect of such interventions on the so–called surrogate endpoints, such as the correction of endothelial function.

REFERENCES

1. C. Andersson, L. van Gaal, I. D. Caterson et al., "Relationship between HbA1c levels and risk of cardiovascular adverse outcomes and all–cause mortality in overweight and obese cardiovascular high–risk women and men with type 2 diabetes," Diabetologia, vol. 55, no. 9, pp. 2348–2355, 2012.
2. S. L. Verhulst, N. Schrauwen, D. Haentjens et al., "Sleep–disordered breathing and the metabolic syndrome in overweight and obese children and adolescents," Journal of Pediatrics, vol. 150, no. 6, pp. 608–612, 2007.
3. A. Sharma, L. M. Grummer–Strawn, K. Dalenius et al., "Obesity prevalence among low–income, preschool–aged children—United States, 1998–2008," The Journal of the American Medical Association, vol. 303, no. 1, pp. 28–30, 2010.

4. S. R. Srinivasan, W. Bao, W. A. Wattigney, and G. S. Berenson, "Adolescent overweight is associated with adult overweight and related multiple cardiovascular risk factors: the Bogalusa heart study," Metabolism, vol. 45, no. 2, pp. 235–240, 1996.
5. A. Tirosh, I. Shai, A. Afek et al., "Adolescent BMI trajectory and risk of diabetes versus coronary disease," The New England Journal of Medicine, vol. 364, no. 14, pp. 1315–1325, 2011.
6. P. W. Franks, R. L. Hanson, W. C. Knowler, M. L. Sievers, P. H. Bennett, and H. C. Looker, "Childhood obesity, other cardiovascular risk factors, and premature death," The New England Journal of Medicine, vol. 362, no. 6, pp. 485–493, 2010.
7. M. Juonala, C. G. Magnussen, G. S. Berenson et al., "Childhood adiposity, adult adiposity, and cardiovascular risk factors," The New England Journal of Medicine, vol. 365, no. 20, pp. 1876–1885, 2011.
8. A. Lerman and A. M. Zeiher, "Endothelial function: cardiac events," Circulation, vol. 111, no. 3, pp. 363–368, 2005.
9. T. Asahara, A. Kawamoto, and H. Masuda, "Concise review: circulating endothelial progenitor cells for vascular medicine," Stem Cells, vol. 29, no. 11, pp. 1650–1655, 2011.
10. A. J. Viera, M. Mooberry, and N. S. Key, "Microparticles in cardiovascular disease pathophysiology and outcomes," Journal of the American Society of Hypertension, vol. 6, no. 4, pp. 243–252, 2012.
11. R. F. Furchgott and J. V. Zawadzki, "The obligatory role of endothelial cells in the relaxation of arterial smooth muscle by acetylcholine," Nature, vol. 288, no. 5789, pp. 373–376, 1980.
12. S. Dimmeler, I. Fleming, B. Fisslthaler, C. Hermann, R. Busse, and A. M. Zeiher, "Activation of nitric oxide synthase in endothelial cells by Akt–dependent phosphorylation," Nature, vol. 399, no. 6736, pp. 601–605, 1999.
13. A. J. Flammer and T. F. Lüscher, "Human endothelial dysfunction: EDRFs," Pflügers Archiv: European Journal of Physiology, vol. 459, no. 6, pp. 1005–1013, 2010.
14. A. Virdis, L. Ghiadoni, and S. Taddei, "Human endothelial dysfunction: EDCFs," Pflügers Archiv: European Journal of Physiology, vol. 459, no. 6, pp. 1015–1023, 2010.
15. G. K. Hansson, "Inflammation, atherosclerosis, and coronary artery disease," The New England Journal of Medicine, vol. 352, no. 16, pp. 1685–1626, 2005.
16. J. Deanfield, A. Donald, C. Ferri et al., "Endothelial function and dysfunction—part I: methodological issues for assessment in the different vascular beds: a statement by the working group on endothelin and endothelial factors of the European society of hypertension," Journal of Hypertension, vol. 23, no. 1, pp. 7–17, 2005.
17. P. M. Vanhoutte, "Endothelial dysfunction—the first step toward coronary arteriosclerosis," Circulation Journal, vol. 73, no. 4, pp. 595–601, 2009.
18. E. Falaschetti, A. D. Hingorani, A. Jones et al., "Adiposity and cardiovascular risk factors in a large contemporary population of pre–pubertal children," European Heart Journal, vol. 31, no. 24, pp. 3063–3072, 2010.
19. D. Shimbo, P. Muntner, D. Mann et al., "Endothelial dysfunction and the risk of hypertension: the multi–ethnic study of atherosclerosis," Hypertension, vol. 55, no. 5, pp. 1210–1216, 2010.

20. M. Juonala, J. S. A. Viikari, T. Rönnemaa, H. Helenius, L. Taittonen, and O. T. Raitakari, "Elevated blood pressure in adolescent boys predicts endothelial dysfunction: the cardiovascular risk in young finns study," Hypertension, vol. 48, no. 3, pp. 424–430, 2006.
21. M. Charakida, A. Jones, E. Falaschetti et al., "Childhood obesity and vascular phenotypes: a population study," Journal of the American College of Cardiology, vol. 60, no. 25, pp. 2643–2650, 2012.
22. J. Koskinen, M. Kähönen, J. S. A. Viikari et al., "Conventional cardiovascular risk factors and metabolic syndrome in predicting carotid intima–media thickness progression in young adults: the cardiovascular risk in young finns study," Circulation, vol. 120, no. 3, pp. 229–236, 2009.
23. M. Juonala, J. S. A. Viikari, T. Laitinen et al., "Interrelations between brachial endothelial function and carotid intima–media thickness in young adults: the cardiovascular risk in young finns study," Circulation, vol. 110, no. 18, pp. 2918–2923, 2004.
24. F. H. Mahmud, D. J. Hill, M. S. Cuerden, and C. L. Clarson, "Impaired vascular function in obese adolescents with insulin resistance," Journal of Pediatrics, vol. 155, no. 5, pp. 678–682, 2009.
25. A. M. Metzig, S. J. Schwarzenberg, C. K. Fox, M. M. Deering, B. M. Nathan, and A. S. Kelly, "Postprandial endothelial function, inflammation, and oxidative stress in obese children and adolescents," Obesity, vol. 19, no. 6, pp. 1279–1283, 2011.
26. A. S. Kelly, A. M. Metzig, K. D. Rudser et al., "Exenatide as a weight–loss therapy in extreme pediatric obesity: a randomized, controlled pilot study," Obesity, vol. 20, no. 2, pp. 364–370, 2012.
27. F. Dangardt, W. Osika, Y. Chen et al., "Omega–3 fatty acid supplementation improves vascular function and reduces inflammation in obese adolescents," Atherosclerosis, vol. 212, no. 2, pp. 580–585, 2010.
28. J. B. Tryggestad, D. M. Thompson, K. C. Copeland, and K. R. Short, "Obese children have higher arterial elasticity without a difference in endothelial function: the role of body composition," Obesity, vol. 20, no. 1, pp. 165–171, 2012.
29. K. Landgraf, D. Friebe, T. Ullrich et al., "Chemerin as a mediator between obesity and vascular inflammation in children," Journal of Clinical Endocrinology and Metabolism, vol. 97, no. 4, pp. E556–E564, 2012.
30. A. E. Loot, J. G. Schreiber, B. Fisslthaler, and I. Fleming, "Angiotensin II impairs endothelial function via tyrosine phosphorylation of the endothelial nitric oxide synthase," Journal of Experimental Medicine, vol. 206, no. 13, pp. 2889–2896, 2009.
31. M. L. Hijmering, E. S. G. Stroes, J. Olijhoek, B. A. Hutten, P. J. Blankestijn, and T. J. Rabelink, "Sympathetic activation markedly reduces endothelium–dependent, flow–mediated vasodilation," Journal of the American College of Cardiology, vol. 39, no. 4, pp. 683–688, 2002.
32. H. M. A. Eid, H. Arnesen, E. M. Hjerkinn, T. Lyberg, and I. Seljeflot, "Relationship between obesity, smoking, and the endogenous nitric oxide synthase inhibitor, asymmetric dimethylarginine," Metabolism, vol. 53, no. 12, pp. 1574–1579, 2004.
33. D. Steinberg and J. L. Witztum, "Is the oxidative modification hypothesis relevant to human atherosclerosis? Do the antioxidant trials conducted to date refute the hypothesis?" Circulation, vol. 105, no. 17, pp. 2107–2111, 2002.

34. P. P. Toth, "Activation of intracellular signaling systems by high–density lipoproteins," Journal of Clinical Lipidology, vol. 4, no. 5, pp. 376–381, 2010.
35. J. R. Nofer, B. Kehrel, M. Fobker, B. Levkau, G. Assmann, and A. V. Eckardstein, "HDL and arteriosclerosis: beyond reverse cholesterol transport," Atherosclerosis, vol. 161, no. 1, pp. 1–16, 2002.
36. P. Schwandt, T. Bertsch, and G. M. Haas, "Anthropometric screening for silent cardiovascular risk factors in adolescents: the PEP family heart study," Atherosclerosis, vol. 211, no. 2, pp. 667–671, 2010.
37. G. Iacobellis, M. C. Ribaudo, A. Zappaterreno, C. V. Iannucci, and F. Leonetti, "Prevalence of uncomplicated obesity in an Italian obese population," Obesity Research, vol. 13, no. 6, pp. 1116–1122, 2005.
38. U. Landmesser, "High density lipoprotein—should we raise it?" Current Vascular Pharmacology, vol. 10, no. 6, pp. 718–719, 2012.
39. R. S. Paffenbarger Jr., R. T. Hyde, A. L. Wing, I. M. Lee, D. L. Jung, and J. B. Kampert, "The association of changes in physical–activity level and other lifestyle characteristics with mortality among men," The New England Journal of Medicine, vol. 328, no. 8, pp. 538–545, 1993.
40. S. N. Blair, H. W. Kohl, C. E. Barlow, R. S. Paffenbarger, L. W. Gibbons, and C. A. Macera, "Changes in physical fitness and all–cause mortality: a prospective study of healthy and unhealthy men," The Journal of the American Medical Association, vol. 273, no. 14, pp. 1093–1098, 1995.
41. M. Nocon, T. Hiemann, F. Müller–Riemenschneider, F. Thalau, S. Roll, and S. N. Willich, "Association of physical activity with all–cause and cardiovascular mortality: a systematic review and meta–analysis," European Journal of Cardiovascular Prevention and Rehabilitation, vol. 15, no. 3, pp. 239–246, 2008.
42. S. G. Trost, L. M. Kerr, D. S. Ward, and R. R. Pate, "Physical activity and determinants of physical activity in obese and non–obese children," International Journal of Obesity, vol. 25, no. 6, pp. 822–829, 2001.
43. I. de Bourdeaudhuij, J. Lefevre, B. Deforche, K. Wijndaele, L. Matton, and R. Philippaerts, "Physical activity and psychosocial correlates in normal weight and overweight 11 to 19 year olds," Obesity Research, vol. 13, no. 6, pp. 1097–1105, 2005.
44. R. A. Abbott, M. A. Harkness, and P. S. W. Davies, "Correlation of habitual physical activity levels with flow–mediated dilation of the brachial artery in 5–10 year old children," Atherosclerosis, vol. 160, no. 1, pp. 233–239, 2002.
45. N. D. Hopkins, G. Stratton, T. M. Tinken et al., "Relationships between measures of fitness, physical activity, body composition and vascular function in children," Atherosclerosis, vol. 204, no. 1, pp. 244–249, 2009.
46. K. Pahkala, O. J. Heinonen, H. Lagström et al., "Vascular endothelial function and leisure–time physical activity in adolescents," Circulation, vol. 118, no. 23, pp. 2353–2359, 2008.
47. K. Pahkala, O. J. Heinonen, O. Simell et al., "Association of physical activity with vascular endothelial function and intima–media thickness," Circulation, vol. 124, no. 18, pp. 1956–1963, 2011.

48. T. C. Otto and M. D. Lane, "Adipose development: from stem cell to adipocyte," Critical Reviews in Biochemistry and Molecular Biology, vol. 40, no. 4, pp. 229–242, 2005.
49. E. E. Kershaw and J. S. Flier, "Adipose tissue as an endocrine organ," Journal of Clinical Endocrinology and Metabolism, vol. 89, no. 6, pp. 2548–2556, 2004.
50. L. F. van Gaal, I. L. Mertens, and C. E. de Block, "Mechanisms linking obesity with cardiovascular disease," Nature, vol. 444, no. 7121, pp. 875–880, 2006.
51. H. R. Lijnen, "Angiogenesis and obesity," Cardiovascular Research, vol. 78, no. 2, pp. 286–293, 2008.
52. G. R. Hajer, T. W. van Haeften, and F. L. J. Visseren, "Adipose tissue dysfunction in obesity, diabetes, and vascular diseases," European Heart Journal, vol. 29, no. 24, pp. 2959–2971, 2008.
53. V. Christiaens and H. R. Lijnen, "Angiogenesis and development of adipose tissue," Molecular and Cellular Endocrinology, vol. 318, no. 1–2, pp. 2–9, 2010.
54. J. M. Friedman, "The function of leptin in nutrition, weight, and physiology," Nutrition Reviews, vol. 60, no. 10, part 2, pp. S1–S14, 2002.
55. C. Vecchione, A. Maffei, S. Colella et al., "Leptin effect on endothelial nitric oxide is mediated through Akt–endothelial nitric oxide synthase phosphorylation pathway," Diabetes, vol. 51, no. 1, pp. 168–173, 2002.
56. H. Y. Park, H. M. Kwon, H. J. Lim et al., "Potential role of leptin in angiogenesis: leptin induces endothelial cell proliferation and expression of matrix metalloproteinases in vivo and in vitro," Experimental and Molecular Medicine, vol. 33, no. 2, pp. 95–102, 2001.
57. M. Singh, U. S. Bedi, P. P. Singh, R. Arora, and S. Khosla, "Leptin and the clinical cardiovascular risk," International journal of cardiology, vol. 140, no. 3, pp. 266–271, 2010.
58. A. Singhal, S. Farooqi, T. J. Cole et al., "Influence of leptin on arterial distensibility: a novel link between obesity and cardiovascular disease?" Circulation, vol. 106, no. 15, pp. 1919–1924, 2002.
59. M. Torigoe, H. Matsui, Y. Ogawa et al., "Impact of the high–molecular–weight form of adiponectin on endothelial function in healthy young men," Clinical Endocrinology, vol. 67, no. 2, pp. 276–281, 2007.
60. Z. V. Wang and P. E. Scherer, "Adiponectin, cardiovascular function, and hypertension," Hypertension, vol. 51, no. 1, pp. 8–14, 2008.
61. R. Ouedraogo, X. Wu, S. Q. Xu et al., "Adiponectin suppression of high–glucose–induced reactive oxygen species in vascular endothelial cells: evidence for involvement of a cAMP signaling pathway," Diabetes, vol. 55, no. 6, pp. 1840–1846, 2006.
62. V. Adams, R. Höllriegel, E. B. Beck et al., "Adiponectin promotes the migration of circulating angiogenic cells through p38–mediated induction of the CXCR4 receptor," International Journal of Cardiology, 2012.
63. R. Shibata, N. Ouchi, S. Kihara, K. Sato, T. Funahashi, and K. Walsh, "Adiponectin stimulates angiogenesis in response to tissue ischemia through stimulation of AMP–activated protein kinase signaling," The Journal of Biological Chemistry, vol. 279, no. 27, pp. 28670–28674, 2004.
64. P. Eren, S. Camus, G. Matrone et al., "Adiponectinemia controls pro–angiogenic cell therapy," Stem Cells, vol. 27, no. 11, pp. 2712–2721, 2009.

65. D. R. Faber, Y. van der Graaf, J. Westerink, and F. L. J. Visseren, "Increased visceral adipose tissue mass is associated with increased C–reactive protein in patients with manifest vascular diseases," Atherosclerosis, vol. 212, no. 1, pp. 274–280, 2010.
66. P. M. Ridker, C. H. Hennekens, J. E. Buring, and N. Rifai, "C–reactive protein and other markers of inflammation in the prediction of cardiovascular disease in women," The New England Journal of Medicine, vol. 342, no. 12, pp. 836–843, 2000.
67. S. Fichtlscherer, G. Rosenberger, D. H. Walter, S. Breuer, S. Dimmeler, and A. M. Zeiher, "Elevated C–reactive protein levels and impaired endothelial vasoreactivity in patients with coronary artery disease," Circulation, vol. 102, no. 9, pp. 1000–1006, 2000.
68. J. S. Yudkin, M. Kumari, S. E. Humphries, and V. Mohamed–Ali, "Inflammation, obesity, stress and coronary heart disease: is interleukin–6 the link?" Atherosclerosis, vol. 148, no. 2, pp. 209–214, 2000.
69. S. Wassmann, M. Stumpf, K. Strehlow et al., "Interleukin–6 induces oxidative stress and endothehal dysfunction by overexpression of the angiotensin II type 1 receptor," Circulation Research, vol. 94, no. 4, pp. 534–541, 2004.
70. L. V. d'Uscio, S. Milstien, D. Richardson, L. Smith, and Z. S. Katusic, "Long–term vitamin C treatment increases vascular tetrahydrobiopterin levels and nitric oxide synthase activity," Circulation Research, vol. 92, no. 1, pp. 88–95, 2003.
71. U. Landmesser, S. Dikalov, S. R. Price et al., "Oxidation of tetrahydrobiopterin leads to uncoupling of endothelial cell nitric oxide synthase in hypertension," Journal of Clinical Investigation, vol. 111, no. 8, pp. 1201–1209, 2003.
72. Z. Y. Jiang, Y. W. Lin, A. Clemont et al., "Characterization of selective resistance to insulin signaling in the vasculature of obese Zucker (fa/fa) rats," Journal of Clinical Investigation, vol. 104, no. 4, pp. 447–457, 1999.
73. M. A. Potenza, F. L. Marasciulo, D. M. Chieppa et al., "Insulin resistance in spontaneously hypertensive rats is associated with endothelial dysfunction characterized by imbalance between NO and ET–1 production," The American Journal of Physiology, vol. 289, no. 2, pp. H813–H822, 2005.
74. M. Montagnani, I. Golovchenko, I. Kim et al., "Inhibition of phosphatidylinositol 3–kinase enhances mitogenic actions of insulin in endothelial cells," The Journal of Biological Chemistry, vol. 277, no. 3, pp. 1794–1799, 2002.
75. J. A. Kim, M. Montagnani, K. K. Kwang, and M. J. Quon, "Reciprocal relationships between insulin resistance and endothelial dysfunction: molecular and pathophysiological mechanisms," Circulation, vol. 113, no. 15, pp. 1888–1904, 2006.
76. H. Zhang, J. Zhang, Z. Ungvari, and C. Zhang, "Resveratrol improves endothelial function: role of TNFα and vascular oxidative stress," Arteriosclerosis, Thrombosis, and Vascular Biology, vol. 29, no. 8, pp. 1164–1171, 2009.
77. T. J. Guzik, S. Mussa, D. Gastaldi et al., "Mechanisms of increased vascular superoxide production in human diabetes mellitus: role of NAD(P)H oxidase and endothelial nitric oxide synthase," Circulation, vol. 105, no. 14, pp. 1656–1662, 2002.
78. W. S. Cheang, W. T. Wong, X. Y. Tian et al., "Endothelial nitric oxide synthase enhancer reduces oxidative stress and restores endothelial function in db/db mice," Cardiovascular Research, vol. 92, no. 2, pp. 267–275, 2011.

79. M. Morita, S. Yano, T. Yamaguchi, and T. Sugimoto, "Advanced glycation end products–induced reactive oxygen species generation is partly through NF–κB activation in human aortic endothelial cells," Journal of Diabetes and its Complications, vol. 27, no. 1, pp. 11–15, 2013.
80. S. Gogg, U. Smith, and P. A. Jansson, "Increased MAPK activation and impaired insulin signaling in subcutaneous microvascular endothelial cells in type 2 diabetes: the role of endothelin–1," Diabetes, vol. 58, no. 10, pp. 2238–2245, 2009.
81. A. Rafnsson, F. Böhm, M. Settergren, A. Gonon, K. Brismar, and J. Pernow, "The endothelin receptor antagonist bosentan improves peripheral endothelial function in patients with type 2 diabetes mellitus and microalbuminuria: a randomised trial," Diabetologia, vol. 55, no. 3, pp. 600–607, 2012.
82. P. L. Ludmer, A. P. Selwyn, and T. L. Shook, "Paradoxical vasoconstriction induced by acetylcholine in atherosclerotic coronary arteries," The New England Journal of Medicine, vol. 315, no. 17, pp. 1046–1051, 1986.
83. J. P. J. Halcox, W. H. Schenke, G. Zalos et al., "Prognostic value of coronary vascular endothelial dysfunction," Circulation, vol. 106, no. 6, pp. 653–658, 2002.
84. D. S. Celermajer, K. E. Sorensen, V. M. Gooch et al., "Non–invasive detection of endothelial dysfunction in children and adults at risk of atherosclerosis," The Lancet, vol. 340, no. 8828, pp. 1111–1115, 1992.
85. A. J. Flammer, T. Anderson, D. S. Celermajer et al., "The assessment of endothelial function: from research into clinical practice," Circulation, vol. 126, no. 6, pp. 753–767, 2012.
86. D. H. J. Thijssen, M. A. Black, K. E. Pyke et al., "Assessment of flow–mediated dilation in humans: a methodological and physiological guideline," The American Journal of Physiology, vol. 300, no. 1, pp. H2–H12, 2011.
87. S. N. Doshi, K. K. Naka, N. Payne et al., "Flow–mediated dilatation following wrist and upper arm occlusion in humans: the contribution of nitric oxide," Clinical Science, vol. 101, no. 6, pp. 629–635, 2001.
88. D. J. Green, H. Jones, D. Thijssen, N. T. Cable, and G. Atkinson, "Flow–mediated dilation and cardiovascular event prediction: does nitric oxide matter?" Hypertension, vol. 57, no. 3, pp. 363–369, 2011.
89. M. Charakida, S. Masi, T. F. Lüscher, J. J. P. Kastelein, and J. E. Deanfield, "Assessment of atherosclerosis: the role of flow–mediated dilatation," European Heart Journal, vol. 31, no. 23, pp. 2854–2861, 2010.
90. M. E. Otto, A. Svatikova, R. B. de Mattos Barretto et al., "Early morning attenuation of endothelial function in healthy humans," Circulation, vol. 109, no. 21, pp. 2507–2510, 2004.
91. R. A. Vogel, M. C. Corretti, and G. D. Plotnick, "Effect of a single high–fat meal on endothelial function in healthy subjects," The American Journal of Cardiology, vol. 79, no. 3, pp. 350–354, 1997.
92. N. Alexopoulos, C. Vlachopoulos, K. Aznaouridis et al., "The acute effect of green tea consumption on endothelial function in healthy individuals," European Journal of Cardiovascular Prevention and Rehabilitation, vol. 15, no. 3, pp. 300–305, 2008.
93. M. M. Engler, M. B. Engler, M. J. Malloy et al., "Antioxidant vitamins C and E improve endothelial function in children with hyperlipidemia: endothelial assess-

ment of risk from lipids in youth (EARLY) trial," Circulation, vol. 108, no. 9, pp. 1059–1063, 2003.

94. M. B. Engler, M. M. Engler, C. Y. Chen et al., "Flavonoid–rich dark chocolate improves endothelial function and increases plasma epicatechin concentrations in healthy adults," Journal of the American College of Nutrition, vol. 23, no. 3, pp. 197–204, 2004.
95. R. A. Harris, S. K. Nishiyama, D. W. Wray, V. Tedjasaputra, D. M. Bailey, and R. S. Richardson, "The effect of oral antioxidants on brachial artery flow–mediated dilation following 5 and 10 min of ischemia," European Journal of Applied Physiology, vol. 107, no. 4, pp. 445–453, 2009.
96. E. Magen, J. R. Viskoper, J. Mishal, R. Priluk, D. London, and C. Yosefy, "Effects of low–dose aspirin on blood pressure and endothelial function of treated hypertensive hypercholesterolaemic subjects," Journal of Human Hypertension, vol. 19, no. 9, pp. 667–673, 2005.
97. M. Charakida, A. E. Donald, M. Terese et al., "Endothelial dysfunction in childhood infection," Circulation, vol. 111, no. 13, pp. 1660–1665, 2005.
98. L. Ghiadoni, A. E. Donald, M. Cropley et al., "Mental stress induces transient endothelial dysfunction in humans," Circulation, vol. 102, no. 20, pp. 2473–2478, 2000.
99. A. E. Donald, M. Charakida, E. Falaschetti et al., "Determinants of vascular phenotype in a large childhood population: the avon longitudinal study of parents and children (ALSPAC)," European Heart Journal, vol. 31, no. 12, pp. 1502–1510, 2010.
100. D. S. Celermajer, K. E. Sorensen, D. Georgakopoulos et al., "Cigarette smoking is associated with dose–related and potentially reversible impairment of endothelium–dependent dilation in healthy young adults," Circulation, vol. 88, no. 5 I, pp. 2149–2155, 1993.
101. K. Kallio, E. Jokinen, O. T. Raitakari et al., "Tobacco smoke exposure is associated with attenuated endothelial function in 11–year–old healthy children," Circulation, vol. 115, no. 25, pp. 3205–3212, 2007.
102. M. Juonala, C. G. Magnussen, A. Venn et al., "Parental smoking in childhood and brachial artery flow–mediated dilatation in young adults: the cardiovascular risk in young finns study and the childhood determinants of adult health study," Arteriosclerosis, Thrombosis, and Vascular Biology, vol. 32, no. 4, pp. 1024–1031, 2012.
103. M. D. Peterson, D. Liu, H. B. IglayReger, W. A. Saltarelli, P. S. Visich, and P. M. Gordon, "Principal component analysis reveals gender–specific predictors of cardiometabolic risk in 6th graders," Cardiovascular Diabetology, vol. 11, no. 1, article 146, 2012.
104. T. Sarkola, C. Manlhiot, C. Slorach et al., "Evolution of the arterial structure and function from infancy to adolescence is related to anthropometric and blood pressure changes," Arteriosclerosis, Thrombosis, and Vascular Biology, vol. 32, no. 10, pp. 2516–2524, 2012.
105. M. Hashimoto, M. Akishita, M. Eto et al., "Modulation of endothelium–dependent flow–mediated dilatation of the brachial artery by sex and menstrual cycle," Circulation, vol. 92, no. 12, pp. 3431–3435, 1995.
106. M. J. Järvisalo, T. Rönnemaa, I. Volanen et al., "Brachial artery dilatation responses in healthy children and adolescents," The American Journal of Physiology, vol. 282, no. 1 51–1, pp. H87–H92, 2002.

107. L. H. Naylor, D. J. Green, T. W. Jones et al., "Endothelial function and carotid intima–medial thickness in adolescents with type 2 diabetes mellitus," Journal of Pediatrics, vol. 159, no. 6, pp. 971–974, 2011.
108. R. Bhattacharjee, W. H. Alotaibi, L. Kheirandish–Gozal, O. S. Capdevila, and D. Gozal, "Endothelial dysfunction in obese non–hypertensive children without evidence of sleep disordered breathing," BMC Pediatrics, vol. 10, article 8, 2010.
109. R. Bhattacharjee, J. Kim, W. H. Alotaibi, L. Kheirandish–Gozal, O. S. Capdevila, and D. Gozal, "Endothelial dysfunction in children without hypertension: potential contributions of obesity and obstructive sleep apnea," Chest, vol. 141, no. 3, pp. 682–691, 2012.
110. A. E. Caballero, R. Bousquet–Santos, L. Robles–Osorio et al., "Overweight latino children and adolescents have marked endothelial dysfunction and subclinical vascular inflammation in association with excess body fat and insulin resistance," Diabetes Care, vol. 31, no. 3, pp. 576–582, 2008.
111. M. M. Ciccone, V. Miniello, R. Marchioli et al., "Morphological and functional vascular changes induced by childhood obesity," European Journal of Cardiovascular Prevention and Rehabilitation, vol. 18, no. 6, pp. 831–835, 2011.
112. N. J. Farpour–Lambert, Y. Aggoun, L. M. Marchand, X. E. Martin, F. R. Herrmann, and M. Beghetti, "Physical activity reduces systemic blood pressure and improves early markers of atherosclerosis in pre–pubertal obese children," Journal of the American College of Cardiology, vol. 54, no. 25, pp. 2396–2406, 2009.
113. S. Kapiotis, G. Holzer, G. Schaller et al., "A proinflammatory state is detectable in obese children and is accompanied by functional and morphological vascular changes," Arteriosclerosis, Thrombosis, and Vascular Biology, vol. 26, no. 11, pp. 2541–2546, 2006.
114. L. Karpoff, A. Vinet, I. Schuster et al., "Abnormal vascular reactivity at rest and exercise in obese boys," European Journal of Clinical Investigation, vol. 39, no. 2, pp. 94–102, 2009.
115. A. A. Meyer, G. Kundt, M. Steiner, P. Schuff–Werner, and W. Kienast, "Impaired flow–mediated vasodilation, carotid artery intima–media thickening, and elevated endothelial plasma markers in obese children: the impact of cardiovascular risk factors," Pediatrics, vol. 117, no. 5, pp. 1560–1567, 2006.
116. E. Mimoun, Y. Aggoun, M. Pousset et al., "Association of arterial stiffness and endothelial dysfunction with metabolic syndrome in obese children," Journal of Pediatrics, vol. 153, no. 1, pp. 65.e1–70.e1, 2008.
117. A. S. Peña, D. P. Belobrajdic, E. Wiltshire, R. Gent, C. Hirte, and J. Couper, "Adiponectin relates to smooth muscle function and folate in obese children," International Journal of Pediatric Obesity, vol. 5, no. 2, pp. 185–191, 2010.
118. A. S. Peña, E. Wiltshire, K. MacKenzie et al., "Vascular endothelial and smooth muscle function relates to body mass index and glucose in obese and nonobese children," Journal of Clinical Endocrinology and Metabolism, vol. 91, no. 11, pp. 4467–4471, 2006.
119. P. Tounian, Y. Aggoun, B. Dubern et al., "Presence of increased stiffness of the common carotid artery and endothelial dysfunction in severely obese children: a prospective study," The Lancet, vol. 358, no. 9291, pp. 1400–1404, 2001.

120. K. S. Woo, P. Chook, C. W. Yu et al., "Overweight in children is associated with arterial endothelial dysfunction and intima–media thickening," International Journal of Obesity, vol. 28, no. 7, pp. 852–857, 2004.
121. M. M. Yilmazer, V. Tavli, O. U. Carti et al., "Cardiovascular risk factors and noninvasive assessment of arterial structure and function in obese Turkish children," European Journal of Pediatrics, vol. 169, no. 10, pp. 1241–1248, 2010.
122. W. Zhu, X. Huang, J. He, M. Li, and H. Neubauer, "Arterial intima–media thickening and endothelial dysfunction in obese Chinese children," European Journal of Pediatrics, vol. 164, no. 6, pp. 337–344, 2005.
123. J. Liu, J. Wang, Y. Jin, H. J. Roethig, and M. Unverdorben, "Variability of peripheral arterial tonometry in the measurement of endothelial function in healthy men," Clinical Cardiology, vol. 32, no. 12, pp. 700–704, 2009.
124. R. J. Widmer, M. A. Freund, A. J. Flammer et al., "Beneficial effects of polyphenol–rich olive oil in patients with early atherosclerosis," European Journal of Nutrition, 2012.
125. L. D. P. Nogueira, M. P. Knibel, M. R. S. G. Torres, J. F. Nogueira Neto, and A. F. Sanjuliani, "Consumption of high–polyphenol dark chocolate improves endothelial function in individuals with stage 1 hypertension and excess body weight," International Journal of Hypertension, vol. 2012, Article ID 147321, 9 pages, 2012.
126. R. J. Miller, K. G. Jackson, T. Dadd et al., "The impact of the catechol–O–methyltransferase genotype on vascular function and blood pressure after acute green tea ingestion," Molecular Nutrition and Food Research, vol. 56, no. 6, pp. 966–975, 2012.
127. Y. Chen, F. Dangardt, W. Osika, K. Berggren, E. Gronowitz, and P. Friberg, "Age– and sex–related differences in vascular function and vascular response to mental stress. Longitudinal and cross–sectional studies in a cohort of healthy children and adolescents," Atherosclerosis, vol. 220, no. 1, pp. 269–274, 2012.
128. N. M. Hamburg, M. J. Keyes, M. G. Larson et al., "Cross–sectional relations of digital vascular function to cardiovascular risk factors in the Framingham heart study," Circulation, vol. 117, no. 19, pp. 2467–2474, 2008.
129. A. Nohria, M. Gerhard–Herman, M. A. Creager, S. Hurley, D. Mitra, and P. Ganz, "Role of nitric oxide in the regulation of digital pulse volume amplitude in humans," Journal of Applied Physiology, vol. 101, no. 2, pp. 545–548, 2006.
130. T. Osanai, N. Fujita, N. Fujiwara et al., "Cross talk of shear–induced production of prostacyclin and nitric oxide in endothelial cells," The American Journal of Physiology, vol. 278, no. 1, pp. H233–H238, 2000.
131. H. Shimokawa, H. Yasutake, K. Fujii et al., "The importance of the hyperpolarizing mechanism increases as the vessel size decreases in endothelium–dependent relaxations in rat mesenteric circulation," Journal of Cardiovascular Pharmacology, vol. 28, no. 5, pp. 703–711, 1996.
132. Y. B. Sverrisdóttir, L. M. Jansson, U. Hägg, and L. M. Gan, "Muscle sympathetic nerve activity is related to a surrogate marker of endothelial function in healthy individuals," PLoS ONE, vol. 5, no. 2, Article ID Article numbere9257, 2010.
133. C. E. McCrea, A. C. Skulas–Ray, M. Chow, and S. G. West, "Test–retest reliability of pulse amplitude tonometry measures of vascular endothelial function: implications for clinical trial design," Vascular Medicine, vol. 17, no. 1, pp. 29–36, 2012.

134. E. S. S. Tierney, J. W. Newburger, K. Gauvreau et al., "Endothelial pulse amplitude testing: feasibility and reproducibility in adolescents," Journal of Pediatrics, vol. 154, no. 6, pp. 901–905, 2009.
135. M. Dhindsa, S. M. Sommerlad, A. E. DeVan et al., "Interrelationships among noninvasive measures of postischemic macro– and microvascular reactivity," Journal of Applied Physiology, vol. 105, no. 2, pp. 427–432, 2008.
136. J. T. Kuvin, A. R. Patel, K. A. Sliney et al., "Assessment of peripheral vascular endothelial function with finger arterial pulse wave amplitude," The American Heart Journal, vol. 146, no. 1, pp. 168–174, 2003.
137. S. Onkelinx, V. Cornelissen, K. Goetschalckx, T. Thomaes, P. Verhamme, and L. Vanhees, "Reproducibility of different methods to measure the endothelial function," Vascular Medicine, vol. 17, no. 2, pp. 79–84, 2012.
138. T. Radtke, K. Khattab, P. Eser, S. Kriemler, H. Saner, and M. Wilhelm, "Puberty and microvascular function in healthy children and adolescents," Journal of Pediatrics, vol. 161, no. 5, pp. 887–891, 2012.
139. A. Bhangoo, S. Sinha, M. Rosenbaum, S. Shelov, and S. Ten, "Endothelial function as measured by peripheral arterial tonometry increases during pubertal advancement," Hormone Research in Paediatrics, vol. 76, no. 4, pp. 226–233, 2011.
140. S. P. Duckles and V. M. Miller, "Hormonal modulation of endothelial NO production," Pflügers Archiv: European Journal of Physiology, vol. 459, no. 6, pp. 841–851, 2010.
141. T. Asahara, T. Murohara, A. Sullivan et al., "Isolation of putative progenitor endothelial cells for angiogenesis," Science, vol. 275, no. 5302, pp. 964–966, 1997.
142. M. C. Yoder, L. E. Mead, D. Prater et al., "Redefining endothelial progenitor cells via clonal analysis and hematopoietic stem/progenitor cell principals," Blood, vol. 109, no. 5, pp. 1801–1809, 2007.
143. S. Wassmann, N. Werner, T. Czech, and G. Nickenig, "Improvement of endothelial function by systemic transfusion of vascular progenitor cells," Circulation Research, vol. 99, no. 8, pp. E74–E83, 2006.
144. D. A. Ingram, L. E. Mead, H. Tanaka et al., "Identification of a novel hierarchy of endothelial progenitor cells using human peripheral and umbilical cord blood," Blood, vol. 104, no. 9, pp. 2752–2760, 2004.
145. F. Timmermans, F. van Hauwermeiren, M. de Smedt et al., "Endothelial outgrowth cells are not derived from CD133+ Cells or CD45+ hematopoietic precursors," Arteriosclerosis, Thrombosis, and Vascular Biology, vol. 27, no. 7, pp. 1572–1579, 2007.
146. J. Case, L. E. Mead, W. K. Bessler, et al., "Human CD34+AC133+VEGFR–2+ cells are not endothelial progenitor cells but distinct, primitive hematopoietic progenitors," Experimental Hematology, vol. 35, pp. 1109–1118, 2007.
147. B. Heissig, K. Hattori, S. Dias et al., "Recruitment of stem and progenitor cells from the bone marrow niche requires MMP–9 mediated release of Kit–ligand," Cell, vol. 109, no. 5, pp. 625–637, 2002.
148. A. Aicher, C. Heeschen, C. Mildner–Rihm et al., "Essential role of endothelial nitric oxide synthase for mobilization of stem and progenitor cells," Nature Medicine, vol. 9, no. 11, pp. 1370–1376, 2003.
149. E. M. van Craenenbroeck and V. M. Conraads, "Endothelial progenitor cells in vascular health: focus on lifestyle," Microvascular Research, vol. 79, no. 3, pp. 184–192, 2010.

150. M. C. Yoder, "Defining human endothelial progenitor cells," Journal of Thrombosis and Haemostasis, vol. 7, no. 1, pp. 49–52, 2009.
151. E. M. F. van Craenenbroeck, V. M. A. Conraads, D. R. van Bockstaele et al., "Quantification of circulating endothelial progenitor cells: a methodological comparison of six flow cytometric approaches," Journal of Immunological Methods, vol. 332, no. 1–2, pp. 31–40, 2008.
152. C. Schmidt–Lucke, S. Fichtlscherer, A. Aicher et al., "Quantification of circulating endothelial progenitor cells using the modified ISHAGE Protocol," PLoS ONE, vol. 5, no. 11, Article ID e13790, 2010.
153. E. M. van Craenenbroeck, A. H. van Craenenbroeck, S. van Ierssel et al., "Quantification of circulating CD34+/KDR+/CD45dim endothelial progenitor cells: analytical considerations," International Journal of Cardiology, 2012.
154. J. O. den Buijs, M. Musters, T. Verrips, J. A. Post, B. Braam, and N. van Riel, "Mathematical modeling of vascular endothelial layer maintenance: the role of endothelial cell division, progenitor cell homing, and telomere shortening," The American Journal of Physiology, vol. 287, no. 6, pp. H2651–H2658, 2004.
155. N. Werner, S. Kosiol, T. Schiegl et al., "Circulating endothelial progenitor cells and cardiovascular outcomes," The New England Journal of Medicine, vol. 353, no. 10, pp. 999–1007, 2005.
156. C. Schmidt–Lucke, L. Rössig, S. Fichtlscherer et al., "Reduced number of circulating endothelial progenitor cells predicts future cardiovascular events: proof of concept for the clinical importance of endogenous vascular repair," Circulation, vol. 111, no. 22, pp. 2981–2987, 2005.
157. N. Werner, S. Wassmann, P. Ahlers et al., "Endothelial progenitor cells correlate with endothelial function in patients with coronary artery disease," Basic Research in Cardiology, vol. 102, no. 6, pp. 565–571, 2007.
158. L. Sibal, A. Aldibbiat, S. C. Agarwal et al., "Circulating endothelial progenitor cells, endothelial function, carotid intima–media thickness and circulating markers of endothelial dysfunction in people with type 1 diabetes without macrovascular disease or microalbuminuria," Diabetologia, vol. 52, no. 8, pp. 1464–1473, 2009.
159. W. Kim, H. J. Myung, H. C. Suk et al., "Effect of green tea consumption on endothelial function and circulating endothelial progenitor cells in chronic smokers," Circulation Journal, vol. 70, no. 8, pp. 1052–1057, 2006.
160. C. Murphy, G. S. Kanaganayagam, B. Jiang et al., "Vascular dysfunction and reduced circulating endothelial progenitor cells in young healthy UK South Asian men," Arteriosclerosis, Thrombosis, and Vascular Biology, vol. 27, no. 4, pp. 936–942, 2007.
161. J. Müller–Ehmsen, D. Braun, T. Schneider et al., "Decreased number of circulating progenitor cells in obesity: beneficial effects of weight reduction," European Heart Journal, vol. 29, no. 12, pp. 1560–1568, 2008.
162. M. Pirro, G. Schillaci, C. Menecali et al., "Reduced number of circulating endothelial progenitors and HOXA9 expression in CD34+ cells of hypertensive patients," Journal of Hypertension, vol. 25, no. 10, pp. 2093–2099, 2007.
163. A. Oliveras, M. J. Soler, O. M. Martínez–Estrada et al., "Endothelial progenitor cells are reduced in refractory hypertension," Journal of Human Hypertension, vol. 22, no. 3, pp. 183–190, 2008.

164. F. Rossi, C. Bertone, F. Montanile et al., "HDL cholesterol is a strong determinant of endothelial progenitor cells in hypercholesterolemic subjects," Microvascular Research, vol. 80, no. 2, pp. 274–279, 2010.
165. G. P. Fadini, M. Miorin, M. Facco et al., "Circulating endothelial progenitor cells are reduced in peripheral vascular complications of type 2 diabetes mellitus," Journal of the American College of Cardiology, vol. 45, no. 9, pp. 1449–1457, 2005.
166. C. G. Egan, R. Lavery, F. Caporali et al., "Generalised reduction of putative endothelial progenitors and CXCR4–positive peripheral blood cells in type 2 diabetes," Diabetologia, vol. 51, no. 7, pp. 1296–1305, 2008.
167. C. Jung, N. Fischer, M. Fritzenwanger et al., "Endothelial progenitor cells in adolescents: impact of overweight, age, smoking, sport and cytokines in younger age," Clinical Research in Cardiology, vol. 98, no. 3, pp. 179–188, 2009.
168. C. Arnold, D. Wenta, J. Mller–Ehmsen, N. Sreeram, and C. Graf, "Progenitor cell number is correlated to physical performance in obese children and young adolescents," Cardiology in the Young, vol. 20, no. 4, pp. 381–386, 2010.
169. R. J. Medina, C. L. O'Neill, M. Sweeney et al., "Molecular analysis of endothelial progenitor cell (EPC) subtypes reveals two distinct cell populations with different identities," BMC Medical Genomics, vol. 3, no. 1, article 18, 2010.
170. R. J. Medina, C. L. O'Neill, T. M. O'Doherty et al., "Myeloid angiogenic cells act as alternative M2 macrophages and modulate angiogenesis through interleukin–8," Molecular Medicine, no. 9–10, pp. 1045–1055, 2011.
171. E. M. van Craenenbroeck, V. Y. Hoymans, P. J. Beckers et al., "Exercise training improves function of circulating angiogenic cells in patients with chronic heart failure," Basic Research in Cardiology, vol. 105, no. 5, pp. 665–676, 2010.
172. M. R. Hoenig, C. Bianchi, and F. W. Sellke, "Hypoxia inducible factor–1α, endothelial progenitor cells, monocytes cardiovascular risk, wound healing, cobalt and hydralazine: a unifying hypothesis," Current Drug Targets, vol. 9, no. 5, pp. 422–435, 2008.
173. E. M. van Craenenbroeck, P. J. Beckers, N. M. Possemiers et al., "Exercise acutely reverses dysfunction of circulating angiogenic cells in chronic heart failure," European Heart Journal, vol. 31, no. 15, pp. 1924–1934, 2010.
174. C. Heiss, S. Keymel, U. Niesler, J. Ziemann, M. Kelm, and C. Kalka, "Impaired progenitor cell activity in age–related endothelial dysfunction," Journal of the American College of Cardiology, vol. 45, no. 9, pp. 1441–1448, 2005.
175. O. J. MacEneaney, E. J. Kushner, G. P. van Guilder, J. J. Greiner, B. L. Stauffer, and C. A. DeSouza, "Endothelial progenitor cell number and colony–forming capacity in overweight and obese adults," International Journal of Obesity, vol. 33, no. 2, pp. 219–225, 2009.
176. N. M. Heida, J. P. Müller, I. F. Cheng et al., "Effects of obesity and weight loss on the functional properties of early outgrowth endothelial progenitor cells," Journal of the American College of Cardiology, vol. 55, no. 4, pp. 357–367, 2010.
177. C. Heiss, A. Schanz, N. Amabile et al., "Nitric oxide synthase expression and functional response to nitric oxide are both important modulators of circulating angiogenic cell response to angiogenic stimuli," Arteriosclerosis, Thrombosis, and Vascular Biology, vol. 30, no. 11, pp. 2212–2218, 2010.

178. I. L. Williams, S. Wheatcroft, A. M. Shah, and M. T. Kearney, "Obesity, atherosclerosis and the vascular endothelium: mechanisms of reduced nitric oxide bioavailability in obese humans," International Journal of Obesity, vol. 26, no. 6, pp. 754–764, 2002.
179. F. Dignat–George and C. M. Boulanger, "The many faces of endothelial microparticles," Arteriosclerosis, Thrombosis, and Vascular Biology, vol. 31, no. 1, pp. 27–33, 2011.
180. C. M. Boulanger, N. Amabile, and A. Tedgui, "Circulating microparticles: a potential prognostic marker for atherosclerotic vascular disease," Hypertension, vol. 48, no. 2, pp. 180–186, 2006.
181. S. V. Brodsky, F. Zhang, A. Nasjletti, and M. S. Goligorsky, "Endothelium–derived microparticles impair endothelial function in vitro," The American Journal of Physiology, vol. 286, no. 5, pp. H1910–H1915, 2004.
182. A. S. Leroyer, T. G. Ebrahimian, C. Cochain et al., "Microparticles from ischemic muscle promotes postnatal vasculogenesis," Circulation, vol. 119, no. 21, pp. 2808–2817, 2009.
183. S. F. Mause and C. Weber, "Microparticles: protagonists of a novel communication network for intercellular information exchange," Circulation Research, vol. 107, no. 9, pp. 1047–1057, 2010.
184. A. M. Weerheim, A. M. Kolb, A. Sturk, and R. Nieuwland, "Phospholipid composition of cell–derived microparticles determined by one–dimensional high–performance thin–layer chromatography," Analytical Biochemistry, vol. 302, no. 2, pp. 191–198, 2002.
185. J. Huber, A. Vales, G. Mitulovic et al., "Oxidized membrane vesicles and blebs from apoptotic cells contain biologically active oxidized phospholipids that induce monocyte–endothelial interactions," Arteriosclerosis, Thrombosis, and Vascular Biology, vol. 22, no. 1, pp. 101–107, 2002.
186. R. Lacroix, C. Judicone, P. Poncelet et al., "Impact of pre–analytical parameters on the measurement of circulating microparticles: towards standardization of protocol," Journal of Thrombosis and Haemostasis, vol. 10, no. 3, pp. 437–446, 2012.
187. S. H. van Ierssel, E. M. van Craenenbroeck, V. M. Conraads et al., "Flow cytometric detection of endothelial microparticles (EMP): effects of centrifugation and storage alter with the phenotype studied," Thrombosis Research, vol. 125, no. 4, pp. 332–339, 2010.
188. K. Esposito, M. Ciotola, B. Schisano et al., "Endothelial microparticles correlate with endothelial dysfunction in obese women," Journal of Clinical Endocrinology and Metabolism, vol. 91, no. 9, pp. 3676–3679, 2006.
189. S. H. van Ierssel, V. Y. Hoymans, E. M. van Craenenbroeck et al., "Endothelial microparticles (EMP) for the assessment of endothelial function: an in vitro and in vivo study on possible interference of plasma lipids," PLoS ONE, vol. 7, no. 2, Article ID e31496, 2012.
190. Z. Siklar, G. Öçal, M. Berberoğlu et al., "Evaluation of hypercoagulability in obese children with thrombin generation test and microparticle release: effect of metabolic parameters," Clinical and Applied Thrombosis/Hemostasis, vol. 17, no. 6, pp. 585–589, 2011.

191. Z. Gündüz, I. Dursun, S. Tülpar et al., "Increased endothelial microparticles in obese and overweight children," Journal of Pediatric Endocrinology and Metabolism, vol. 25, no. 11–12, pp. 1111–1117, 2012.
192. S. E. Barlow, "Expert committee recommendations regarding the prevention, assessment, and treatment of child and adolescent overweight and obesity: summary report," Pediatrics, vol. 120, supplement 4, pp. S164–S192, 2007.
193. S. B. Sondike, N. Copperman, and M. S. Jacobson, "Effects of a low–carbohydrate diet on weight loss and cardiovascular risk factors in overweight adolescents," Journal of Pediatrics, vol. 142, no. 3, pp. 253–258, 2003.
194. T. Reinehr and W. Andler, "Changes in the atherogenic risk factor profile according to degree of weight loss," Archives of Disease in Childhood, vol. 89, no. 5, pp. 419–422, 2004.
195. C. L. Kaufman, D. R. Kaiser, A. S. Kelly, J. L. Dengel, J. Steinberger, and D. R. Dengel, "Diet revision in overweight children: effect on autonomic and vascular function," Clinical Autonomic Research, vol. 18, no. 2, pp. 105–108, 2008.
196. K. S. Woo, P. Chook, C. W. Yu et al., "Effects of diet and exercise on obesity–related vascular dysfunction in children," Circulation, vol. 109, no. 16, pp. 1981–1986, 2004.
197. K. Watts, P. Beye, A. Siafarikas et al., "Exercise training normalizes vascular dysfunction and improves central adiposity in obese adolescents," Journal of the American College of Cardiology, vol. 43, no. 10, pp. 1823–1827, 2004.
198. A. A. Meyer, G. Kundt, U. Lenschow, P. Schuff–Werner, and W. Kienast, "Improvement of early vascular changes and cardiovascular risk factors in obese children after a six–month exercise program," Journal of the American College of Cardiology, vol. 48, no. 9, pp. 1865–1870, 2006.
199. A. E. Tjønna, T. O. Stølen, A. Bye et al., "Aerobic interval training reduces cardiovascular risk factors more than a multitreatment approach in overweight adolescents," Clinical Science, vol. 116, no. 4, pp. 317–326, 2009.
200. A. B. R. Maggio, Y. Aggoun, X. E. Martin, L. M. Marchand, M. Beghetti, and N. J. Farpour–Lambert, "Long–term follow–up of cardiovascular risk factors after exercise training in obese children," International Journal of Pediatric Obesity, vol. 6, no. 2, pp. e603–e610, 2011.
201. R. Hambrecht, V. Adams, S. Erbs et al., "Regular physical activity improves endothelial function in patients with coronary artery disease by increasing phosphorylation of endothelial nitric oxide synthase," Circulation, vol. 107, no. 25, pp. 3152–3158, 2003.
202. V. Adams, A. Linke, N. Kränkel et al., "Impact of regular physical activity on the NAD(P)H oxidase and angiotensin receptor system in patients with coronary artery disease," Circulation, vol. 111, no. 5, pp. 555–562, 2005.
203. G. S. L. Ferrari and C. K. B. Ferrari, "Exercise modulation of total antioxidant capacity (TAC): towards a molecular signature of healthy aging," Frontiers in Life Science, vol. 5, no. 3–4, pp. 81–90, 2011.
204. J. D. Widder, W. Chen, L. Li et al., "Regulation of tetrahydrobiopterin biosynthesis by shear stress," Circulation Research, vol. 101, no. 8, pp. 830–838, 2007.
205. K. Sonnenschein, T. Horváth, M. Mueller et al., "Exercise training improves in vivo endothelial repair capacity of early endothelial progenitor cells in subjects with metabolic syndrome," European Journal of Cardiovascular Prevention and Rehabilitation, vol. 18, no. 3, pp. 406–414, 2011.

206. J. H. Park, M. Miyashita, Y. C. Kwon et al., "A 12–week after–school physical activity programme improves endothelial cell function in overweight and obese children: a randomised controlled study," BMC Pediatrics, vol. 12, article 111, 2012.
207. C. Walther, L. Gaede, V. Adams et al., "Effect of increased exercise in school children on physical fitness and endothelial progenitor cells: a prospective randomized trial," Circulation, vol. 120, no. 22, pp. 2251–2259, 2009.
208. R. Christensen, P. K. Kristensen, E. M. Bartels, H. Bliddal, and A. Astrup, "Efficacy and safety of the weight–loss drug rimonabant: a meta–analysis of randomised trials," The Lancet, vol. 370, no. 9600, pp. 1706–1713, 2007.
209. W. P. T. James, I. D. Caterson, W. Coutinho et al., "Effect of sibutramine on cardiovascular outcomes in overweight and obese subjects," The New England Journal of Medicine, vol. 363, no. 10, pp. 905–917, 2010.
210. S. Czernichow, C. M. Y. Lee, F. Barzi et al., "Efficacy of weight loss drugs on obesity and cardiovascular risk factors in obese adolescents: a meta–analysis of randomized controlled trials," Obesity Reviews, vol. 11, no. 2, pp. 150–158, 2010.
211. R. D. Brook, R. L. Bard, L. Glazewski et al., "Effect of short–term weight loss on the metabolic syndrome and conduit vascular endothelial function in overweight adults," The American Journal of Cardiology, vol. 93, no. 8, pp. 1012–1016, 2004.
212. C. Vitale, G. Mercuro, A. Cornoldi, M. Fini, M. Volterrani, and G. M. C. Rosano, "Metformin improves endothelial function in patients with metabolic syndrome," Journal of Internal Medicine, vol. 258, no. 3, pp. 250–256, 2005.
213. C. L. Clarson, F. H. Mahmud, J. E. Baker et al., "Metformin in combination with structured lifestyle intervention improved body mass index in obese adolescents, but did not improve insulin resistance," Endocrine, vol. 36, no. 1, pp. 141–146, 2009.

CHAPTER 9

INSULIN SENSITIVITY, SERUM LIPIDS, AND SYSTEMIC INFLAMMATORY MARKERS IN SCHOOL-AGED OBESE AND NONOBESE CHILDREN

JINKWAN KIM, RAKESH BHATTACHARJEE,
LEILA KHEIRANDISH-GOZAL, ABDELNABY KHALYFA,
OSCAR SANS CAPDEVILA, RIVA TAUMAN, and DAVID GOZAL

9.1 INTRODUCTION

Childhood obesity is a serious and progressively increasing public health problem that has reached epidemic proportions and in the United States disproportionately affects low–income and minority children [1–3]. Metabolic and cardiovascular complications of obesity in childhood, while less common than in adulthood, may nevertheless include insulin resistance and type 2 diabetes. Body mass index (BMI) tracks from childhood to adulthood and as such, overweight and obese children are at greater risk of developing not only hypertension, metabolic, and cardiovascular diseases, but also asthma and sleep apnea in later life [4–8].

In adults, obesity is associated with increases in systemic inflammatory markers, as evidenced by studies documenting the association of BMI and visceral obesity with circulating levels of cytokines and acute–phase reactants [9–11]. In children, the presence of obesity also appears to be associated with increased levels of high–sensitivity CRP (hsCRP) [12], as well as other inflammatory mediators [13–17], that promote the development of endothelial and metabolic dysfunction [18–21]. A recent review on this topic [22] concluded that although there appears to be sufficient evidence to support the existence of an association between obesity and increased hsCRP along with decreased adiponectin levels, the circulating levels found in the majority of the studies published in the literature are generally within the normative range, and could therefore underestimate the concentrations of these mediators at the tissue level. Tam et al. further recommended additional studies measuring IL–6 and TNF–α, as well as other interleukins and chemokines in young children [22].

Concurrent with such understanding of the published literature and in agreement with the recommendations by Tam and colleagues [22], we hypothesized that by examining the concentration of fasting morning plasma inflammatory mediators obtained from community–based obese children, we would expect to find variable expression of specific inflammatory mediators in this cohort that have previously been shown to increase the cardiovascular and/or metabolic disease risk.

9.2 METHODS

9.2.1 SUBJECTS

The study was approved by the University of Louisville Human Research Committee, and informed consent was obtained from the legal caregiver of each participant. Consecutive children between the ages of 5 to 8 years attending public schools in Jefferson County, Louisville, KY, were invited to participate in the study, after they underwent a school–based health screening, which included height and weight measurements. Based on such screening, children were identified when their BMI *z* score was ≥1.65 and age– (within 3–6 months), gender–, ethnicity–, and area–of–residence–matched children with BMI scores <1.65 were then identified and recruited to serve as controls. Of note, all children were otherwise healthy, were recruited from the community via the Jefferson County Public School Health Screening Program, and were representative of the demographic characteristics of the general population of the city of Louisville (http://ksdc.louisville.edu/sdc/census2000/cityprofiles/LouisvilleDP.

pdf). Children were excluded if they had known diabetes or prediabetes (http://www.diabetes.org/pre–diabetes/pre–diabetes–symptoms.jsp), any defined genetic abnormality or underlying systemic disease including hypertension (as defined by a systolic or diastolic blood pressure exceeding the 95th percentile for age, gender, and height using population data obtained by the National Heart Lung and Blood Institute), or if they were within a month from any acute infectious process.

9.2.2 ANTHROPOMETRY

To verify the school–health screening initial reports, children were weighed using a calibrated scale to the nearest 0.1 kg and height (to 0.1 cm) was measured with a stadiometer (Holtain, Crymych, UK). Body mass index (BMI) was calculated and BMI z–score was computed using CDC 2000 growth standards (http://www.cdc.gov/growthcharts/) and online software (http://www.cdc.gov/epiinfo/). A BMI z score $\geq$1.65 was considered as fulfilling the criteria for obesity.

9.2.3 BLOOD BASED ASSAYS

Fasting blood samples were drawn by venipuncture in the morning. Blood samples were immediately centrifuged, and plasma was frozen at –80°C until assay. Plasma insulin levels were measured using a commercially available radioimmunoassay kit (Coat–A–Count Insulin; Diagnostic Products Inc). This method has a detection level of 1.2 μIU/mL and exhibits linear behavior up to 350 μIU/mL, with intra–assay and interassay coefficients of variability of 3.1% and 4.9%, respectively. Plasma glucose level was measured using a commercial kit based on the hexokinase–glucose–6–phosphate dehydrogenase method (Flex Reagent Cartridges; Dade Behring, Newark, DE). Insulin resistance was assessed using a widely validated mathematical formula, the homeostasis model assessment (HOMA) equation (fasting insulin × fasting glucose/22.5) [23].

Plasma hsCRP levels were measured within 2–3 hours after collection using the Flex Reagent Cartridge (Date Behring, Newark, DE). This method has a detection level of 0.05 mg/dL and exhibits linear behavior up to 255 mg/dL, with intra–assay and interassay coefficients of variability of 9% and 18%, respectively, for hsCRP. Serum lipids including total cholesterol, high–density lipoprotein (HDL) cholesterol, calculated low–density lipoprotein cholesterol (LDL), and triglycerides (TG) were also assessed using Flex Reagent Cartridges (Dade Behring).

Plasma IL–6, IL–20, monocyte chemotactic protein (MCP), retinol–binding protein 4 (RBP4), and tumor necrosis factor (TNF–α) levels were measured using commercial ELISA kits (R&D systems, Minneapolis, MN). Plasma apolipoprotein B (ApoB), myeloid–related protein (MRP) 8/14, and macrophage inhibitory factor (MIF) levels were also measured using commercial ELISA kits (ALPCO Diagnostics, Salem, NH) following the manufacturer's instructions. Circulating levels of ICAM–1 and P–selectin were measured with commercially available kits (R&D System, Abington, UK). For ICAM–1, the sensitivity was 0.35 ng/mL and the intra– and interassay coefficients of variation were 2.5 and 1.8%, respectively. For P–selectin, the sensitivity was 0.5 ng/mL and the intra– and interassay coefficients of variation were 3.6 and 6.9%, respectively. All assays were performed in duplicate, and a calibration curve was included in each assay.

9.2.4 STATISTICAL ANALYSIS

Data were expressed as mean ± SD. Significant differences within groups were analyzed using ANOVA followed by post–hoc tests with Bonferroni corrections for multiple comparisons for continuous variables and chi–square tests for categorical variables. Spearman's correlation analyses were conducted to examine potential associations between BMI and plasma concentrations of the various inflammatory mediators. Statistical analyses were performed using SPSS software (version 16.0; SPPS Inc., Chicago, Ill.). All *P* values reported are 2–tailed with statistical significance set at <.05.

9.3 RESULTS

A total of 354 obese children and 350 age–, gender–, and ethnicity–matched nonobese children were recruited between May 2004 and October 2008. The demographic characteristics of this cohort are shown in Table 1 and are virtually identical to the published demographics of the city of Louisville, Kentucky. It should be pointed out that not every child could have all of the inflammatory markers assayed, as dictated by the limited amounts of plasma, such that, the number of inflammatory marker assays varied from one child to another.

TABLE 1: Demographic characteristics of obese and nonobese children.

	Obese (BMI $Z \geq 1.65$)	**Nonobese (BMI $Z < 1.65$)**	***P*–value***
Age (years)	7.1 ± 0.4 (n = 354)	7.1 ± 0.4 (n = 350)	
Gender (% male)	51.2	51.1	
Ethnicity			
White Caucasian % (n)	77.9 (n = 277)	78.0 (n = 273)	
African American % (n)	18.1 (n = 64)	18.0 (n= 63)	
BMI z score	2.21 ± 0.14	1.06 ± 0.21	<0.00001
Systolic blood pressure (mmHg)	112.4 ± 9.7 (n = 354)	107.1 ± 10.9 (n = 350)	<0.01
Diastolic blood pressure (mmHg)	67.1 ± 8.2 (n = 354)	62.7 ± 7.9 (n = 350)	<0.01
Triglycerides	84.1 ± 5.1	73.5 ± 2.8	<0.001
HDL	42.5 ± 1.8	62.9 ± 1.8	<0.00001
VLDL	19.1 ± 2.9	14.7 ± 0.9	<0.001
LDL	98.7 ± 2.9	93.3 ± 1.8	<0.001
Glucose	81.3 ± 1.8	78.2 ± 1.7	<0.04
Insulin	14.9 ± 1.7	6.7 ± 1.0	<0.001
HOMA–IR	2.3 ± 0.3	1.2 ± 0.3	<0.001

**Statistical significance determined using ANOVA test.*

As anticipated, obese children had higher HOMA values, indicative of insulin resistance, and also exhibited higher LDL, VLDL, and TG levels and lower HDL concentrations compared to nonobese children (Table 1).

Obese children also had significantly higher levels of hsCRP, IL–6, MRP 8/14, P–selectin, ICAM–1, IL–20, RBP4, MIF, and TNF–α compared to nonobese children (Table 2). Only ApoB and MCP showed similar levels among the 2 groups. Among the obese children, 57% showed one, 34% two, 19% three, and 15% four or more cardiometabolic biomarkers with elevated plasma levels.

TABLE 2: Fasting morning plasma concentrations of inflammatory biomarkers in obese and nonobese children.

	Obese (BMI $Z \geq 1.65$)	Nonobese (BMI $Z <$ 1.65)	*P*–values*
MRP 8/14 (ug/mL)	1.5 ± 0.9 (n = 128)	0.9 ± 0.8 (n = 176)	0.001
Apo B (mg/dL)	77.8 ± 19.6 (n = 65)	73.7 ± 16.3 (n = 51)	0.233
MCP (pg/mL)	132.0 ± 29.6 (n = 77)	120. 1 ± 18.8 (n = 68)	0.494
RBP4 (ng/mL)	33.4 ± 16.3 (n = 80)	18.2 ± 14.0 (n = 80)	0.01
IL–20 (pg/mL)	4.0 ± 5.4 (n = 67)	2.3 ± 4.5 (n = 44)	0.02
IL–6 (pg/mL)	1.8 ± 0.9 (n = 132)	1.0 ± 0.6 (n = 134)	0.01
MIF (pg/mL)	188.6 ± 103.1 (n = 132)	131.7 ± 86.4 (n = 106)	0.02
hsCRP (mg/dL)	2.3 ± 1.4 (n = 354)	0.5 ± 0.3 (n = 350)	<0.0001
TNF–α (pg/mL)	598.5 ± 62.2 (n = 172	317.3 ± 42.2 (n = 176)	0.0001
P selectin (ng/mL)	94.3 ± 22.2 (n = 56)	37.7 ± 12.9(n = 90)	0.0001
ICAM–1 (ng/mL)	297.4 ± 83.5 (n = 56)	247.4 ± 53.5 (n = 90)	0.0001

**Statistical significance determined using ANOVA test.*

Regression analyses between each of the inflammatory markers and actual BMI or BMI z score revealed significant correlations for the majority of these markers (Table 3). In addition, hsCRP was also significantly associated with IL–6 ($r = 0.35$; $n = 145$; $P < 0.01$), MRP 8/14 ($r = 0.67$; $n = 236$; $P < 0.01$), and with TNF–α ($r = 0.46$; $n = 122$; $P < 0.01$). Furthermore, hsCRP was also positively correlated with ICAM–1 ($r = 0.29$; $n = 98$; $P < 0.03$) and with RBP4 ($r = 0.32$; $n = 80$; $P < 0.05$). Similarly, significant correlations emerged between RBP4, MRP 8/14, TNF–α, hsCRP, and HOMA (Table 3). In addition MRP 8/14, IL–6, hsCRP, MIF, P–selectin, ICAM–1, and TNF–α levels showed significant associations with LDL/HDL (Table 3).

TABLE 3: Correlation coefficients between BMI and BMI score and plasma levels of several inflammatory markers in children.

	BMI *r*–value	**BMI *z* score *r*–value**	**HOMA *r*–value**	**LDL/HDL ratio *r*–value**
MRP 8/14 (ug/mL) (*n* = 249)	0.354*	0.375*	0.332*	0.227*
Apo B (mg/dL) (*n* = 116)	0.143	0.160	0.02	0.10
MCP (pg/mL) (*n* = 143)	0.00	0.00	0.00	0.12
RBP4 (ng/mL) (*n* = 160)	0.213*	0.317*	0.376*	0.09
IL–20 (pg/mL) (*n* = 111)	–0.054	–0.036	0.01	0.03
IL–6 (pg/mL) (*n* = 266)	0.161*	0.177*	0.132*	0.134*
MIF (pg/mL) (*n* = 108)	0.043	0.02	0.110*	0.176*
hsCRP (mg/dL) (*n* = 704)	0.472*	0.367*	0.276*	0.365*
TNF–α (pg/mL) (*n* = 348)	0.298*	0.336*	0.187*	0.215*
P selectin (ng/mL) (*n* = 146)	0.310*	0.376*	0.087	0.387*
ICAM–1 (ng/mL) (*n* = 146)	0.254*	0.276*	0.104	0.413*

In parenthesis in the number of observations for which data were available,
*P < *0.05. Statistical significance determined using Spearman's correlational analysis.*

9.4 DISCUSSION

This study shows that systemic inflammatory processes are activated in otherwise asymptomatic, community–dwelling, obese, and school–age prepubertal children. Interestingly, the degree and nature of the activation of these inflammatory processes varied from child to child, and only a minority of obese children exhibited extensive derangements across multiple cardiometabolic inflammatory biomarkers. Furthermore, there were significant associations between specific subsets of the inflammatory markers and the degree of obesity, insulin resistance, or hyperlipidemia. However, while there was some degree of overlap among the inflammatory markers associated with each of these variables, the more remarkable finding was that even if a particular marker was elevated and correlated with end–organ dysfunction, this did not necessarily imply that all other markers were affected as well. The pathophysiology of end–organ morbidities that are traditionally perceived as resulting from the long–term effects of obesity on health has been postulated to involve low–grade activation of multiple pathways of inflammation. Based on the current understanding of the roles played by these inflammatory mediators, it becomes imperative to explore the role of genetic, environmental, and lifestyle influences on the modulation of the inflammatory responses in the context of obesity, and its consequences [24, 25]. Similarly, it will be critical to assess in the future the effect of interventions such as dietary changes and physical activity on the reversibility of these inflammatory responses and on the progression of obesity–related morbidities [26–30].

Before we discuss any further the potential implications of our findings, some technical and methodological approaches deserve comment. We selected a wide net array of known biomarkers for both cardiovascular and metabolic dysfunction, in an attempt to characterize the variability and the potential associations of these markers in the context of pediatric obesity. However, since hsCRP has been extensively used in past studies, we aimed to include this measure in all children. We also restricted the age range of our cohort to a narrow time span that is associated with the initial 3 years of attendance in the public school system, a period during which changes in eating patterns are now well described [31–34]. We also selected our population based on a representative community sample for the city of Louisville, Kentucky, and identified this cohort in the school system itself, rather than using a clinical referral cohort. As such, we also selected a closely matching nonobese child, in the same school, to control for and all of the potential confounders that could be introduced in the process of cohort allocation. However, neither pubertal status nor the presence of hepatic steatosis was assessed.

Globally, the findings from this study indicate the presence of a substantial inflammatory burden in obese prepubertal children along with a high risk for insulin resistance and increased serum lipids. Therefore, our study is in close agreement with multiple other studies in children that have examined a selected number of inflammatory mediators in the context of obesity [12, 35–46]. Of note, in a study by Nagel and colleagues, these investigators did not find any evidence for a significant association between increased body weight in children and ApoB levels, and our current findings are in close concordance with such report in German children [44]. Similarly, our findings concur with the increased plasma levels of adhesion molecules in a pediatric overweight cohort from Mexico, suggesting evidence of endothelial dysfunction in a substantial proportion of obese children [47]. Of note, the inflammatory markers that were presently associated with either BMI, BMI *z* score, HOMA, and hypercholesterolemia varied, suggesting different and coordinated biological roles for clusters of inflammatory modulators. Indeed, although there was some degree of overlap among the inflammatory markers and their respective significant associations, only a portion of the markers were correlated with any given end–organ dysfunction.

In summary, this study clearly shows that obesity in childhood carries a substantial inflammatory burden that is strongly, yet selectively, associated with specific functional alterations, such as insulin sensitivity or lipid homeostasis. With the emergence of multiplexed ELISA assays that exhibit improved sensitivities, we would advocate that future community–based intervention or correlational studies on childhood obesity [48] should explore more expansive panels of inflammatory markers along with functional phenotypes, and should also consider incorporation of genomic variance assessments using recently developed cardiovascular or metabolic gene–centric polymorphism arrays [49].

REFERENCES

1. D. W. Haslam and W. P. T. James, "Obesity," Lancet, vol. 366, no. 9492, pp. 1197–1209, 2005.
2. Y. Wang and M. A. Beydoun, "The obesity epidemic in the United States—gender, age, socioeconomic, racial/ethnic, and geographic characteristics: a systematic review and meta–regression analysis," Epidemiologic Reviews, vol. 29, no. 1, pp. 6–28, 2007.
3. C. L. Ogden, M. D. Carroll, and K. M. Flegal, "High body mass index for age among US children and adolescents, 2003–2006," Journal of the American Medical Association, vol. 299, no. 20, pp. 2401–2405, 2008.
4. D. S. Freedman, D. A. Patel, S. R. Srinivasan et al., "The contribution of childhood obesity to adult carotid intima–media thickness. The Bogalusa Heart Study," International Journal of Obesity, vol. 32, no. 5, pp. 749–756, 2008.
5. D. S. Freedman, P. T. Katzmarzyk, W. H. Dietz, S. R. Srinivasan, and G. S. Berenson, "Relation of body mass index and skinfold thicknesses to cardiovascular disease risk factors in children. The Bogalusa Heart Study," American Journal of Clinical Nutrition, vol. 90, no. 1, pp. 210–216, 2009.
6. D. S. Freedman, W. H. Dietz, S. R. Srinivasan, and G. S. Berenson, "Risk factors and adult body mass index among overweight children. The Bogalusa Heart Study," Pediatrics, vol. 123, no. 3, pp. 750–757, 2009.

7. N. Mattsson, T. Rönnemaa, M. Juonala, J. S. A. Viikari, and O. T. Raitakari, "Childhood predictors of the metabolic syndrome in adulthood. The cardiovascular risk in Young Finns study," Annals of Medicine, vol. 40, no. 7, pp. 542–552, 2008.
8. American Academy of Pediatrics, "Policy statement: prevention of pediatric overweight and obesity," Pediatrics, vol. 112, pp. 424–430, 2003.
9. U. N. Das, "Is obesity an inflammatory condition?" Nutrition, vol. 17, no. 11–12, pp. 953–966, 2001.
10. J. Sacheck, "Pediatric obesity: an inflammatory condition?" Journal of Parenteral and Enteral Nutrition, vol. 32, no. 6, pp. 633–637, 2008.
11. C. Maffeis, D. Silvagni, R. Bonadonna, A. Grezzani, C. Banzato, and L. Tatò, "Fat cell size, insulin sensitivity, and inflammation in obese children," Journal of Pediatrics, vol. 151, no. 6, pp. 647–652, 2007.
12. M. Visser, L. M. Bouter, G. M. McQuillan, M. H. Wener, and T. B. Harris, "Low–grade systemic inflammation in overweight children," Pediatrics, vol. 107, no. 1, p. E13, 2001.
13. D. Nemet, P. Wang, T. Funahashi et al., "Adipocytokines, body composition, and fitness in children," Pediatric Research, vol. 53, no. 1, pp. 148–152, 2003.
14. M. Halle, U. Korsten–Reck, B. Wolfarth, and A. Berg, "Low–grade systemic inflammation in overweight children: impact of physical fitness," Exercise Immunology Review, vol. 10, pp. 66–74, 2004.
15. F. S. Ezgü, A. Hasanoğlu, L. Tümer, F. Özbay, C. Aybay, and M. Gündüz, "Endothelial activation and inflammation in prepubertal obese Turkish children," Metabolism, vol. 54, no. 10, pp. 1384–1389, 2005.
16. M. Juonala, J. S. A. Viikari, T. Rönnemaa, L. Taittonen, J. Marniemi, and O. T. Raitakari, "Childhood C–reactive protein in predicting CRP and carotid intima–media thickness in adulthood. The Cardiovascular Risk in Young Finns Study," Arteriosclerosis, Thrombosis, and Vascular Biology, vol. 26, no. 8, pp. 1883–1888, 2006.
17. A. R. Brasil, R. C. Norton, M. B. Rossetti, E. Leão, and R. P. Mendes, "C–reactive protein as an indicator of low intensity inflammation in children and adolescents with and without obesity," Jornal de Pediatria, vol. 83, no. 5, pp. 477–480, 2007.
18. A. Sbarbati, F. Osculati, D. Silvagni et al., "Obesity and inflammation: evidence for an elementary lesion," Pediatrics, vol. 117, no. 1, pp. 220–223, 2006.
19. T. Reinehr, W. Kiess, G. De Sousa, B. Stoffel–Wagner, and R. Wunsch, "Intima media thickness in childhood obesity: relations to inflammatory marker, glucose metabolism, and blood pressure," Metabolism, vol. 55, no. 1, pp. 113–118, 2006.
20. M. Valle Jiménez, R. M. Estepa, R. MA. M. Camacho, R. C. Estrada, F. G. Luna, and F. B. Guitarte, "Endothelial dysfunction is related to insulin resistance and inflammatory biomarker levels in obese prepubertal children," European Journal of Endocrinology, vol. 156, no. 4, pp. 497–502, 2007.
21. S. Lee, F. Bacha, N. Gungor, and S. Arslanian, "Comparison of different definitions of pediatric metabolic syndrome: relation to abdominal adiposity, insulin resistance, adiponectin, and inflammatory biomarkers," Journal of Pediatrics, vol. 152, no. 2, pp. 177–184, 2008.

22. C. S. Tam, K. Clément, L. A. Baur, and J. Tordjman, "Obesity and low–grade inflammation: a paediatric perspective," Obesity Reviews, vol. 11, no. 2, pp. 118–126, 2010.
23. D. R. Matthews, J. P. Hosker, and A. S. Rudenski, "Homeostasis model assessment: insulin resistance and β–cell function from fasting plasma glucose and insulin concentrations in man," Diabetologia, vol. 28, no. 7, pp. 412–419, 1985.
24. D. M. Stringer, E. A. C. Sellers, L. L. Burr, and C. G. Taylor, "Altered plasma adipokines and markers of oxidative stress suggest increased risk of cardiovascular disease in First Nation youth with obesity or type 2 diabetes mellitus," Pediatric Diabetes, vol. 10, no. 4, pp. 269–277, 2009.
25. G. S. Berenson, "Cardiovascular risk begins in childhood. A time for action," American Journal of Preventive Medicine, vol. 37, supplement 1, pp. S1–S2, 2009.
26. T. Reinehr, B. Stoffel–Wagner, C. L. Roth, and W. Andler, "High–sensitive C–reactive protein, tumor necrosis factor α, and cardiovascular risk factors before and after weight loss in obese children," Metabolism, vol. 54, no. 9, pp. 1155–1161, 2005.
27. N. E. Thomas and D. R. R. Williams, "Inflammatory factors, physical activity, and physical fitness in young people: review," Scandinavian Journal of Medicine and Science in Sports, vol. 18, no. 5, pp. 543–556, 2008.
28. A. L. Carrel, J. J. McVean, R. R. Clark, S. E. Peterson, J. C. Eickhoff, and D. B. Allen, "School–based exercise improves fitness, body composition, insulin sensitivity, and markers of inflammation in non–obese children," Journal of Pediatric Endocrinology and Metabolism, vol. 22, no. 5, pp. 409–415, 2009.
29. R. Kelishadi, M. Hashemi, N. Mohammadifard, S. Asgary, and N. Khavarian, "Association of changes in oxidative and proinflammatory states with changes in vascular function after a lifestyle modification trial among obese children," Clinical Chemistry, vol. 54, no. 1, pp. 147–153, 2008.
30. A. A. Meyer, G. Kundt, U. Lenschow, P. Schuff–Werner, and W. Kienast, "Improvement of early vascular changes and cardiovascular risk factors in obese children after a six–month exercise program," Journal of the American College of Cardiology, vol. 48, no. 9, pp. 1865–1870, 2006.
31. N. D. Ernst and E. Obarzanek, "Child health and nutrition: obesity and high blood cholesterol," Preventive Medicine, vol. 23, no. 4, pp. 427–436, 1994.
32. B. M. Popkin and P. Gordon–Larsen, "The nutrition transition: worldwide obesity dynamics and their determinants," International Journal of Obesity, vol. 28, supplement 3, pp. S2–S9, 2004.
33. M. Chopra, S. Galbraith, and I. Darnton–Hill, "A global response to a global problem: the epidemic of overnutrition," Bulletin of the World Health Organization, vol. 80, no. 12, pp. 952–958, 2002.
34. M. B. Zimmermann and I. Aeberli, "Dietary determinants of subclinical inflammation, dyslipidemia and components of the metabolic syndrome in overweight children: a review," International Journal of Obesity, vol. 32, supplement 6, pp. S11–S18, 2008.
35. E. S. Ford, D. A. Galuska, C. Gillespie, J. C. Will, W. H. Giles, and W. H. Dietz, "C–reactive protein and body mass index in children: findings from the 3rd national

health and nutrition examination survey, 1988–1994," Journal of Pediatrics, vol. 138, no. 4, pp. 486–492, 2001.
36. D. G. Cook, M. A. Mendall, P. H. Whincup et al., "C–reactive protein concentration in children: relationship to adiposity and other cardiovascular risk factors," Atherosclerosis, vol. 149, no. 1, pp. 139–150, 2000.
37. A. D. Aygun, S. Gungor, B. Ustundag, M. K. Gurgoze, and Y. Sen, "Proinflammatory cytokines and leptin are increased in serum of prepubertal obese children," Mediators of Inflammation, vol. 2005, no. 3, pp. 180–183, 2005.
38. A. A. Meyer, G. Kundt, M. Steiner, P. Schuff–Werner, and W. Kienast, "Impaired flow–mediated vasodilation, carotid artery intima–media thickening, and elevated endothelial plasma markers in obese children: the impact of cardiovascular risk factors," Pediatrics, vol. 117, no. 5, pp. 1560–1567, 2006.
39. S. D. De Ferranti, K. Gauvreau, D. S. Ludwig, J. W. Newburger, and N. Rifai, "Inflammation and changes in metabolic syndrome abnormalities in US adolescents: findings from the 1988–1994 and 1999–2000 National Health and Nutrition Examination Surveys," Clinical Chemistry, vol. 52, no. 7, pp. 1325–1330, 2006.
40. E. V. Economou, A. V. Malamitsi–Puchner, C. P. Pitsavos, E. E. Kouskouni, I. Magaziotou–Elefsinioti, and G. Creatsas, "Low–grade systemic inflammation profile, unrelated to homocysteinemia, in obese children," Mediators of Inflammation, vol. 2005, no. 6, pp. 337–342, 2005.
41. G. Akinci, B. Akinci, S. Coskun, P. Bayindir, Z. Hekimsoy, and B. Özmen, "Evaluation of markers of inflammation, insulin resistance and endothelial dysfunction in children at risk for overweight," Hormones, vol. 7, no. 2, pp. 156–162, 2008.
42. J. Y. Shin, S. Y. Kim, M. J. Jeung et al., "Serum adiponectin, C–reactive protein and TNF–α levels in obese Korean children," Journal of Pediatric Endocrinology and Metabolism, vol. 21, no. 1, pp. 23–29, 2008.
43. M. E. Atabek, "Obese related effects of inflammatory markers and insulin resistance on increased carotid intima–media thickness in pre–pubertal children," Atherosclerosis, vol. 200, no. 2, p. 446, 2008.
44. G. Nagel, K. Rapp, M. Wabitsch et al., "Prevalence and cluster of cardiometabolic biomarkers in overweight and obese schoolchildren: results from a large survey in Southwest Germany," Clinical Chemistry, vol. 54, no. 2, pp. 317–325, 2008.
45. J. R. Ruiz, F. B. Ortega, J. Warnberg, and M. Sjöström, "Associations of low–grade inflammation with physical activity, fitness and fatness in prepubertal children. The European Youth Heart Study," International Journal of Obesity, vol. 31, no. 10, pp. 1545–1551, 2007.
46. S. Kapiotis, G. Holzer, G. Schaller et al., "A proinflammatory state is detectable in obese children and is accompanied by functional and morphological vascular changes," Arteriosclerosis, Thrombosis, and Vascular Biology, vol. 26, no. 11, pp. 2541–2546, 2006.
47. A. E. Caballero, R. Bousquet–Santos, L. Robles–Osorio et al., "Overweight latino children and adolescents have marked endothelial dysfunction and subclinical vascular inflammation in association with excess body fat and insulin resistance," Diabetes Care, vol. 31, no. 3, pp. 576–582, 2008.

48. C. A. Pratt, J. Stevens, and S. Daniels, "Childhood obesity prevention and treatment. Recommendations for future research," American Journal of Preventive Medicine, vol. 35, no. 3, pp. 249–252, 2008.
49. B. J. Keating, S. Tischfield, S. S. Murray et al., "Concept, design and implementation of a cardiovascular gene–centric 50 K SNP array for large–scale genomic association studies," PLoS ONE, vol. 3, no. 10, article e3583, 2008.

This chapter was originally published under the Creative Commons License. Kim, J., Bhattacharjee, R., Kheirandish-Gozal, L., Khalyfa, A., Sans Capdevila, O., Tauman, R., and Gozal, D. Insulin Sensitivity, Serum Lipids, and Systemic Inflammatory Markers in School-Aged Obese and Nonobese Children. International Journal of Pediatrics, 2010; 2010: 846098. doi:10.1155/2010/846098.

CHAPTER 10

IS CHILDHOOD OBESITY ASSOCIATED WITH BONE DENSITY AND STRENGTH IN ADULTHOOD?

KIRSTI UUSI–RASI, PEKKA KANNUS, MATTI PASANEN, and HARRI SIEVÄNEN

10.1 INTRODUCTION

Compared to normal–weight adults, overweight and obese persons have not only greater bone mineral density (BMD), but they lose bone at a slower pacc, and may even have a reduced risk of fragility fractures [1, 2]. As regards the effect of childhood obesity on bone mass, findings are inconsistent [3–5]. Some studies have attributed greater bone mass to excess body weight in the growing years [5–8] but it has also been suggested that obese children have lower bone mass for a given weight [9–11].

In principle, the greater bone mass in obese subjects could simply be a consequence of increased body mass. While the weight load per se can stimulate bone formation, obese subjects also have moderate increase in muscle mass [12, 13], which is an important determinant of bone mass and strength [14]. In absolute terms bone mass and strength have been shown to be greater in obese women but reduced relative to body weight varying in proportion to lean mass, not to body weight or fat mass [15]. Interestingly, some studies have suggested that fat mass, the other major component of body weight, may also stimulate bone accrual in growing children, but these results have remained inconsistent showing both positive [16, 17] and negative associations [11, 13, 18].

Despite the apparent simplicity, the relationship between body weight and bone mass and strength is not straightforward. While obesity is represented by increased body weight, and thus the mass needed to move during habitual activities, the incident muscular contractions, particularly the magnitude of force, the rate of force production, and the total amount of contractions play a more important role than body weight. Body weight alone imposes relatively small and static load on bones (corresponding to Earth's gravity, G), whereas the load can be substantially amplified in different activities by muscle actions [19]. Theoretically, a physically active obese individual has stronger muscles than a sedentary, similarly obese person. Consequently, the greater muscle force would permit more bone remodeling and thus result in increased bone mass and strength in the former. On the other hand, if an obese individual is sedentary or becomes inactive and loses muscle mass, incident muscle forces imposed on bones decline and osteopenia could ensue no matter how obese the individual might be [12]. However, it is not yet known whether the influence of increased body weight on bones would be stronger in childhood and persist until adulthood.

The conventional paradigm suggests that obesity results in greater bone mass as a consequence of greater body weight and could thus protect against osteoporosis. In this cross–sectional study we hypothesized that those women who have been obese since childhood would show some benefit in bone traits compared to women who have gained extra weight not earlier than in adulthood. This hypothesis was based on observations that the period of adolescent growth provides the window of opportunity for efficient bone accrual [20]. In addition, we evaluated whether statistical adjusting for different anthropometric variables affects the results, and if so, to what extent.

10.2 METHODS

10.2.1 SUBJECTS

Sixty–two obese (BMI > 30) clinically healthy premenopausal women with the age range of 25 to 45 years participated in the study [21]. According to self–reports, 12 of these women had been obese since childhood, and 50 had gained excess weight after maturity. None of the subjects had evidence for any disease, prior injuries, or drug use (hormonal contraceptives were allowed) that would have affected their skeleton. The study protocol was approved by the Ethics Committee of The Pirkanmaa Hospital District, Tampere, Finland, and each participant gave her written informed consent.

10.2.2 HEALTH HISTORY QUESTIONNAIRE

Information on self–reported health, injuries, medication, diseases, diet, nonpregnant weight at the age of 25, history of weight cycling, weight loss attempts, menstrual status, and lifestyle factors such as current physical activity, smoking, and consumption of alcohol was obtained with a questionnaire. The questionnaire was completed in an interview.

10.2.3 MEASUREMENTS

10.2.3.1 PHYSICAL ACTIVITY

History of physical activity was assessed in an interview, and current daily walking steps were measured with a pedometer (Omron HL–112–E, Omron Healthcare Europe) on six days.

10.2.3.2 MUSCLE PERFORMANCE

The maximal isometric leg extension force (N/kg) was measured by a strain gauge dynamometer at a knee angle of 110 degrees (Tamtron, Tampere, Finland). The maximal take–off force (N/kg) and power (W/kg) during a vertical counter–movement jump were measured on a force–plate (Kistler Ergojump 1.04, Kistler Instrumente AG, Winterthur, Switzerland). Functional agility (s) was evaluated by a figure–8 running test and the grip strength of both forearms (kg) with a standard grip strength meter.

10.2.3.3 ANTHROPOMETRY AND BODY COMPOSITION

Body height was measured to the nearest 0.1 cm and body weight to the nearest 0.1 kg with a high–precision scale with the participants wearing only their underwear. Waist circumference was measured midway between the lowest rib and the iliac crest and hip circumference at the tip of the greater trochanter. Body composition (fat mass and lean mass) and the android (trunk) and gynoid (hip) fat–% were assessed with dual–energy X–ray absorptiometry (DXA, Lunar Prodigy Advance, GE Lunar, Madison, WI, USA). In our laboratory, the in vivo precision (coefficient variation, CV%) based on repeated scans of 27 subjects with repositioning is 1.3% for the fat mass and 0.8% for the lean mass (Sievänen, unpublished data).

10.2.3.4 BONE MASS AND STRUCTURE

Bone mineral content (BMC, g) of the total body, BMC, and bone mineral density (BMD, g/cm2) of the left proximal femur and lumbar spine were assessed with DXA (Lunar Prodigy Advance). Femoral neck structure was assessed using the Advanced Hip Analysis (AHA) which provided data on cross–sectional area occupied by bone mineral (CSA, mm2, an index of bone strength against compression), section modulus (Z, mm3, an index of bone strength against bending), and outer diameter (Width, mm). In our laboratory, the in vivo precision is 1.4% for the total body BMC, about 1% for the lumbar spine BMC, 1.5% for the femoral neck BMC, and about 3% for the trochanter BMC (Sievänen 2005, unpublished data). The in vivo precision for CSA, Z, and Width is 2.3%, 3.8%, and 1.2%, respectively.

In addition to the DXA measurements, the left radius and tibia were measured with pQCT (Norland/Stratec XCT 3000, Pforzheim, Germany).

The tomographic slices were taken from the shaft and distal part of the tibia (50% and 5% from the distal endplate of the tibia, resp.) and of the radius (30% and 4% from the distal endplate of the radius, resp.) according to our standard procedures [22]. For the shaft regions, the analyzed bone traits were total cross–sectional area (ToA, mm2), cortical density (CoD, mg/cm3), and density–weighted polar section modulus (SSI, mm3, an index of bone strength against bending). For the distal parts of the radius and tibia the traits were ToA, trabecular density (TrD, mg/cm3), and SSI. In our laboratory, the in vivo precision of the pQCT traits is between 0.7% (CoD of the tibial shaft) and 7.7% (SSI of the distal radius) [22].

10.2.3.5 STATISTICAL ANALYSES

Mean and standard deviations (SD) were used as descriptive statistics. Between–group differences were evaluated by an analysis of covariance, and as possible confounding factors body height, fat, and lean mass were used as covariates. Since the body mass index (BMI) is a commonly used covariate in the pertinent literature, the analyses were adjusted for BMI for comparison. Linear regression analyses were used to find the predictors for bone traits; body height, lean mass, fat mass, and childhood obesity were included as the dichotomous variable (ObC = 0; ObA = 1) in the regression models.

10.3 RESULTS

The group characteristics (mean, SD) are given in Table 1. The ObC–group was on average 5.2 cm taller and compared to the ObA–group they had 2.5 kg and 3.5 kg more lean and fat mass, respectively. However, difference in mean relative body fat and lean mass was small indicating similar body composition in both groups for given body height in adulthood. As shown in Table 1, the muscle strength, functional agility, and steps/day were similar between the two groups.

TABLE 1: Group characteristics (SD) among women who have been obese since childhood (ObC) and women who have become obese in adulthood.

	ObC, *n* =12	**ObA, *n* = 50**	**P**
Height, cm	170.0 (8.2)	164.8 (5.6)	0.053
Weight, kg	98.2 (16.6)	92.1 (14.0)	0.25
Weight at the age of 25, kg	81.9 (14.9)	69.5 (8.6)	0.016
BMI	33.8 (3.4)	33.9 (4.8)	0.93
Age, yrs	37.8 (5.9)	40.5 (4.8)	0.17
Fat mass, kg	45.4 (10.0)	41.9 (9.5)	0.29
Lean mass, kg	49.1 (8.4)	46.6 (5.5)	0.35
Fat, %	47.9 (4.5)	47.0 (4.4)	0.53
Waist circumference, cm	109 (15)	107 (13)	0.75
Hip circumference, cm	119 (9)	117 (10)	0.55
Android fat, %	53.1 (6.2)	53.0 (5.0)	0.94
Gynoid fat, %	51.0 (5.4)	51.0 (5.6)	1.00
Grip strength, kg	37.4 (5.2)	35.7 (4.4)	0.31
Time of figure–8 run, s	15.6 (1.6)	15.6 (1.4)	0.92
Isometric leg extension, N/kg	25.8 (6.9)	25.9 (8.2)	0.94
Jumping force, N/kg	20.2 (2.9)	21.5 (3.3)	0.22
Jumping power, W/kg	31.7 (5.6)	31.5 (5.5)	0.95
Steps/day	8,590 (3,486)	7,687 (2,939)	0.42

The DXA–based bone mass and strength were greater in the ObC–group than that those in the ObA–group (Table 2). Adjusting for BMI had practically no effect on the mean differences, whereas the adjustment for height, fat, and lean mass affected the results, which were no more statistically significant.

Although the ObC–group tended to have greater TrD both at the distal radius and at the distal tibia, cortical density was similar in both groups. There was also a tendency for greater total bone cross–sectional area in ObC–group especially in the weight–bearing tibia. However, this difference disappeared after adjusting for body height and fat and lean tissue. In the nonweight–bearing radius, cross–sectional bone area was similar in both groups, and adjusting for anthropometric variables did not affect the results.

TABLE 2: Unadjusted bone values (SD), BMI–adjusted, and height, fat, and lean mass–adjusted between–group (ObC versus ObA) mean differences (95% CI).

	ObC, *n* = 12	**ObA, *n* = 50**	**unadjusted mean difference, %**	**BMI adjusted mean difference, %**	**Height, fat, and lean adjusted mean difference, %**
Total body BMC, g	3233 (536)	2904 (386)	8.4 (0.1 to 17.4)	11.0 (1.4 to 21.5)	3.4 (–4.6 to 12.2)
Lumbar spine BMC	80.12 (17.57)	69.30 (12.10)	14.8 (2.3 to 28.9)	14.8 (2.2 to 29.0)	6.2 (–4.8 to 18.4)
Femoral neck BMC	5.72 (0.93)	5.09 (0.76)	12.1 (1.8 to 23.3)	12.1 (1.9 to 23.4)	6.0 (–3.2 to 16.1)
Trochanter BMC	12.34 (2.84)	11.43 (2.57)	7.8 (–6.4 to 24.2)	7.9 (–6.4 to 24.3)	–0.7 (–13.3 to 13.9)
Femoral neck Z, mm^3	776 (176)	660 (113)	16.7 (4.2 to 30.6)	16.7 (4.2 to 30.7)	8.6 (–2.0 to 20.4)
Distal radius					
TrD, mg/cm^3	224.1 (35.5)	203.9 (31.1)	9.8 (–0.9 to 21.7)	9.8 (–1.0 to 21.9)	7.9 (–3.4 to 20.4)
ToA, mm^2	277.5 (41.0)	286.0 (44.7)	–2.8 (–12.5 to 8.0)	3.0 (–10.4 to 18.4)	–5.8 (–14.7 to 4.1)
SSI, mm^3	376.9 (86.0)	365.1 (74.0)	0.5 (–11.5 to 14.2)	3.0 (–10.4 to 18.4)	4.0 (–9.5 to 19.5)
Radial shaft					
CoD, mg/cm^3	1201.6 (15.7)	1202.4 (19.5)	–0.1 (–1.1 to 1.0)	–0.1 (–1.1 to 1.0)	0.0 (–1.1 to 1.2)
ToA, mm^2	105.9 (12.0)	105.8 (14.3)	0.5 (–8.0 to 9.8)	0.7 (–7.7 to 10.0)	–2.2 (–10.2 to 6.6)
SSI, mm^3	215.1 (32.3)	215.8 (43.1)	0.5 (–11.5 to 14.2)	0.8 (–11.2 to 14.5)	–3.5 (–15.0 to 9.5)
Distal tibia					
TrD, mg/cm^3	254.0 (27.0)	238.0 (27.6)	6.8 (–0.7 to 15.0)	6.9 (–0.7 to 15.0)	5.4 (–2.5 to 13.9
ToA, mm^2	897.6 (112.3)	348.4 (105.4)	5.8 (–2.4 to 14.7)	5.9 (–2.1 to 14.6)	0.5 (–5.8 to 7.1)
SSI, mm^3	1523.9 (385.7)	1402.2 (288.7)	7.5 (–6.6 to 23.7)	7.6 (–6.3 to 23.6)	2.4 (–11.2 to 18.0)
Tibial shaft					
CoD, mg/cm^3	1145.5 (13.0)	1143.8 (22.1)	0.2 (–1.0 to 1.3)	0.2 (–1.0 to 1.3)	0.3 (–0.9 to 1.6)
ToA, mm^2	501.5 (41.3)	471.0 (53.3)	6.8 (–0.4 to 14.5)	6.9 (0.0 to 14.3)	2.6 (–3.4 to 9.0(
SSI, mm^3	2138.3 (213.6)	1931.1 (331.0)	11.8 (0.8 to 24.0)	11.9 (1.3 to 23.8)	5.9 (–3.9 to 16.7)

The most significant predictors for bone traits were height and lean mass (Tables 3, 4, and 5). Height was positively associated with bone mass and strength at the axial skeleton, while lean mass was the strongest correlate for most bone traits at the appendicular skeleton.

TABLE 3: Multiple linear regressions showing the simultaneous effects of height, lean, and fat mass and age of weight gain on bone traits in the axial skeleton.

Axial skeleton	β	SE	P
Total body BMC, g			
Height, cm	40.7	9.4	<0.001
Fat mass, g	6.7	6.0	0.27
Lean mass, g	–11.6	11.1	0.30
Age of weight gain (ObC = 0; ObA = 1)	–119.8	121.3	0.33
Femoral neck BMC, g			
Height, cm	0.046	0.019	0.019
Fat mass, g	–0.001	0.012	0.97
Lean mass, g	0.026	0.022	0.25
Age of weight gain (ObC = 0; ObA = 1)	–0.322	0.243	0.19
Trochanter BMC, g			
Height, cm	0.151	0.063	0.021
Fat mass, g	0.003	0.040	0.95
Lean mass, g	0.063	0.074	0.40
Age of weight gain (ObC = 0; ObA = 1)	0.043	0.816	0.96
Lumbar spine BMC, g			
Height, cm	1.15	0.31	0.001
Fat mass, g	0.042	0.198	0.83
Lean mass, g	–0.14	0.37	0.71
Age of weight gain (ObC = 0; ObA = 1)	–4.97	4.02	0.22
Femoral Neck Z, mm^3			
Height, cm	8.70	2.88	0.004
Fat mass, g	–1.64	1.83	0.38
Lean mass, g	4.94	3.39	0.15
Age of weight gain (ObC = 0; ObA = 1)	–63.77	37.17	0.092

TABLE 4: Multiple linear regressions showing the simultaneous effects of height, lean, and fat mass and age of weight gain on bone traits in the appendicular nonweight–bearing skeleton.

Appendicular nonweight–bearing skeleton	β	SE	P
Distal radius TrD, mg/cm³			
Height, cm	0.64	0.86	0.46
Fat mass, g	0.44	0.55	0.42
Lean mass, g	–0.73	1.06	0.5
Age of weight gain (ObC = 0; ObA = 1)	–16.88	11.37	0.14
Distal radius SSI, mm³			
Height, cm	–0.7	1.94	0.72
Fat mass, g	–2.25	1.25	0.076
Lean mass, g	6.18	2.41	0.013
Age of weight gain (ObC = 0; ObA = 1)	–14.75	25.7	0.57
Distal radius ToA, mm²			
Height, cm	1.57	1.05	0.14
Fat mass, g	–0.58	0.67	0.39
Lean mass, g	2.96	1.3	0.026
Age of weight gain (ObC = 0; ObA = 1)	17.12	13.87	0.22
Radial shaft CoD, mg/cm³			
Height, cm	0.079	0.5	0.88
Fat mass, g	–0.106	0.32	0.74
Lean mass, g	–0.77	0.62	0.22
Age of weight gain (ObC = 0; ObA = 1)	0.03	6.58	0.996
Radial shaft SSI, mm³			
Height, cm	1.6	1.04	0.13
Fat mass, g	0.29	0.67	0.67
Lean mass ,g	1.12	1.29	0.39
Age of weight gain (ObC = 0; ObA = 1)	10.01	13.78	0.47
Radial shaft ToA, mm²			
Height, cm	0.42	0.34	0.17
Fat mass, g	0.01	0.22	0.85
Lean mass, g	0.65	0.42	0.13
Age of weight gain (ObC = 0; ObA = 1)	2.79	4.51	0.54

TABLE 5: Multiple linear regressions showing the simultaneous effects of height, lean, and fat mass and age of weight gain on bone traits in the appendicular weight–bearing skeleton.

Appendicular weight–bearing skeleton	β	SE	P
Distal tibia TrD, mg/cm^3			
Height, cm	0.52	0.73	0.47
Fat mass, g	0.76	0.46	0.1
Lean mass, g	–0.89	0.86	0.3
Age of weight gain (ObC = 0; ObA = 1)	–12.76	9.39	0.18
Distal tibia SSI, mm^3			
Height, cm	6.04	7.75	0.44
Fat mass, g	0.56	4.93	0.91
Lean mass, g	14.49	9.11	0.12
Age of weight gain (ObC = 0; ObA = 1)	–52.17	99.95	0.6
Distal tibia ToA, mm^2			
Height, cm	5.24	2.12	0.016
Fat mass, g	–1.64	1.35	0.23
Lean mass, g	9.26	2.49	<0.001
Age of weight gain (ObC = 0; ObA = 1)	–4.43	27.31	0.87
Tibial shaft CoD, mg/cm^3			
Height, cm	0.026	0.549	0.96
Fat mass, g	0.11	0.35	0.76
Lean mass, g	–0.87	0.65	0.19
Age of weight gain (ObC = 0; ObA = 1)	–3.38	7.08	0.64
Tibial shaft SSI, mm^3			
Height, cm	10.47	7.23	0.15
Fat mass, g	2.77	4.6	0.55
Lean mass ,g	16.4	8.5	0.059
Age of weight gain (ObC = 0; ObA = 1)	–102.09	93.25	0.28
Tibial shaft ToA, mm^2			
Height, cm	1.75	1.09	0.11
Fat mass, g	0.06	0.69	0.93
Lean mass, g	3.93	1.28	0.003
Age of weight gain (ObC = 0; ObA = 1)	–11.33	14.02	0.42

10.4 DISCUSSION

According to our findings, women who had been obese since childhood were taller than women who had gained weight in adulthood. Obese children tend to be tall for their age and advanced in sexual and skeletal development, but not necessarily in terms of adult height, which was found in the present study. Since body size is strongly associated with bone mass and strength, and taller people have more massive skeleton than their smaller counterparts, body height and weight, or their aggregate measure BMI, are commonly used in literature as covariates to statistically control for the between–group differences in anthropometric variables. Besides the conventional variables, we measured both fat and lean tissue (the main components of body weight) in this study, and both of these variables were used as covariates instead of pure body weight. According to our findings, body height is a much stronger predictor of bone mass and strength than body weight, while lean mass turned out to be a stronger predictor for pQCT–assessed appendicular bone traits than height.

Taller persons usually have bigger bone cross–sections, and in line with this notion the total CSA of the weight–bearing tibia was greater in the ObC group than in the ObA group. In this respect, it is recalled that BMI (by definition = weight /height^2 in meters) does not totally correct differences in body size, which was also shown by our results. Thus we argue that BMI is not an appropriate variable in adjusting for different body sizes. In general, we found that BMI–adjusted DXA– and pQCT traits were similar to unadjusted crude differences. However, after adjusting for body height, lean, and fat mass the mean differences declined and were no statistically significant. With regard to bone traits which are basically size–independent, such as cortical density, the adjustment did not affect the results. To sum up, if it is not possible to separate lean, and fat mass from each other, use of total body weight, in addition to height, better takes into account the differences in body size than BMI, and the former approach is recommended.

Increased biomechanical loading of the skeleton due to increased body weight and increased lean mass (muscle forces) may contribute to greater bone mass and size. Obese persons (both children and adults) have to move greater body mass in habitual physical activity compared to the persons with normal body weight, which implies greater loading on bones provided that the intensity and the amount of exercise are similar. Some previous DXA–based studies, but not all, have shown that obese children have normal or increased bone mass relative to healthy weight peers [4, 6]. In contrast, Goulding et al. found decreased bone mass in obese children relative to bone size and body weight [9]. Also a Canadian cohort showed that the greater bone strength in obese children was appropriate for their lean mass, not their fat mass [8]. In a recently published pQCT–study the obese children had similar cortical density but greater radial and trabecular density than normal weight children [7], which gives support to our finding. Some previous studies have also suggested that fat mass may modulate periosteal bone growth in childhood [16, 23], but we did not find such associations in adulthood, the weak correlation with the bone strength index at the distal radius excluded.

Due to cross–sectional study design, we cannot definitely say whether the ObC and ObA groups differed essentially in their habits of physical activity in childhood (commuting to school, playing, and related activities). However, at the time of the study the participants were 40 years old on average, which means that when they were school kids in 1970s, a decade when children in Finland commonly walked or cycled to school, played outdoor games, spent less time watching TV, and had no internet games or computers. This being the apparent case, their bones received versatile dynamic loading in childhood and thus an adequate mechanical stimulus. In contrast, it is well possible that when the present–day children are grown–up, the situation is different. Nowadays children are more and more transported to school by car or bus, and instead of playing outdoor games, they are engaged in computer games or internet in their leisure activities. The obesity epidemic is associated with sedentary life style, which means substantially less dynamic loading on bones and most likely more fragile bone in adulthood.

Similarly we can only speculate about hormonal effects on bone development in childhood and adolescence. However, extra weight in the form of fat mass, besides its potential mechanical influence on bones [12, 14], may increase production of many hormones, such as estrogen, leptin, and insulin [13, 24]. Adipose tissue is known to express aromatase enzymes that convert steroid precursors to estrogen, which has been reported to stimulate and suppress periosteal bone growth in childhood [25]. Obese children may enter puberty earlier than normal–weight children, and they also achieve similar estradiol levels and bone ages at significantly earlier chronological ages than nonobese children [26, 27]. Hormonal actions may thereby potentially contribute to the increased linear growth and skeletal mass observed in childhood obesity [28]. Further studies are, however, needed to elucidate effects of adipose tissue and adipose–modulated hormones on adult bone mass and strength.

The strength of this study was the use of pQCT besides the conventional DXA, which allowed not only the differentiation of trabecular and cortical bone but also information on bone geometry. In addition, bone traits both at the weight–bearing tibia and nonweight–bearing radius were evaluated, and the muscle performance of the subjects was assessed. This study also had limitations the most important being the small sample size. Only twelve women reported that they had been obese since childhood. Obviously this rendered the study underpowered to statistically uncover meaningful differences, such as the properly adjusted 5 to 8% differences in trabecular density. Second, our study group represented a convenience sample and focused only on premenopausal women. Third, the analyses were based on a self–reported retrospective data, and we do not have exact body weight data from childhood over the adolescent period. Fourth, the ObC group was about 5 cm taller, which may be just a coincidence. Childhood obesity has been shown to be associated with greater height–for–age, advanced maturation for age, and greater lean mass for height, although the advantage in growth gradually decreases and probably does not account for taller height in adulthood [4]. It is, however, recalled that the difference in body height was statistically controlled for and it did not affect the present conclusions.

In conclusion, childhood obesity seemed to be slightly associated with denser trabecular bone in adulthood in both weight–bearing and non-weight–bearing bones among obese premenopausal women, but not with cortical density, neither was childhood obesity associated with total cross–sectional bone area. It is recalled that the study sample was small and the between–group mean differences did not reach statistical significance. Nevertheless, the results bring up an interesting question, whether childhood obesity can result in denser trabecular bone in adulthood. Finally, in this kind of research attention must be paid on proper adjustment of the data.

REFERENCES

1. C. De Laet, J. A. Kanis, and J. A. Kanis, "Body mass index as a predictor of fracture risk: a meta–analysis," Osteoporosis International, vol. 16, no. 11, pp. 1330–1338, 2005.
2. A. C. Looker, K. M. Flegal, and L. J. Melton III, "Impact of increased overweight on the projected prevalence of osteoporosis in older women," Osteoporosis International, vol. 18, no. 3, pp. 307–313, 2007.
3. P. Manzoni, P. Brambilla, A. Pietrobelli, L. Beccaria, A. Bianchessi, S. Mora, and G. Chiumello, "Influence of body composition on bone mineral content in children and adolescents," American Journal of Clinical Nutrition, vol. 64, no. 4, pp. 603–607, 1996.
4. N. Stettler, R. I. Berkowtiz, J. L. Cronquist, J. Shults, T. A. Wadden, B. S. Zemel, and M. B. Leonard, "Observational study of bone accretion during successful weight loss in obese adolescents," Obesity, vol. 16, no. 1, pp. 96–101, 2008.
5. M. A. Petit, T. J. Beck, J. Shults, B. S. Zemel, B. J. Foster, and M. B. Leonard, "Proximal femur bone geometry is appropriately adapted to lean mass in overweight children and adolescents," Bone, vol. 36, no. 3, pp. 568–576, 2005.
6. M. B. Leonard, J. Shults, B. A. Wilson, A. M. Tershakovec, and B. S. Zemel, "Obesity during childhood and adolescence augments bone mass and bone dimensions," The American Journal of Clinical Nutrition, vol. 80, no. 2, pp. 514–523, 2004.
7. G. Ducher, S. L. Bass, G. A. Naughton, P. Eser, R. D. Telford, and R. M. Daly, "Overweight children have a greater proportion of fat mass relative to muscle mass in the upper limbs than in the lower limbs: implications for bone strength at the distal forearm," American Journal of Clinical Nutrition, vol. 90, no. 4, pp. 1104–1111, 2009.
8. R. J. Wetzsteon, M. A. Petit, H. M. Macdonald, J. M. Hughes, T. J. Beck, and H. A. McKay, "Bone structure and volumetric BMD in overweight children: a longitudinal study," Journal of Bone and Mineral Research, vol. 23, no. 12, pp. 1946–1953, 2008.

9. A. Goulding, R. W. Taylor, I. E. Jones, K. A. McAuley, P. J. Manning, and S. M. Williams, "Overweight and obese children have low bone mass and area for their weight," International Journal of Obesity, vol. 24, no. 5, pp. 627–632, 2000.
10. A. Goulding, R. W. Taylor, I. E. Jones, P. J. Manning, and S. M. Williams, "Spinal overload: a concern for obese children and adolescents?" Osteoporosis International, vol. 13, no. 10, pp. 835–840, 2002.
11. M. A. Petit, T. J. Beck, J. M. Hughes, H.–M. Lin, C. Bentley, and T. Lloyd, "Proximal femur mechanical adaptation to weight gain in late adolescence: a six–year longitudinal study," Journal of Bone and Mineral Research, vol. 23, no. 2, pp. 180–188, 2008.
12. H. M. Frost, "Obesity, and bone strength and "mass": a tutorial based on insights from a new paradigm," Bone, vol. 21, no. 3, pp. 211–214, 1997.
13. N. K. Pollock, E. M. Laing, C. A. Baile, M. W. Hamrick, D. B. Hall, and R. D. Lewis, "Is adiposity advantageous for bone strength? A peripheral quantitative computed tomography study in late adolescent females," American Journal of Clinical Nutrition, vol. 86, no. 5, pp. 1530–1538, 2007.
14. H. M. Frost, "Bone's mechanostat: a 2003 update," Anatomical Record—Part A, vol. 275, no. 2, pp. 1081–1101, 2003.
15. T. J. Beck, M. A. Petit, G. Wu, M. S. LeBoff, J. A. Cauley, and Z. Chen, "Does obesity really make the femur stronger? BMD, geometry, and fracture incidence in the women's health initiative–observational study," Journal of Bone and Mineral Research, vol. 24, no. 8, pp. 1369–1379, 2009.
16. E. M. Clark, A. R. Ness, and J. H. Tobias, "Adipose tissue stimulates bone growth in prepubertal children," Journal of Clinical Endocrinology and Metabolism, vol. 91, no. 7, pp. 2534–2541, 2006.
17. M. Lorentzon, C. Swanson, N. Andersson, D. Mellström, and C. Ohlsson, "Free testosterone is a positive, whereas free estradiol is a negative, predictor of cortical bone size in young Swedish men: the GOOD study," Journal of Bone and Mineral Research, vol. 20, no. 8, pp. 1334–1341, 2005.
18. A. Janicka, T. A. L. Wren, M. M. Sanchez, F. Dorey, P. S. Kim, S. D. Mittelman, and V. Gilsanz, "Fat mass is not beneficial to bone in adolescents and young adults," Journal of Clinical Endocrinology and Metabolism, vol. 92, no. 1, pp. 143–147, 2007.
19. B. K. Weeks and B. R. Beck, "The BPAQ: a bone–specific physical activity assessment instrument," Osteoporosis International, vol. 19, no. 11, pp. 1567–1577, 2008.
20. P. Kannus, H. Haapasalo, and H. Haapasalo, "Effect of starting age of physical activity on bone mass in the dominant arm of tennis and squash players," Annals of Internal Medicine, vol. 123, no. 1, pp. 27–31, 1995.
21. K. Uusi–Rasi, A. Rauhio, P. Kannus, et al., "Three–month weight reduction does not compromise bone strength in obese premenopausal women," Bone, vol. 46, no. 5, pp. 1286–1293, 2010.
22. H. Sievänen, V. Koskue, A. Rauhio, P. Kannus, A. Heinonen, and I. Vuori, "Peripheral quantitative computed tomography in human long bones: evaluation of in vitro and in vivo precision," Journal of Bone and Mineral Research, vol. 13, no. 5, pp. 871–882, 1998.

23. N. J. Timpson, A. Sayers, G. Davey–Smith, and J. H. Tobias, "How does body fat influence bone mass in childhood? a mendelian randomization approach," Journal of Bone and Mineral Research, vol. 24, no. 3, pp. 522–533, 2009.
24. K. A. Larmore, D. O'Connor, T. I. Sherman, V. L. Funanage, S. G. Hassink, and K. O. Klein, "Leptin and estradiol as related to change in pubertal status and body weight," Medical Science Monitor, vol. 8, no. 3, pp. CR206–CR210, 2002.
25. E. Schoenau, C. M. Neu, F. Rauch, and F. Manz, "The development of bone strength at the proximal radius during childhood and adolescence," Journal of Clinical Endocrinology and Metabolism, vol. 86, no. 2, pp. 613–618, 2001.
26. K. O. Klein, K. A. Larmore, E. De Lancey, J. M. Brown, R. V. Considine, and S. G. Hassink, "Effect of obesity on estradiol level, and its relationship to leptin, bone maturation, and bone mineral density in children," Journal of Clinical Endocrinology and Metabolism, vol. 83, no. 10, pp. 3469–3475, 1998.
27. H. A. McKay, D. A. Bailey, R. L. Mirwald, K. S. Davison, and R. A. Faulkner, "Peak bone mineral accrual and age at menarche in adolescent girls: a 6– year longitudinal study," Journal of Pediatrics, vol. 133, no. 5, pp. 682–687, 1998.
28. G. Maor, M. Rochwerger, Y. Segev, and M. Phillip, "Leptin acts as a growth factor on the chondrocytes of skeletal growth centers," Journal of Bone and Mineral Research, vol. 17, no. 6, pp. 1034–1043, 2002.

CHAPTER 11

OVERWEIGHT, OBESITY AND UNDERWEIGHT IS ASSOCIATED WITH ADVERSE PSYCHOSOCIAL AND PHYSICAL HEALTH OUTCOMES AMONG 7-YEAR-OLD CHILDREN: THE 'BE ACTIVE, EAT RIGHT' STUDY

AMY VAN GRIEKEN, CARRY M. RENDERS, ANNE I. WIJTZES, REMY A. HIRASING, and HEIN RAAT

11.1 INTRODUCTION

The prevalence of overweight and obesity is increasing worldwide and has become a public health challenge [1]. The tracking of childhood overweight and associated health consequences into adulthood is of concern [2], [3], [4]. Several serious physical conditions are associated with overweight and, especially obesity, among children including asthma, sleep problems, cardiovascular diseases and type–2 diabetes [4], [5]. Also psychosocial conditions such as lower self–esteem, depressive feelings and body dissatisfaction [6], [7], [8], [9], [10], [11], [12], [13] are associated with overweight or obesity in childhood and adolescence.

A decrease in self–esteem or more depressive symptoms may be caused by weight–related teasing or bullying that may start in childhood and continue into adolescence [10], [11], [13], [14], [15], [16], [17]. Children can experience different types and varying amounts of weight–related teasing [6], [8], [10], [12], [15], [16], [17], [18]. For example, teasing may be peer–related or parent–related and may be daily or sporadic. Moreover, girls and boys report differences in the type and amount of teasing they experience [11].

Most studies have evaluated psychosocial outcomes of overweight and obesity in adolescents and older children (7–12 years) [10], [11], [13], [14], [15], [16], [17] whereas few have evaluated the association between overweight and health outcomes in younger children aged e.g. 5–8 years [18], [19], [20]. Studying the association among younger children, and evaluating changes in weight and potential changes in health outcomes with longitudinal studies, may help to develop appropriate (preventive) interventions for children and parents to deal with both overweight and associated psychosocial/physical consequences.

Similarly, the association between underweight and health outcomes has rarely been assessed, and seldom among younger underweight children [2], [4], [7]. Moreover, ambiguous results are reported on young underweight children and the associations with health outcomes. For example, one study reported decreases in physical health for underweight children compared to normal weight or obese children [4] whereas another reported diminishing associations after controlling for confounding variables [7].

This study aimed to evaluate two questions 1) what are the associations between childhood underweight, overweight and obesity at age 5 years and physical and psychosocial health outcomes at age 7 years, and 2) what is the association between change in body mass index (BMI) measured at age 5 and 7 years and health outcomes at age 7 years.

11.2 METHODS

11.2.1 STUDY DESIGN

In order to answer both study questions two types of study design were used. For the first study question a prospective design was used, investigating the association between determinants at age 5 years and parent–reported health outcome at age 7 years. For the second study question longitudinal data, obtained at age 5 and age 7, was summarized into one variable defining weight trajectories at age 7 years. Subsequently, a cross–sectional design was used to study the association between weight trajectories and health outcomes both at the age of 7 years. Data to evaluate the study questions originated from the 'Be active, eat right' study.

11.2.2 STUDY POPULATION

Data was obtained from the 'Be active, eat right' study, a cluster randomized controlled trial which aims to assess the effects of an overweight prevention protocol, described in detail elsewhere [21]. The Medical Ethics Committee of Erasmus MC University Medical Center Rotterdam approved the study protocol (reference number MEC–2007–163). A total of 13,638 parents visiting one of the 44 participating youth health care (YHC) centers for their 5–year–old child's regular preventive health visit between 2007 and 2008 were invited to participate in the study. Of the parents who were present at their child's health visit, 64.4% provided informed consent (n = 8,784) and 98.9% (n = 8,683) returned a first questionnaire. At 2–year follow–up a questionnaire was distributed among all parents that had given informed consent, for which the response rate was 62.9%.

For this specific study we excluded study participants that were allocated to the intervention condition (n = 4,842) to prevent interference of the intervention with regard to the associations under study. Records with missing data on child height, weight, gender and age at baseline (n = 30) were excluded. In addition, records with missing data for all of the outcomes of interest (n = 1,540) were excluded; these are records from parents that did not return the questionnaire at 2–year follow–up which included the health outcome measures.

Two study populations were created. The first study question, was answered using the population of n = 2,372 children that remained after excluding abovementioned missing data. To evaluate the second study question children with missing data on height, weight, age and gender at age 7 years were additionally excluded; a study population of n = 1,995 remained for the analysis.

11.2.3 WEIGHT STATUS OF THE CHILD

During the regular preventive health visit, each child's weight, height and waist circumference was measured by YHC professionals using standardized methods [22]. At 2–year follow–up, research assistants measured child height and weight according to the same methods. BMI was calculated by the researchers as weight in kilograms divided by height in meters squared. Children were categorized into one out of four weight categories according to their age– and gender–specific BMI: underweight, normal weight, overweight (not obesity) and obesity [23], [24].

Children were categorized in weight trajectories to describe changes in weight status. Based on the child's age– and gender–specific categorization at 5–years and 7–years, children were allocated to one of four trajectories (n = 1,995): low–stable, increasing, high–stable and decreasing. The first trajectory (low–stable) consisted of children maintaining a categorization of normal or underweight BMI, and children moving from the underweight category to the normal weight category or vice versa (n = 1,732). The second trajectory (increasing) consisted of children changing from a normal weight or overweight BMI categorization to the overweight or obesity BMI category (n = 115). The third trajectory (high stable) consisted of children maintaining the categorization of an overweight or obesity BMI (n = 103). The final trajectory (decreasing) consisted of children changing from an obesity or overweight BMI category to an overweight or normal weight BMI category (n = 45). Children changing from an underweight to an overweight or obese BMI category or vice versa were excluded (n = 1).

11.2.4 HEALTH OUTCOMES

Data on children's health outcomes (age 7 years) were obtained by questionnaires completed by parents. Table 1 presents the items assessing the indicators of psychosocial health outcomes. The items measuring psychosocial health outcomes were based on existing studies and developed by the researchers to fit the population under study [25], [26], [27].

TABLE 1: Overview of items providing an indication of psychosocial health.

Item name	Question/assessment	Scoring
Happiness	How often, in the past four weeks, was your child happy?	1 = always to 5 = never
Insecurity	I sometimes feel that my child feels insecure due to his/her weight	1 = totally disagree to 5 = totally agree
Adverse treatment	I sometimes feel that my child is treated adversely due to his/her weight (for example, being teased, left behind, or ignored)	1 = totally disagree to 5 = totally agree
Parental concern	Sometimes, I am concerned about my child's weight and the consequences thereof for his/her health	1 = totally disagree to 5 = totally agree

Parents were invited to fill in the number of visits to the general practitioner (GP) due to issues with the child's weight in the past two years, which was categorized as none versus one or more visits. The presence of common health conditions was assessed by the item: "Does your child have one of the following conditions?". Parents could choose 'yes' or 'no' to each of the following conditions: hearing difficulties, seeing difficulties, abdominal pain, headaches or migraine, allergies, and eczema. The presence of asthma symptoms was assessed with the wheezing and dyspnea questions from the International Study of Asthma and Allergies in Childhood (ISAAC) questionnaire [28]: "Has your child suffered from wheezing or a whistling noise in the chest/shortness of breath or breathlessness in the past 12 months?" and "How often in the past 12 months has your child suffered from wheezing or a whistling noise in the chest/shortness of breath or breathlessness?" Answers were dichotomized into 'symptoms' versus 'no symptoms'.

Health–related quality of life of the child was assessed with the General Health scale from the Child Health Questionnaire Parent Form 28 (CHQ–PF28) [29]. The General Health scale was dichotomized [mean: 86.40 (sd: 15.43)] into low scoring (<71.97) or average to high scoring (≥71.97).

11.2.5 CHILD AND MATERNAL CHARACTERISTICS

Information on child gender, age (years) and ethnic background (Dutch, non–Dutch) was obtained at enrolment. Child ethnicity was determined by the parents' country of birth: if both parents were born in the Netherlands the child was classified as 'Dutch', otherwise the child was classified as 'non–Dutch' [30].

In our study sample most of the questionnaires were completed by mothers (90.3%). Information on maternal age (years), height (meters), weight (kilograms), country of birth (the Netherlands, other) and educational level was obtained at enrolment. Maternal BMI was calculated as weight in kilograms divided by height in meters squared. Maternal level of education consisted of four levels: low (no education, primary school, or ≤3 years of general secondary school), mid–low (>3 years of general secondary school), mid–high (higher vocational training, undergraduate programs, or bachelor's degree), and high (higher academic education) [31].

11.2.6 ANALYSES

Normal weight, overweight, obese and underweight children were compared with regard to baseline child (gender, age, ethnic background, BMI) and maternal characteristics (age, country of birth, education level and BMI) by means of one–way analysis of variance and chi–square tests.

The non–normal distribution of the data required ordinal regression analyses to be performed for the items on happiness, insecurity, adverse treatment and parental concern. Logistic regression models were fitted for all other health outcomes: GP visits, ISAAC items, health conditions, and general health.

For both study questions regression models were fitted. First a model with with BMI in categories at age 5 years as independent variable was fitted; secondly a model with weight trajectories as independent variable was fitted. Underweight, overweight and obese children were compared with normal weight children (reference category) and the high–stable, increasing and decreasing weight trajectories were compared with the low–stable weight trajectory (reference category). All regression models were corrected for potential confounding variables gender, ethnic background of the child and education level of the mother [12].

The ordinal regression model provides the estimated odds ratio (OR) with 95% confidence interval (95%CI) for having a higher score on the outcome variable, if the independent variable would increase with one unit. An OR is estimated for each category of the independent variable compared to the reference category; the model assumes that all other factors stay constant. Tests of parallel lines were performed to check the use of the ordinal regression model. A multinomial regression model was fitted when the parallel lines test was significant; coefficients of the ordinal and multinomial regression models were compared. All ordinal regression models coefficients were in line with the coefficients of the multinomial models.

Effect modification was explored for gender, ethnic background and maternal education level: an interaction term with weight category was added to the above– described models [12]. Interaction terms were evaluated at p<0.10. Wardle et al. [12] recommended presenting results stratified.

Demographic characteristics of mothers (age, country of birth, education level and BMI) with missing data on the outcomes (n = 1,540) were compared with characteristics of mothers with no missing data (n = 2,372) by means of descriptive statistics.

Analyses were performed using SPSS version 20.0 for Windows (International Business Machines (IBM) Corp., SPSS statistics, version 20.0, Armonk, New York, USA).

11.3 RESULTS

The mean age of the included children was 5.8 (sd: 0.4) years, 50.0% were boys and 89.9% were Dutch (Table 2). There was a significant difference in the distribution of boys (p<0.001) and non–Dutch children (p<0.001) across the weight categories; more girls were categorized having overweight and more non–Dutch children were categorized as having obesity. Table S1 (available in the supplemental material) presents the baseline distribution of health conditions. Compared to mothers with no missing outcome data, the mothers of children with missing outcome data were younger [mean: 35.6 (sd: 4.6) years vs. 36.7 (sd: 4.2) years; p<0.001], more often non–Dutch (56.1% vs. 37.4%; p<0.05), less often higher educated (32.5% vs. 67.5%, p<0.001) and had a higher BMI [mean: 24.0 (sd: 4.4) vs. 23.7 (sd: 3.7); p<0.01].

TABLE 2: General characteristics of the study population, stratified by children's weight status (n = 2372).

	Total (n =2732)	**Under–weightᵗ (n = 355)**	**Normal weightᵗ (n = 1830)**	**Overweightᵗ (n = 148)**	**Obesityᵗ (n = 39)**	**p–value***
Child characterisics						
Age in years, mean (SD) [missing n=0]	5.8 (0.4)	5.8 (0.4)	5.7 (0.4)	5.8 (0.4)	5.8 (0.6)	0.396
Gender, % boys [missing n = 0]	50.0	47.9	51.5	37.2	43.6	0.005
Ethic background, % Dutch [missing n= 34]	89.9	89.2	90.6	87.8	68.4	<0.001
BMI, mean (SD) [missing n=0]	15.4 (1.5)	13.4 (0.5)	15.4 (0.9)	18.2 (0.7)	21.1 (1.2)	<0.001
Characteristics of the mother						
Age in years, mean (SD) [missing n =251]	36.7 (4.2)	36.6 (4.1)	36.7 (4.2)	36.8 (4.4)	37.6 (5.7)	0.581
Born in the Netherlands, % yes [missing n =2]	92.9	93.0	93.6	88.5	74.4	<0.001
Educational level [missing n =6]						
% low	3.0	2.0	3.1	3.4	5.1	0.050
% mid–low	16.7	17.2	16.0	20.9	28.2	
% mid–high	45.6	45.4	45.2	52.0	46.2	
% high	34.7	35.5	35.8	23.6	20.5	
BMI mean (SD) [missing n = 64]	23.7 (3.7)	23.0 (3.9)	23.6 (3.5)	24.8 (4.5)	27.9 (5.6)	<0.001

ᵗCategories based on international age– and gender–specific BMI cut–off values.
**p–value from Chi–square tests for categorical variables and ANOVA for continuous variables, comparing general characteristics across weight categories.*

11.3.1 ASSOCIATIONS BETWEEN UNDERWEIGHT, OVERWEIGHT OR OBESITY AT AGE 5 YEARS AND HEALTH OUTCOMES AT AGE 7 YEARS

Table 3 presents the results of the regressions analysis predicting health outcomes at age 7 years with BMI at age 5 years as independent variable.

TABLE 3: Health outcomes at age 7 years, predicted by BMI status at age 5 years.

		Underweight (n = 355)'	**Normal weight (n=1830)** '	**Overweight (n = 148)**'	**Obesity (n = 39)**'
	n	OR (95% CI)	OR (95% CI)	OR (95% CI)	OR (95% CI)
One or more visits to GP (yes)[1]	2316	**2.64 (1.37; 5.00)****	1.00	**3.41 (1.51; 7.69)****	**13.39 (5.43; 33.03)*****
Respiratory symptoms (yes)–wheezing[1]	2328	1.25 (0.79; 1.99)	1.00	1.57 (0.84; 2.94)	2.40 (0.91; 6.36)
Respiratory symptoms (yes)–dyspnea[1]	2322	1.06 (0.67; 1.67)	1.00	0.96 (0.47; 1.93)	1.74 (0.60; 5.04)
Hearing difficulties (yes)[1]	2301	1.14 (0.73; 1.78)	1.00	1.14 (0.60; 2.17)	1.18 (0.35; 3.94)
Seeing difficulties (yes)[1]	2295	1.32 (0.53; 3.26)	1.00	1.63 (0.48; 5.54)	4.48 (0.99; 20.34)
Abdominal pain (yes)[1]	2298	**1.44 (1.02; 2.03)***	1.00	1.00 (0.57; 1.76)	1.00 (0.35; 2.89)
Headaches or migraines (yes)[1]	2296	1.25 (0.57; 2.74)	1.00	1.51 (0.52; 4.35)	1.40 (0.18; 10.71)
Allergies (yes)[1]	2301	**1.81 (1.32; 2.49)*****	1.00	1.22 (0.71; 2.07)	1.07 (0.37; 3.08)
Eczema (yes)[1]	2292	0.65 (0.42; 1.00)	1.00	1.15 (0.68; 1.97)	0.72 (0.22; 2.38)
Lower score on general health[1]	2333	**1.47 (1.12; 1.94)****	1.00	1.31 (0.87; 1.98)	**2.60 (1.32; 5.13)****
Lower score on happiness[2]	2306	0.99 (0.77; 1.26)	1.00	1.21 (0.84; 1.75)	0.80 (0.41; 1.55)
Higher score on feeling insecure[2]	2330	1.12 (0.85; 1.47)	1.00	**6.37 (4.62; 8.78)*****	**23.81 (13.05; 43.42)*****
Higher score on adverse treatment[2]	2329	**1.39 (1.05; 1.84)***	1.00	**5.70 (4.10; 7.92)*****	**35.34 (19.16;65.17)*****
Parental concern[2]	2331	**1.84 (1.48; 2.30)*****	1.00	**7.22 (5.31; 9.87)*****	**26.10 (14.21; 47.94)*****

1 Odds ratio (OR) and 95% confidence interval (95% CI) from logistic regression analysis.
2 Odds ratio (OR) and 95% confidence interval (95% CI from ordinal regression analysis.
'Categories based on international age– and gender–specific BMI cut–off values. Note: all models are corrected for confounding by gender and ethnic background of the child and education level of the mother. Numbers printed in bold represent significant OR and asterisks represent significance level:
$*p<0.05$.
$**p<0.01$
$***p<0.001$

The OR for visiting the GP once or more was 2.64 (95% CI: 1.37 to 5.00) for underweight children. Compared to normal weight children, underweight children had an OR of 1.39 (95% CI: 1.05 to 1.84) for being treated adversely according to their parents. Their parents had an OR of 1.84 (95% CI: 1.48 to 2.30) for being concerned (Table 3).

For overweight children the OR for visiting the GP once or more was 3.41 (95% CI: 1.51 to 7.69). Compared to normal weight children, overweight children had an OR of 6.37 (95% CI: 4.62 to 8.78) for feeling insecure and 5.70 (95% CI: 4.10 to 7.92) for being treated adversely, as reported by their parents. Compared to normal weight children, the OR for parental concern was 7.22 (95% CI: 5.31 to 9.87) for overweight children (Table 3).

The OR for visiting the GP once or more was 13.39 (95% CI: 5.43 to 33.03) for obese children. Compared to the reports of parents of normal weight children, obese children had an OR of 35.34 (95% CI: 19.16 to 65.17) for being treated adversely. Compared to normal weight children, the OR was 26.10 (95% CI: 14.21 to 47.94) for parental concern about obese children (Table 3).

11.3.2 ASSOCIATIONS BETWEEN WEIGHT TRAJECTORIES AND HEALTH OUTCOMES AT AGE 7 YEARS

Table 4 presents the results for the regression model predicting health outcomes at age 7 years using the weight trajectories as independent variable.

Compared to children with a low stable weight, parents of children with an increasing weight trajectory reported significantly higher ORs for their child feeling insecure (OR: 5.61, 95% CI: 3.88 to 8.11) and being treated adversely (OR: 4.72, 95% CI: 3.23 to 6.90). Compared to parents of children with a low stable BMI, parents of children with an increasing weight trajectory reported a higher OR for concern (OR: 5.55, 95% CI: 3.91 to 7.89) (Table 4).

Compared to children with a low stable trajectory, children with a high–stable weight trajectory had a significantly higher OR for visiting the

TABLE 4: Health outcomes at age 7 years, predicted by weight trajectory between the age of 5 and 7 years.

		Low stable (n = 1732)	**Increasing (n=115)**	**High stable (n = 103)**	**Decreasing (n = 45)**
	n	OR (95% CI)	OR (95% CI)	OR (95% CI)	OR (95% CI)
One or more visits to GP (yes)[1]	1951	1.00	0.88 (0.21; 3.74)	**3.87 (1.75; 8.54)****	1.86 (0.42; 8.17)
Respiratory symptoms (yes)–wheezing[1]	1959	1.00	0.88 (0.38; 2.07)	1.93 (0.99; 3.76)	0.78 (0.19; 3.27)
Respiratory symptoms (yes)–dyspnea[1]	1953	1.00	0.37 (0.11; 1.17)	0.99 (0.45; 2.20)	0.65 (0.16; 2.73)
Hearing difficulties (yes)[1]	1939	1.00	1.67 (0.89; 3.15)	1.52 (0.77; 3.04)	0.32 (0.04; 2.33)
Seeing difficulties (yes)[1]	1934	1.00	0.00 (0.00; –)	2.56 (0.74; 1.99)	1.86 (0.29; 14.21)
Abdominal pain (yes)[1]	1937	1.00	1.07 (0.58; 1.95)	1.06 (0.56; 1.99)	0.81 (0.29; 2.31)
Headaches or migraines (yes)[1]	1935	1.00	1.46 (0.44; 4.88)	2.23 (0.76; 6.54)	1.26 (0.11; 9.49)
Allergies (yes)[1]	1940	1.00	1.05 (0.58; 1.92)	1.24 (0.68; 2.28)	0.84 (0.30; 2.39)
Eczema (yes)[1]	1935	1.00	0.99 (0.52; 1.89)	0.98 (0.50; 1.94)	1.57 (0.68; 3.59)
Lower score on general health[1]	1964	1.00	1.09 (0.52; 2.29)	1.28 (0.79; 2.07)	1.14 (0.71; 1.84)
Lower score on happiness[2]	1940	1.00	1.10 (0.73; 1.67)	1.19 (0.76; 1.85)	0.76 (0.41; 1.40)
Higher score on feeling insecure[2]	1961	1.00	**5.61 (3.88; 8.11)*****	**16.35 (11.08; 24.36)*****	**4.38 (2.48; 7.73)*****
Higher score on adverse treatment[2]	1961	1.00	**4.72 (3.23; 6.90)*****	**15.94 (10.75;23.64)*****	**2.82 (1.53; 5.20)*****
Parental concern[2]	1963	1.00	**5.55 (3.91; 7.89)*****	**14.54 (9.91; 21.33)*****	**3.61 (2.10; 6.18)*****

1 Odds ratio (OR) and 95% confidence interval (95% CI) from logistic regression analysis.
2 Odds ratio (OR) and 95% confidence interval (95%CI) from ordinal regression analysis
Note: all models are corrected for confounding by gender and ethnic background of the child and education level of the mother. Numbers printed in bold represent significant OR and asterisks represent significance level:
$*p<0.05$
$**p<0.01$
$***p<0.001$

GP once or more (OR: 3.87, 95% CI: 1.75 to 8.54). Children with a high stable weight trajectory had higher ORs for feeling insecure (OR: 16.35, 95% CI: 11.08 to 24.36) and being treated adversely (OR: 15.94, 95% CI: 10.75 to 23.64), according to their parents. Compared to children with a low stable weight trajectory, parents of children with a high–stable weight trajectory showed a higher OR for parental concern (OR: 14.54, 95% CI: 9.91 to 21.33) (Table 4).

Compared to parents of children with a low stable weight trajectory, parents of children with a decreasing weight trajectory reported their child having a higher OR for feeling insecure (OR: 4.38, 95% CI: 2.48 to 7.73) and being treated adversely (OR: 2.82, 95% CI: 1.53 to 5.20). Compared to low stable weight children, parents of children with a decreasing weight reported more concern (OR: 3.61; 95% CI: 2.10 to 6.18) (Table 4).

11.4 DISCUSSION

This study first prospectively evaluated the association between the weight status of children aged 5 years and parent–reported health outcomes at age 7 years. The study shows that overweight and obese children visit the GP more often; there was no indication that overweight or obese children have more physical health conditions such as asthma symptoms or allergies. According to their parents, overweight and obese children experience more insecurity and are more often being treated adversely. Underweight children also appeared to experience more adverse treatment and lower general health. Parents of both underweight and overweight or obese children reported more concern about their child's weight compared to parents of normal weight children.

Secondly this study evaluated the association of weight status trajectory, based on the weight status of the child at age 5 years and at age 7 years, with parent–reported health outcomes at child age 7 years. Overweight or obesity at 5 and 7 years of age was associated with more insecurity and adverse treatment compared to children with a normal weight at both ages.

11.4.1 ASSOCIATIONS BETWEEN OVERWEIGHT AND OBESITY, AND HEALTH OUTCOMES

According to their parents, overweight and obese children had a higher risk of feeling insecure or being treated adversely due to their weight. This concurs with other studies on children of a similar age (5–9 years) [18], [19], [20]. For example, one study that also evaluated the association between weight status and wellbeing in a longitudinal sample of general population children (but did not report on adverse treatment experienced by children), reported significantly higher odds for 4–5 year old overweight/obese children to have emotional and peer problems at age 8–9 years [19].

Parents of children with overweight and obesity reported more concern with regard to their child's weight. Although our study did not indicate that overweight and obese children experience more health conditions (e.g. asthma symptoms, allergies), several health conditions are reported to be potentially associated with overweight and obesity (e.g. type 2 diabetes, sleep problems) [32]. Hypothetically, parents and children may have visited the GP for conditions unmeasured in the current study. Also, the smaller number of children with specific conditions may have created a lack of power to detect an effect of weight status in the current study. Nevertheless, other researchers have reported that parents are more aware and likely to identify the overweight of their 6–year–old child, compared to parents with younger children, and therefore concern with regard to the child's weight may have increased among these parents [33].

11.4.2 ASSOCIATIONS BETWEEN UNDERWEIGHT AND HEALTH OUTCOMES

Underweight children had slightly higher odds for being treated adversely compared to normal weight children. Two studies reported that adverse treatment was associated with both underweight and overweight [13] [15]. Although their results indicate that the odds for experiencing adverse treatment are much greater for overweight/obese children compared to the odds for underweight children, future studies will need to include underweight as a separate subgroup to further explore the associations with health outcomes.

Parents of underweight children reported slightly higher odds for higher levels of concern compared to parents of normal weight children. Also, a higher frequency of GP visits and lower scoring on the general health scale of the Child Health Questionnaire at age 7 years was observed. Hypothetically, underweight children may be more prone to seasonal diseases (such as influenza or a cold) which may partly explain the increased risk for visits to the GP and the overall lower scores on general health.

Although we observed some interesting associations between underweight in children and parent–reported health outcomes, these associations are to be interpreted with caution. We, for example, did not measure whether these children had specific diseases during preschool. Because children with relative underweight may develop a normal weight when they grow older [34], longitudinal data needs to provide more insight in weight patterns of these children. Health care practitioners may be attentive to health problems associated with childhood underweight so that appropriate advice can be given; however, more research is needed before reliable advice with regard to counseling for underweight children and their parents can be given.

11.4.3 ASSOCIATIONS BETWEEN WEIGHT STATUS TRAJECTORIES AND HEALTH OUTCOMES

Our study indicates that children with an increasing BMI between the age of 5 and 7 years have higher odds for being treated adversely and feeling insecure, as also reported by other studies [18], [19], [20]. Weight patterns have been associated with lower school functioning among elementary school–aged children [35]. The association between weight patterns and lower school functioning has been found to be mediated by internalizing factors (e.g. loneliness, low self–esteem) [35], [36]. This emphasizes the need to develop and evaluate appropriate interventions for overweight/ obese children at young ages to prevent further decreases in school performance, social participation, health outcomes and quality of life. Also, the pathways and environmental characteristics through which health outcomes are affected by overweight or obesity need further clarification; qualitative studies are required to gain more insight into these mechanisms. Combining multiple resources, such as child, parent report and teacher reports, or performing observational studies, may help to elucidate the association between weight and health outcomes.

Based on the methods used in other studies we categorized children in weight status trajectories using the international cutoff values at age 5 and age 7 years, which may result in a relatively crude categorization [35], [36]. Children may decrease or increase within a weight category, but not reach the criterion to be categorized in another weight category. We explored whether gain in BMI was associated with higher risk for adverse psychosocial outcomes (data not shown). Children that gained BMI had a higher risk for being treated adversely and feeling insecure at age 7, as reported by their parents. Also, parents reported more concern with regard to their child's weight. Considering the physical outcomes, children had a higher parent–reported OR for having one or more health conditions or visits to the GP when they gained BMI between age 5 and 7. This is in line with the results we observed using the trajectories approach for high stable and increasing weight status. Longitudinal studies having access to multiple BMI measures may be able to create individual pathways of BMI development using statistical models [37], [38], [39]. These longitudinal trajectories or developmental pathways may reveal more distinct patterns of for example, late or early onset BMI gain, and can be related to health outcomes [37], [38], [39].

11.4.4 METHODOLOGICAL CONSIDERATIONS

Strengths of this study include the large sample size, the ability to create subgroups based on the international cut–off values for BMI, inclusion of a large group of underweight children and the availability of data at child age 5 years and child age 7 years.

Limitations include the missing data at child age 7 years and parents self–report of the children's health outcomes. Also, mothers of children with complete outcome measures differ from mothers with missing outcome data; however, this does not necessarily influence the associations under study. With regard to the items used to measure psychosocial health outcomes of the child, these have not been examined with regard to validity and reliability. Additional analyses (data not shown) were performed to gain insight in the validity of the items used. These analyses showed that normal weight mothers reported a higher OR for their overweight or obese child to be treated adversely and feel insecure, normal weight mothers also reported more concern for their overweight or obese child compared to normal weight children (data not shown). Nevertheless, we recommend future research to evaluate the validity and reliability of the items measuring psychosocial health outcomes.

The use of parent self–report may have led to over– or underestimation of the children's health outcomes and needs to be taken into account when interpreting the findings. Measures of depression and self–esteem of the child were not included in the questionnaire because of the already reasonably high respondent burden. Although child report may have provided more accurate estimates of consequences on health outcomes, measuring concepts such as self–esteem and depression is known to be challenging among young children [40]. Also, at younger age self–concept indicators, such as teasing and insecurity, may be more informative compared to self–esteem questionnaires due to the developmental stage of the children [40].

11.5 CONCLUSIONS

In conclusion, parents reported their overweight, obese and underweight children to be more often treated adversely or feel insecure due to their weight. Parents of overweight, obese and underweight children expressed more concern about health outcomes associated with their child's overweight or underweight. These concerns seem to be reflected in the more frequent parent–reported visits to the GP of children with overweight, obesity and underweight.

Future studies need to follow–up on the associations between weight status and health outcomes when children develop and reach adolescence and adulthood. Also, underlying mechanisms and pathways associated with weight status and health outcomes need to be assessed in, preferably, longitudinal research.

In the meantime, we recommend that healthcare providers be alert to early signs of adverse treatment and insecure feelings in both overweight and underweight children. Appropriate counseling for teasing and insecure feelings should be offered in addition to, or as part of, interventions aiming at a positive change in weight status.

REFERENCES

1. World Health Organization (WHO). Childhood overweight and obesity. Available: http://www.who.int/dietphysicalactivity/childhood/en/index.html. Accessed 18 November 2012.
2. Reilly JJ, Kelly J (2011) Long–term impact of overweight and obesity in childhood and adolescence on morbidity and premature mortality in adulthood: systematic review. Int J Obes (Lond) 35: 891–898. doi: 10.1038/ijo.2010.222.
3. Wijga AH, Scholtens S, Bemelmans WJ, de Jongste JC, Kerkhof M, et al. (2010) Comorbidities of obesity in school children: a cross–sectional study in the PIAMA birth cohort. BMC Public Health 10: 184. doi: 10.1186/1471–2458–10–184.
4. Wake M, Clifford SA, Patton GC, Waters E, Williams J, et al.. (2012) Morbidity patterns among the underweight, overweight and obese between 2 and 18 years: population–based cross–sectional analyses. Int J Obes (Lond). DOI: 10.1038/ijo.2012.86.
5. Dietz W (1998) Health consequences of obesity in youth: childhood predictors of adult disease. Pediatrics 101: 518–525.
6. Erickson SJ, Robinson TN, Haydel KF, Killen JD (2000) Are overweight children unhappy? Body mass index, depressive symptoms, and overweight concerns in elementary school children. Arch Pediatr Adolesc Med 154: 931–935. doi: 10.1001/archpedi.154.9.931.
7. Drukker M, Wojciechowski F, Feron FJ, Mengelers R, van Os J (2009) A community study of psychosocial functioning and weight in young children and adolescents. Int J Pediatr Obes 4: 91–97. doi: 10.1080/17477160802395442.
8. Strauss RS (2000) Childhood Obesity and Self–Esteem. Pediatrics 105: e15. doi: 10.1542/peds.105.1.e15.
9. Taylor A, Wilson C, Slater A, Mohr P (2012) Self–esteem and body dissatisfaction in young children: Associations with weight and perceived parenting style. Clin Psychol 16: 25–35. doi: 10.1111/j.1742–9552.2011.00038.x.
10. Janssen I, Craig WM, Boyce WF, Pickett W (2004) Associations between overweight and obesity with bullying behaviors in school–aged children. Pediatrics 113: 1187–1194. doi: 10.1542/peds.113.5.1187.
11. Griffiths LJ, Wolke D, Page AS, Horwood JP (2006) Team AS (2006) Obesity and bullying: different effects for boys and girls. Arch Dis Child 91: 121–125.
12. Wardle J, Cooke L (2005) The impact of obesity on psychological well–being. Best Practice Res Clin Endocrinology Metabol 19: 421–440. doi: 10.1016/j.beem.2005.04.006.
13. Eisenberg ME, Neumark – Sztainer D, Story M (2003) Associations of weight–based teasing and emotional well–being among adolescents. Arch Pediatr Adolesc Med 157: 733–738. doi: 10.1001/archpedi.157.8.733.

14. McCormack LA, Laska MN, Gray C, Veblen–Mortenson S, Barr–Anderson D, et al. (2011) Weight–related teasing in a racially diverse sample of sixth–grade children. J Am Diet Assoc 111: 431–436. doi: 10.1016/j.jada.2010.11.021.
15. Neumark–Sztainer D, Falkner N, Story M, Perry C, Hannan PJ, et al. (2002) Weight–teasing among adolescents: correlations with weight status and disordered eating behaviors. Int J Obes Relat Metab Disord 26: 123–131. doi: 10.1038/sj.ijo.0801853.
16. Young–Hyman D, Tanofsky–Kraff M, Yanovski SZ, Keil M, Cohen ML, et al. (2006) Psychological status and weight–related distress in overweight or at–risk–for–overweight children. Obesity (Silver Spring) 14: 2249–2258. doi: 10.1038/oby.2006.264.
17. Goldfield G, Moore C, Henderson K, Buchholz A, Obeid N, et al. (2010) The relation between weight–based teasing and psychological adjustment in adolescents. Paediatr Child Health 15: 283–288.
18. Davison K, Birch LL (2002) Processes linking weight status and self–concept among girls from ages 5 to 7 years. Develop Psychol 38: 735–748. doi: 10.1037/0012–1649.38.5.735.
19. Sawyer MG, Harchak T, Wake M, Lynch J (2011) Four–year prospective study of BMI and mental health problems in young children. Pediatrics 128: 677–684. doi: 10.1542/peds.2010–3132.
20. Shunk JA, Birch LL (2004) Girls at risk for overweight at age 5 are at risk for dietary restraint, disinhibited overeating, weight concerns, and greater weight gain from 5 to 9 years. J Am Diet Assoc 104: 1120–1126. doi: 10.1016/j.jada.2004.04.031.
21. Veldhuis L, Struijk MK, Kroeze W, Oenema A, Renders CM, et al. (2009) 'Be active, eat right', evaluation of an overweight prevention protocol among 5–year–old children: design of a cluster randomised controlled trial. BMC Public Health 9: 177. doi: 10.1186/1471–2458–9–177.
22. Bulk–Bunschoten AMW, Renders CM, Van Leerdam FJM, HiraSing RA (2005) Signaleringsprotocol overgewicht in de jeugdgezondheidszorg [Youth Health Care Overweight–detection–protocol]. Woerden: Platform Jeugdgezondheidszorg.
23. Cole TJ, Bellizzi MC, Flegal KM, Dietz WH (2000) Establishing a standard definition for child overweight and obesity worldwide: international survey. BMJ 320: 1240–1243. doi: 10.1136/bmj.320.7244.1240.
24. Cole TJ, Flegal KM, Nicholls D, Jackson AA (2007) Body mass index cut offs to define thinness in children and adolescents: international survey. BMJ 335: 194. doi: 10.1136/bmj.39238.399444.55.
25. Lyubomirsky S, Lepper H (1999) A measure of subjective happiness: Preliminary reliability and construct validation. Social Indicators Research 46: 137–155. doi: 10.1023/a:1006824100041.
26. Hills P, Argyle M (2002) The Oxford happiness questionnaire: A compact scale for the measurement of psychological well–being. Personality and Individual Differences 33: 1071–1082. doi: 10.1016/s0191–8869(01)00213–6.
27. Griffin RS, Gross AM (2004) Childhood bullying: Current empirical findings and future directions for research. Aggression and Violent Behavior 9: 379–400. doi: 10.1016/s1359–1789(03)00033–8.
28. The international study of asthma and allergies in childhood. Available: http://isaac.auckland.ac.nz/index.html. Accessed: 18 November 2012.

29. Landgraf J, Abetz L, Ware JE (1996) The CHQ's user manual. Boston: The Health Institute, New England Medical Center.
30. Statistics Netherlands. Definition of migrants. Available: http://www.cbs.nl/nl–NL/menu/methoden/begrippen/default.htm?conceptid=37. Accessed 18 November 2012.
31. Statistics Netherlands. Dutch Standard Classification of Education. Available: http://www.cbs.nl/nl–NL/menu/methoden/classificaties/overzicht/soi/default.htm. Accessed; 18 November 2012.
32. Wabitsch M (2000) Overweight and obesity in European children: definition and diagnostic procedures, risk factors and consequences for later health outcome. Eur J Pediatr 159: S8–S13. doi: 10.1007/pl00014368.
33. Eckstein KC, Mikhail LM, Ariza AJ, Thomson JS, Millard SC, et al. (2006) Parents' perceptions of their child's weight and health. Pediatrics 117: 681–690. doi: 10.1542/peds.2005–0910.
34. Luigi G, Chris P, Catherine P (1995) Adult outcome of normal children who are short or underweight at age 7 years. Bmj 310: 696–700. doi: 10.1136/bmj.310.6981.696.
35. Datar A, Sturm R (2006) Childhood overweight and elementary school outcomes. Int J Obes (Lond) 30: 1449–1460. doi: 10.1038/sj.ijo.0803311.
36. Gable S, Krull JL, Chang Y (2012) Boys' and Girls' Weight Status and Math Performance From Kindergarten Entry Through Fifth Grade: A Mediated Analysis. Child Develop 83: 1822–1839. doi: 10.1111/j.1467–8624.2012.01803.x.
37. Lumeng JC, Forrest P, Appugliese DP, Kaciroti N, Corwyn RF, et al. (2010) Weight Status as a Predictor of Being Bullied in Third Through Sixth Grades. Pediatrics 125: 1301–1307. doi: 10.1542/peds.2009–0774.
38. Huang DY, Lanza HI, Wright–Volel K, Anglin MD (2013) Developmental trajectories of childhood obesity and risk behaviors in adolescence. J Adolesc 36: 139–148. doi: 10.1016/j.adolescence.2012.10.005.
39. Bisset S, Fournier M, Pagani L, Janosz M (2013) Predicting academic and cognitive outcomes from weight status trajectories during childhood. Int J Obes (Lond) 37: 154–159. doi: 10.1038/ijo.2012.106.
40. Davis–Kean PE, Sandler HM (2001) A Meta–Analysis of Measures of Self–Esteem for Young Children: A Framework for Future Measures. Child Develop 72: 887–906. doi: 10.1111/1467–8624.00322.

PART III

PREVENTION AND TREATMENT

CHAPTER 12

DETERMINANTS OF WEIGHT LOSS IN AN INTERDISCIPLINARY LONG–TERM CARE PROGRAM FOR CHILDHOOD OBESITY

A. C. DUBUISSON, F. R. ZECH, M. M. DASSY, N. B. JODOGNE, and V. M. BEAULOYE

12.1 INTRODUCTION

Childhood obesity has spread dramatically over the previous decades. To curtail this major health issue, long–term effective weight control programs are essential. In the short term, several studies have shown positive and encouraging outcomes of multidisciplinary approach for childhood obesity. A combination of dietary, physical activity and behavioural interventions compared to standard care or self–help can produce a significant and clinically meaningful reduction in obesity in children and adolescents [1, 2]. This highlights the importance of multidisciplinary programs as the best first–line treatment. However, in most studies, programs are limited to between 6 and 12 months of duration, and beneficial effects are partly lost from 6 to 12 months after completion [3], especially for severely obese children [4]. Long–term follow–up studies of paediatric obesity interventions show a mean 10% reduction in relative weight but also substantial relapse [5, 6]. As obesity is a chronic disease, the question of the need of a chronic care program is raised [7]. Whether the continuation of the program will still be beneficial and how to implement this in a real–life situation remains to be answered.

Moreover, why some children respond differently to obesity treatment remains unclear. Identification of factors associated with better outcomes can help maximize the effectiveness of existing interventions, tailor treatment programs to the specific needs of the patients, and set realistic weight loss goals. Treatment for children presents a unique challenge as nutrition education, physical activity, and behaviour modification must be presented to both the parents/caregiver and child. Parental involvement and individual counselling have been recognised as an important feature of behavioural programs, particularly in preadolescent children [3, 4, 9]. It is thus relevant to examine the impact of family characteristics and psychosocial factors on children's weight loss.

In this paper, we first analyzed the 5–year results of an interdisciplinary long–term care program for childhood obesity. Secondly, we determined the baseline factors (medical, dietary, and psychosocial) which were associated with weight loss.

12.2 SUBJECTS AND METHODS

12.2.1 INTERVENTION STRUCTURE

In 2000, we set up an interdisciplinary approach for the treatment of childhood obesity. Our approach is an individually adapted (specific for each patient) family–based, behavioural lifestyle and dietary intervention program. It consists in joint consultations where each child and his family are seen by both a psychologist and a paediatric endocrinologist at the same time. After a time together, the patient is examined (weight, height, blood pressure, and Tanner stage) by the physician in a separate room which gives the opportunity to the child of having personal time with the paediatrician. During this time, the parents/family/caregivers are seen by the psychologist trained in family therapy. Then, the child and the physician get back to the psychologist and the family for a resume of the situation in order to make some decisions. Thereafter, the child and his family are taken by the psychologist to the dietician. The psychologist gives a summary of the situation, and then all the family is seen by the dietician.

Before each session, the interdisciplinary team reviews the situation of each patient. At the end of the visit a letter, including all the decisions taken with the child and his family, and some personal encouragement for the child is written and sent to the patient one month later. For further analysis, the implementation of those decisions was defined as adherence to the treatment. The patient attends the next appointment, generally every 3 to 6 months. The duration of intervention is determined by the needs of the patient. They participate in our program as long as they want or need. Between two interdisciplinary visits, the patient and his family may meet the dietician or the psychologist individually if needed. Some patients are referred to a specific psychotherapist, and individual physiotherapy can also be prescribed in some situations to reintroduce physical activity especially if joining a sport club is difficult at the beginning of the treatment. The psychotherapy and the physiotherapy are defined in our approach as "adjuvant therapies."

12.2.2 PROGRAM ORIENTATION

Our way to treat obese children is based on a solution–focused therapy [10]. The team develops realistic goals (small step changes) together with the child and his family rather than imposing ideas and assumptions about what they need to do to change their lifestyle [11]. We start the treatment from their questions or their needs (What can we do for you?) in order to stimulate them to be an active player in their own changes. We focus on the development of the confidence and competence of the parents or caregivers and of the children.

Our approach is also in agreement with the evidences published in the BMJ [12] by Edmunds et al. in 2001. We encourage the child to "grow without gaining weight" which decreases BMI slowly. The dietician does not prescribe any specific diet but focuses on healthy eating patterns (decrease exposure to obesogenic foods, designate times for family meals, and allocate individual portions) and on increasing the intake of healthy foods. Adolescents who were educated about better food choices of moderate portion sizes had been described to be more successful in the long term than teenagers who were given a structured meal plan or restrictive diets [13]. We propose that the child joins a sports club or a youth organization. But, we mostly encourage regular daily activities such as riding a bicycle, walking the dog, dancing, gardening, using the stairs instead of elevators, and playing outside with friends who are more easily integrated into a child's lifestyle than participation in organized sport teams. Data suggested that less structured, more flexible lifestyle exercise may be superior to more structured and higher–intensity aerobic exercise for weight control [14]. Recommended activities must be enjoyable and consistent with the child's and his/her family's lifestyle and be rewarding irrespective of the health benefits [14]. A complementary approach is also to reduce sedentary free–time activities. A psychologist is present during the interdisciplinary visits. Obesity may be reactive to an event of life (divorce of their parents, difficulties at school…). Parents are encouraged to not focus on weight loss but address their and the child's internal needs by expressing feelings and nurturing the child emotionally. The parents are targeted as the main agents of change, and they are responsible for inducing this change in the family home [9], not specifically at the target child. Extended family members are included as a means of reaching all people who play a significant role in the child's health.

12.2.3 SUBJECTS AND ASSESSMENTS

Four hundred twenty eight medical files of children who entered the interdisciplinary consultations between 2002 and 2007 were retrospectively reviewed. Inclusion criteria were [1] to have participated in, at least, 2 interdisciplinary visits and [2] to have at least one year of treatment. Children with obesity due to an organic/syndromic cause or with type 2 diabetes were excluded. Among the included patients: 73% were Caucasian, 12% were Hispanic, 10% were Arabian, 2% were African, 2% were Asiatic (representative of the national population). The latest visit was defined as the most recent visit reported in the medical files when they were reviewed between 2007 and 2009. Thus, for some children the latest visit is the last visit before they were no longer monitored. For other children, the latest visit is not the last because they are still monitored by the team.

Weight was measured (patient in socks with no shoes and wearing a light gown) in kilograms to the nearest 0.1 kg using a medical weight scale (SECA nondigital medical scale), zeroed and calibrated before each weight. A stadiometer (Holtain limited, Crymmych, PEMBS. UK), calibrated in 0.1–cm intervals, was used to determine height. BMI (kg/m2) were expressed relative to the Cole population reference data [8]. Weight loss was defined by reduction of the BMI standard deviation score (BMI–z–score) since BMI is gender and age dependent in childhood. BMI–z–score standardizes an individual's size, adjusting for age and sex, and allows comparison between values on an equivalent basis. Puberty development was scored in the adolescents according to Tanner stages [15].

The medical, dietary, and psychosocial factors characterizing the child and his family at baseline were assessed retrospectively by an external consultant (who was not made aware of the patients weight evolution) by reviewing the records (standard home–made questionnaire) filled in by the team at the first visit. As it is a retrospective and not a prospective study, a semiquantitative (low/intermediate/high) approach was used to evaluate each factor. For food consumption, no/low means not every day or never; intermediate (if applicable) means every day but at a low (1–2) level; yes/high means every day at a high (>2) level. Physical activity means that the child attends a sports club or a youth organization at least twice a week (yes/high), once a week (intermediate) or never (no/low). Delayed puberty was considered when a girl was assessed M1 > 13.5 years or a boy was assessed G1 > 14 years. Obesity in the family means the child has at least one of the two parents who is obese (BMI > 30 kg/m^2). Bad quality of sleep was assessed by snoring or short sleep duration (<9 h/night). Recent weight gain means a weight gain within the last year. Family encouragement for the project and child's motivation were assessed according to the involvement of the family and of the child in the project (who had the idea to come here? A doctor, the parents, and/or the child himself?). Social integration was assessed as the participation of the child in out–of–school activities. Family encouragement for leisure activities was assessed as activities realised by all the family out of the home.

12.2.4 STATISTICAL ANALYSIS

We use the quasilikelihood estimation [16] with a linear or a logistic canonical link. For the measures of the same patient, we used an autoregressive correlation matrix and computed the covariance matrix by quasi–least–squares (QLSs) [17].

For the significance of the results, we used the sandwich variance matrix augmented by the correction proposed by Morel et al. [18] which may be evaluated by the normal distribution.

For continuous independent variables, we worked with nonlinear regression by P splines (B splines with penalization [19] of degree 4). Penalisation was chosen to minimize the Akaike information criterion [20].

All the significances were expressed as two sided. Significance was taken at P value <0.05.

12.3 RESULTS

Out of 428 patients seen between 2002 and 2007, 322 patients (75%) were interested in our interdisciplinary treatment and attended a second visit. Of those, 144 children (45%) (59 boys (41.0%), 85 girls (59.0%); mean age: 10.5 ± 3.1 years (range 4–19 years; 105 (73%) <13 years, 39 (26%) >13 years; mean BMI–z–score: 2.73 ± 0.62) had at least a 1 year intervention program and were selected for our study. Mean length of treatment was 2.2 years (1–6.7 years) with an average of 3 ± 1 visits per year. After 24 months, 72 children were still monitored; 14 achieved a 48–month intervention. The length of treatment or assiduity (number of visits) did not depend on the initial weight loss (Δ BMI–z–score) between the first, and the second visit ($P = 0.63$). Sex, age and BMI–z–score at the first visit did also not influence the length of intervention ($P = 0.76$; $P = 0.009$ and $P = 0.43$ resp.).

Table 1 described the percentage of patients where BMI–z–score decreased and was stable or increased at the second visit and at the latest visit. At the second visit (approximately after 3–6 months), BMI–z–score was decreased in 53% of the patients, remained unchanged in 31%, and was increased in 16%. At the latest visit, 77% of patients had a stable or decreased weight.

In fact, weight loss was mainly observed during the first 6 months of treatment (Figures 1(a) and 1(b)) but was sustained long–term. The mean BMI–z–score was decreased by 8% ± 1% (−0.23 ± 0.04 mean BMI–z–score) of the initial mean BMI–z–score after a mean of intervention of 2.2

years and decreased by 12% ± 2%(−0.28 ± 0.06 mean BMI–z–score) for patients with a 48–month treatment. Initial BMI–z–score, age at the first visit, sex (data of boys and girls were thus combined for further analysis) and number of visit(s) per year did not influence these results ($P = 0.73$; $P = 0.27$; $P = 0.95$ and $P = 0.89$ resp.). Furthermore, an additional weight loss was observed between 6 and 48 months of intervention whatever the Δ BMI–z–score between the first and the second visit (Figure 2). Patients with a BMI–z–score reduction ≥0.3 units were 23% (20–27; 95% CI) of the population at 3 months versus 49% (40–58; 95% CI) at 48 months of treatment (Figure 3(a)). However, 16% (13–19; 95% CI) of our patients gained weight after 3 months of treatment. The percentage was increased to 31% (25–38; 95% CI) of the patients reaching 48 months of treatment (Figure 3(b)). The weight change between the first and the second visit was predictive of the additional weight change over the time (Figure 4).

TABLE 1: Evolution of the BMI–z–score of the patients during the intervention.

% of patients where BMI–z–score[a]	↓	=	↑
At the 2nd visit (± after 3–6 months)	53%	31%	16%
At the latest visit[b] (mean = 2.2 y (1–6.7 years))	67%	10%	23%

[a] *BMI–z–score: body mass index–standard deviation score [8].*

[b] *The latest visit is the most recent visit found in medical files when reviewed between 2007 and 2009 (Section 2).*

No evidence of adverse effects on growth, eating disorder pathology or mental health, was found.

We next investigated whether baseline medical, dietary, and psychosocial parameters reported at the first visit could influence the weight change over the time and which one could be associated with weight loss.

Patients who exercised in daily life before joining the interdisciplinary treatment were the most successful in term of weight loss (Table 2). Preexisting regular physical activity had a statistically significant positive influence (−0.42 ± 0.11 of mean BMI–z–score) on the weight evolution of the child, in comparison with those who did not exercise before starting the treatment (−0.18 ± 0.04 of mean BMI–z–score). Having delayed puberty had a negative influence on the evolution of the mean BMI–z–score of the patients (−0.02 ± 0.10 of mean BMI–z–score) in comparison with those who did not (−0.23 ± 0.03 of mean BMI–z–score, $P = 0.046$).

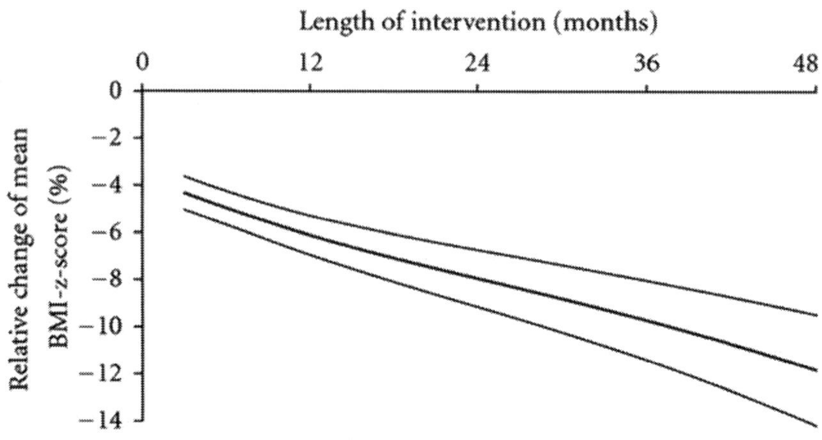

FIGURE 1: (a) Relative change (%) of mean BMI–z–score during intervention and (b) mean BMI–z–score during intervention. Data are expressed as mean ± SEM. BMI–z–score: body mass index–standard deviation score [8].

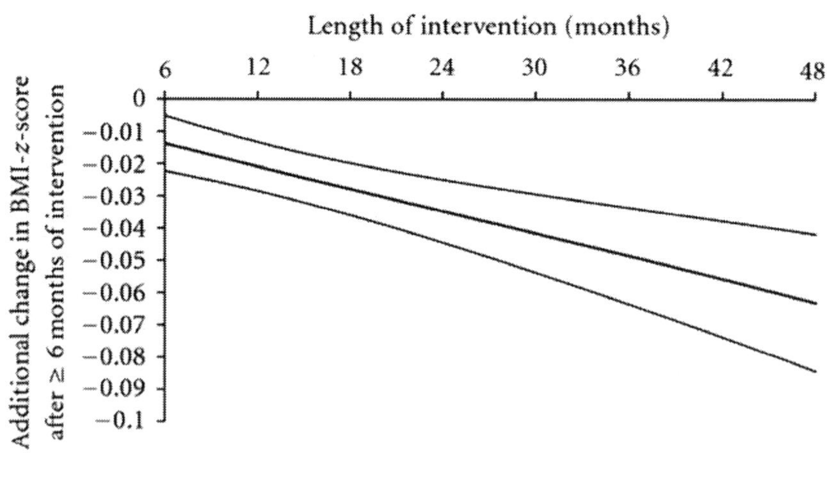

FIGURE 2: Additional change in BMI–z–score observed ≥6–month intervention, controlled for the Δ BMI–z–score between the first and the second visit. Data are expressed as mean ± SEM, bivariate analysis.

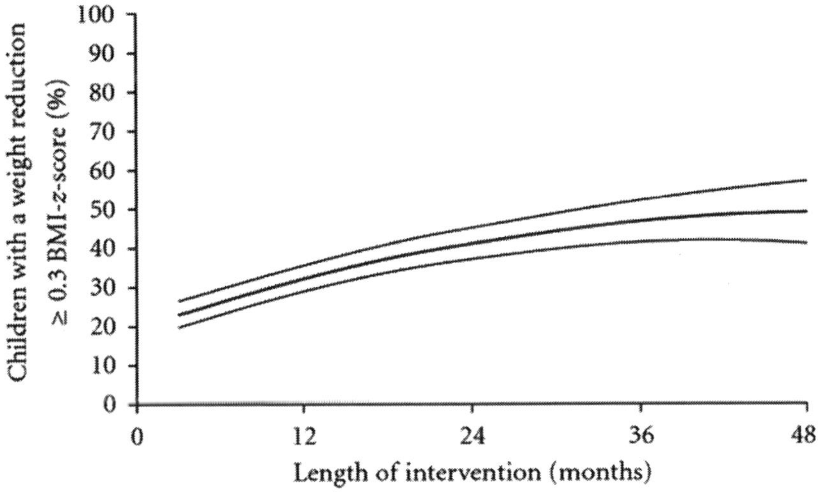

FIGURE 3: (a) Percentage of children with a BMI–z–score reduction ≥0.3 during intervention. Data are expressed as mean (95% CI). (b) Percentage of patients where BMI–z–score increased during the intervention (Δ BMI–z–score ≥0 at time of intervention in comparison with the initial visit). Data are expressed as mean (95% CI).

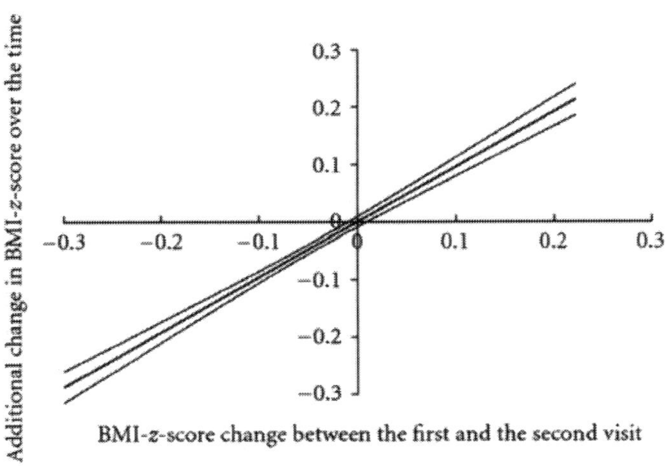

FIGURE 4: Additional change in BMI–z–score obtained in function of the change observed between the first and the second visit, controlled for the length of intervention. Data are expressed as mean ± SEM, bivariate analysis.

TABLE 2: Influence of medical factors assessed at the first visit on the weight change observed at 9 months of intervention.

Factors studied	No/low	Intermediate	Yes/high	P value
Significative				
Physical activity[a]	−0.18 ± 0.04	−0.30 ± 0.05	−0.42 ± 0.11	0.037
Delayed Puberty[b]	−0.23 ± 0.03		−0.02 ± 0.10	0.046
Nonsigificative				
Birthweight (>4000g)	−0.21 ± 0.03		−0.31 ± 0.11	0.37 NS
Gestational diabetes	−0.23 ± 0.04		−0.02 ± 0.14	0.15 NS
Breastfeeding (≥6 months)	−0.20 ± 0.04		−0.30 ± 0.07	0.21 NS
Obesity in the family[c]	−0.21 ± 0.05		−0.18 ± 0.04	0.62 NS
Asthma	−0.21 ± 0.03		−0.37 ± 0.11	0.18 NS
Bad quality of sleep[d]	−0.18 ± 0.05		−0.13 ± 0.07	0.61 NS

Data are expressed as change in mean BMI–z–score ± SEM at 9 months of intervention. $P < 0.05$, significant. NS: not significant.

[a]Physical activity means that the child joins a sport club or a youth organization at least twice a week (yes/high); once a week (intermediate) or never (no/low).

[b]Delayed puberty was considered when a girl was assessed M1 > 13.5 years or a boy was assessed G1 > 14 years.

[c]Obesity in the family means that the child has at least one of the two parents who is obese (BMI > 30 kg/m^2).

[d]Bad quality of sleep was assessed by snoring or short sleep duration (<9 h/night). This information was not available for all the patients (n = 76).

Moreover, baseline daily water intake and daily soda intake had a statistical significant impact on the children's weight outcome ($P = 0.046$ and $P = 0.00006$ resp.) (Table 3).

We then determined whether the psychosocial context of the child at the first visit may influence the weight change observed later on (Table 4). We showed that being a single child, having family encouragement for the project, the child's motivation, the adherence to the treatment, and the compliance to adjuvant therapies had a statistically positive effect on the mean BMI–z–score at 9 months of intervention. The duration of the obesity and dual parent households did not impact the weight changes observed.

TABLE 3: Influence of dietary factors assessed at the first visit on the weight change observed at 9 months of intervention.

Factors studied	No/low[a]	Intermediate[b]	Yes/high[c]	*P* value
Significative				
Daily water intake	–0.16 ± 0.04		–0.25 ± 0.04	0.046
Daily soft drinks intake	–0.38 ± 0.06	–0.15 ± 0.03	+0.08 ± 0.07	0.00006
Nonsignificative				
Daily fruits intake	–0.18 ± 0.04	–0.25 ± 0.04	–0.32 ± 0.08	0.16 NS
Eating breakfast every day	–0.18 ± 0.05		–0.25 ± 0.04	0.22 NS
2 hot meals a day	–0.23 ± 0.04		–0.16 ± 0.06	0.29 NS
Daily juice intake	–0.20 ± 0.05	–0.22 ± 0.03	–0.23 ± 0.04	0.73 NS
Daily vegetable intake	–0.25 ± 0.04	–0.20 ± 0.04	–0.15 ± 0.07	0.31 NS
Daily soup intake	–0.23 ± 0.04	–0.21 ± 0.04	–0.19 ± 0.07	0.68 NS
Daily cookies intake	–0.39 ± 0.11	–0.30 ± 0.06	–0.21 ± 0.03	0.12 NS
Snacker	–0.15 ± 0.05		–0.24 ± 0.04	0.15 NS
Large portions	–0.24 ± 0.05		–0.21 ± 0.04	0.55 NS

Data are expressed as change in mean BMI–z–score ± SEM at 9 months of intervention. P < 0.05, significant. NS: not significant.

[a]*Low means not every day or never.*

[b]*Intermediate means every day but in a low quantity (1–2).*

[c]*High means every day in a high quantity (>2).*

12.4 DISCUSSION

This retrospective real–life study reported the outcomes of a long–term approach for treating childhood obesity and identified baseline predictors of weight changes.

This intervention used interdisciplinary strategies (with effective interaction between the team, not only juxtaposed competences) but had the specificity to be individually adapted with a continuous care program. To our knowledge, this is the first time that the sustained benefit of a chronic intervention is reported and that the feasibility of a long–term intervention in real life is described in obese children. Our results were comparable to results reported by short intervention clinical trials. For example, family–based lifestyle behavioural treatment for obese children with similar clinical characteristics resulted in an average % decrease in overweight of ~7% after 6 months of treatment [3, 4] which is comparable to the 8% decrease observed at 24 months in the current study. However, the duration for the weight change was different. We described only a 4% BMI–z–score decrease at the second visit (3–6 months) but a 12% BMI–z–score decrease at 48 months. In contrast, in the long term, the results of the abovementioned studies were not as promising as they were immediately after completion of the program (~ −3.5% decrease in weight at 12 months, −1% at 18 months). The beneficial effects of short intervention programs (from 3 to 6 months) were partly lost on the follow–up. Even with a 12–month drug (Orlistat) intervention combined to lifestyle, the initial weight loss was not maintained for more than 6 months [21]. With this emphasis on acute short–term intervention, contemporary healthcare may not be well suited to meet the long–term needs of overweight children and their families fighting against this chronic disease. This indicates the need to develop chronic care models to optimize results, especially for severely obese children [4, 7]. The addition of a 4–month maintenance treatment after short–term weight loss treatment resulted in better maintenance of weight loss compared with the no maintenance group (−0.04 versus +0.05 BMI–z–score) but no additional weight loss was obtained over followup [22]. Similar findings were also reported by Reinehr et al. after a 4–year followup [6]. In contrast, our program was still beneficial after 48 months of treatment. Moreover, the percentage of patients with a 0.3 BMI–z–score reduction increased over time. At least 50% of children reaching a ≥36–month intervention presented a 0.3 reduction of BMI–z–score.

Longer treatments create challenges in maintaining participants in the program. In adults, the longer the treatment, the greater the proportion of patients who do not attend [23]. This problem may be magnified with families, who may have more challenges in scheduling than individual adults, and where there are multiple people who may want to drop out of the treatment [9]. In our study, 72 participants were still monitored at 24 months. Mean drop–out rates in the literature are varying from 10% to 60% at 12 months of followup [1]. For example, in an Italian multicentric study of nutritional intervention in obese paediatric patients, drop–out rates ranged from 30–34% to 90–94% after 2 years [24]. According to the literature, the main reasons for dropout are loss of interest, relocation, schedule conflict, transport difficulties, family issues for example, limited time for recurrent group sessions, and even for daily household demands [3, 25]. However, even for those patients, encouraging data recently published by Nemet et al. [26] showed that participants in a 3–month brief multidisciplinary intervention still maintained an increased leisure–time physical activity compared to the control group subjects after 1 year of followup. Even the weight benefit was modest after 1 year of followup; this could help them to better general health in the long term.

TABLE 4: Influence of psychosocial factors assessed at the first visit on the weight change observed at 9 months of intervention.

Factors studied	No/low	Intermediate	Yes/high	*P* value
Significative				
Only Child	–0.19 ± 0.03		–0.36 ± 0.07	0.026
Familial encouragement to the project[a]	–0.12 ± 0.04	–0.26 ± 0.04	–0.39 ± 0.08	0.0035
Child's movivation[a]	–0.05 ± 0.03	–0.30 ± 0.04	–0.55 ± 0.07	0.000000014
Adherence to the treatment[b]	–0.04 ± 0.03	–0.29 ± 0.03	0.54 ± 0.07	0.000000021
Compliance to adjuvant therapies[c]	–0.10 ± 0.04	–0.23 ± 0.03	–0.35 ± 0.06	0.0014
Nonsigificative				
Recent weight gain[d]	–0.17 ± 0.11		–0.22 ± 0.03	0.64 NS
Parents at home after school	–0.22 ± 0.05	–0.22 ± 0.04	–0.21 ± 0.09	0.91 NS
Dual parents households	–0.24 ± 0.05		–0.20 ± 0.04	0.50 NS
Social integration[e]	–0.21 ± 0.04	–0.23 ± 0.04	–0.24 ± 0.08	0.74 NS
Familial encouragement to leisure activities[f]	–0.19 ± 0.04	–0.26 ± 0.05	–0.33 ± 0.09	0.17 NS

Data are expressed as change in mean BMI–z–score ± SEM at 9 months of intervention. P > 0.05, significant. NS: not significant.

[a]*Familial encouragement to the project and child's motivation were assessed by the team using the involvement of the family and of the child in the project (Section 2).*

[b]*Adherence to treatment was assessed by the implementation of the decisions taken together (team and family).*

[c]*Compliance to adjuvant therapies means that the child and his family took part in psychotherapy or in physiotherapy as suggested by the team.*

[d]*Recent weight gain means a weight gain for less than 1 year.*

[e]*Social integration was assessed by the team using the participation of the child in extrascholar activities.*

[f]*Familial encouragement to leisure activities was assessed by the team according to activities realised by all the family out of home.*

The fall in BMI observed in our study may be clinically relevant as demonstrated by many studies, even though not analysed here. Short–term family–based treatment which combined nutrition education, behavioural modification and exercise was shown to improve body composition, lipids profile, blood pressure, and insulin resistance [4, 25, 26]. Many of the obesity–associated complications can be reversed with a 5% decrease in age–adjusted BMI percentile [27]. In adults with impaired glucose tolerance, the Diabetes Prevention Program demonstrated that an intensive lifestyle program that reduced body weight by 7% delayed or prevented the development of type 2 diabetes [28]. Moreover, Savoye et al. [13] reported that obese adolescents with impaired glucose tolerance who were able to limit increases in BMI reverted to normal glucose tolerance 2 years later. Thus, the BMI changes observed over time in our study are likely to be clinically significant as those changes were sustained over the longer term.

We determined parameters characterizing the families and children at baseline conditions which were associated with a better weight control. Indeed, for those less or not responsive patients, new research studies should try to devise new treatments to optimize long–term weight benefits. We demonstrated that those patients could be identified quickly according to the initial weight change observed between the first and the second visits. Tanaka et al. [29] also reported that a greater weight loss between the first and the second visits was a predictive factor in the success of the treatment. Goldschmidt et al. [30] also reported that early weight change seems to be related to treatment response through to the end of the treatment and also the 2–year followup. Identification of factors that promote early weight change is critical because modification of these factors before initiation of the treatment may promote a better early response.

In our study, similarly to Reinehr et al. [6], reduction of overweight was independent of initial BMI–z–score, age at the first visit, and sex. Preexisting regular physical activity contributed significantly to the early treatment response. It is well know that physical activity is related to long–term weight maintenance [31] but, as suggested in another study [32], our data supports its role before the initiation of the treatment. Children with large birth weight, gestational diabetes, no or short–term breastfeeding, parental obesity, asthma, and short sleep duration were described as having an increased risk of obesity [33, 34]. Our analysis suggested that those factors were not determinants for weight loss.

Baseline daily water and soda intake seemed to be a good predictor of early weight change. Consumption of sugar sweetened drinks by adolescents is an independent variable associated with increasing BMI [35], but its role on early weight loss was never examined. Healthy eating habits as eating breakfast and participating in programmed exercises were described to be correlated to healthful BMI, suggesting that these factors may be potentially protective against obesity in 12–16–year old adolescents. Our study extends those results by showing that prevention policy may also be helpful even for children who have to lose weight.

Our results demonstrated that motivated children given family encouragement were more likely to succeed in our treatment. Interestingly, recent reports suggested greater weight loss in obese children when parents alone are targeted for intervention [9], which emphasizes the role of the parents in the child's weight reduction. Moreover, some studies have analyzed the parent's weight changes during the treatment. Larger reductions in adult BMI were associated with more successful results, which indicates that working to enhance the adult role in child weight control programs may improve results [4]. Data from Rhee et al. [36] suggested that parents having an older child, believing that they themselves were overweight, perceiving that their child had a health problem were associated with greater parental readiness to make changes. Emerging research also indicates that overweight children with psychosocial problems or the occurrence of maternal psychopathology are less responsive to weight–control intervention over the longer term [5, 37]. There have been significant lifestyle changes in the family during the previous decades. The divorce rate has increased as well as the number of families with both parents working. Our data suggested that the dual parent households did not affect weight changes observed at 9 months [38]. This is in contrast with a recent study [39] which showed a relationship between single–parents status and excess weight in children. Further studies are needed to explore the dynamics of single–parent households and its influence on childhood diet and obesity. Interestingly, our study showed that a family with an only child may expect a greater weight loss. Other factors [4] such as higher incomes and higher level of education for the mother were also reported to be associated with better results but were not analyzed in the current study.

In conclusion, this study was a first step in determining whether weight loss was achievable with our interdisciplinary approach and highlighted potential success of a continuous care weight control program to lower BMI. An early weight change seems to be a marker for children's long–term treatment response. Preexisting regular physical activity, normal timing of puberty, baseline daily water and soda intake, motivation and some family characteristics predict the early response to the treatment. Better prevention policy and parental support may thus improve the success of the childhood obesity treatment. Our data may provide a better understanding of the factors involved in better weight control and may help to optimize/adapt our strategies for participants who do not benefit from treatment.

REFERENCES

1. H. Oude Luttikhuis, L. Baur, H. Jansen et al., "Interventions for treating obesity in children," Cochrane Database of Systematic Reviews, no. 1, Article ID CD001872, 2009.
2. E. P. Whitlock, E. A. O'Connor, S. B. Williams, T. L. Beil, and K. W. Lutz, "Effectiveness of weight management interventions in children: a targeted systematic review for the USPSTF," Pediatrics, vol. 125, no. 2, pp. e396–e418, 2010.
3. M. P. Kalavainen, M. O. Korppi, and O. M. Nuutinen, "Clinical efficacy of group–based treatment for childhood obesity compared with routinely given individual counseling," International Journal of Obesity, vol. 31, no. 10, pp. 1500–1508, 2007.
4. M. A. Kalarchian, M. D. Levine, S. A. Arslanian et al., "Family–based treatment of severe pediatric obesity: randomized, controlled trial," Pediatrics, vol. 124, no. 4, pp. 1060–1068, 2009.
5. E. Moens, C. Braet, and M. Van Winckel Myriam, "An 8–year follow–up of treated obese children: children's, process and parental predictors of successful outcome," Behaviour Research and Therapy, vol. 48, no. 7, pp. 626–633, 2010.
6. T. Reinehr, M. Temmesfeld, M. Kersting, G. De Sousa, and A. M. Toschke, "Four–year follow–up of children and adolescents participating in an obesity intervention program," International Journal of Obesity, vol. 31, no. 7, pp. 1074–1077, 2007.
7. L. H. Epstein and R. R. Wing, "Behavioral treatment of childhood obesity," Psychological Bulletin, vol. 101, no. 3, pp. 331–342, 1987.

8. T. J. Cole, J. V. Freeman, and M. A. Preece, "British 1990 growth reference centiles for weight, height, body mass index and head circumference fitted by maximum penalized likelihood," Statistics in Medicine, vol. 17, no. 4, pp. 407–429, 1998.
9. M. Golan, V. Kaufman, and D. R. Shahar, "Childhood obesity treatment: targeting parents exclusively v. parents and children," British Journal of Nutrition, vol. 95, no. 5, pp. 1008–1015, 2006.
10. H. Klar and W. L. Coleman, "Brief solution–focused strategies for behavioral pediatrics," Pediatric Clinics of North America, vol. 42, no. 1, pp. 131–141, 1995.
11. Y. Peterson, "Family therapy treatment: working with obese children and their families with small steps and realistic goals," Acta Paediatrica, International Journal of Paediatrics, Supplement, vol. 94, no. 448, pp. 42–44, 2005.
12. L. Edmunds, E. Waters, and E. J. Elliott, "Evidence based paediatrics: evidence based management of childhood obesity," BMJ, vol. 323, no. 7318, pp. 916–919, 2001.
13. M. Savoye, D. Berry, J. Dziura et al., "Anthropometric and psychosocial changes in obese adolescents enrolled in a Weight Management Program," Journal of the American Dietetic Association, vol. 105, no. 3, pp. 364–370, 2005.
14. L. H. Epstein, M. D. Myers, H. A. Raynor, and B. E. Saelens, "Treatment of pediatric obesity," Pediatrics, vol. 101, no. 3, pp. 554–570, 1998.
15. J. M. Tanner and R. H. Whitehouse, "Clinical longitudinal standards for height, weight, height velocity, weight velocity, and stages of puberty," Archives of Disease in Childhood, vol. 51, no. 3, pp. 170–179, 1976.
16. P. McCullagh, "Quasi–likelihood functions," The Annals of Statistics, vol. 11, pp. 59–67, 1983.
17. N. R. Chaganty and J. Shults, "On eliminating the asymptotic bias in the quasi–least squares estimate of the correlation parameter," Journal of Statistical Planning and Inference, vol. 76, no. 1–2, pp. 145–161, 1999.
18. J. G. Morel, M. C. Bokossa, and N. K. Neerchal, "Small sample correction for the variance of GEE estimators," Biometrical Journal, vol. 45, no. 4, pp. 395–409, 2003.
19. P. H. C. Eilers and B. D. Marx, "Flexible smoothing with B–splines and penalties," Statistical Science, vol. 11, no. 2, pp. 89–121, 1996.
20. M. P. Wand and J. T. Ormerod, "On semiparametric regression with O'Sullivan penalized splines," Australian and New Zealand Journal of Statistics, vol. 50, no. 2, pp. 179–198, 2008.
21. J. P. Chanoine, S. Hampl, C. Jensen, M. Boldrin, and J. Hauptman, "Effect of orlistat on weight and body composition in obese adolescents: a randomized controlled trial," JAMA, vol. 293, no. 23, pp. 2873–2883, 2005.
22. D. E. Wilfley, R. I. Stein, B. E. Saelens et al., "Efficacy of maintenance treatment approaches for childhood overweight: a randomized controlled trial," JAMA, vol. 298, no. 14, pp. 1661–1673, 2007.
23. R. R. Wing, E. Blair, M. Marcus, L. H. Epstein, and J. Harvey, "Year–long weight loss treatment for obese patients with type II diabetes: does including an intermittent very–low–calorie diet improve outcome?" American Journal of Medicine, vol. 97, no. 4, pp. 354–362, 1994.

24. L. Pinelli, N. Elerdini, M. S. Faith et al., "Childhood obesity: results of a multicenter study of obesity treatment in Italy," Journal of Pediatric Endocrinology and Metabolism, vol. 12, no. 3, pp. 795–799, 1999.
25. M. Savoye, M. Shaw, J. Dziura et al., "Effects of a weight management program on body composition and metabolic parameters in overweight children: a randomized controlled trial," JAMA, vol. 297, no. 24, pp. 2697–2704, 2007.
26. D. Nemet, S. Barkan, Y. Epstein, O. Friedland, G. Kowen, and A. Eliakim, "Short– and long–term beneficial effects of a combined dietary–behavioral– physical activity intervention for the treatment of childhood obesity," Pediatrics, vol. 115, no. 4, pp. e443–e449, 2005.
27. T. Reinehr and W. Andler, "Changes in the atherogenic risk factor profile according to degree of weight loss," Archives of Disease in Childhood, vol. 89, no. 5, pp. 419–422, 2004.
28. W. C. Knowler, E. Barrett–Connor, S. E. Fowler et al., "Reduction in the incidence of type 2 diabetes with lifestyle intervention or metformin," The New England Journal of Medicine, vol. 346, no. 6, pp. 393–403, 2002.
29. S. Tanaka, M. Yoshinaga, K. Sameshima et al., "Predictive factors in the success of intervention to treat obesity in elementary school children," Circulation Journal, vol. 69, no. 2, pp. 232–236, 2005.
30. A. B. Goldschmidt, R. I. Stein, B. E. Saelens, K. R. Theim, L. H. Epstein, and D. E. Wilfley, "Importance of early weight change in a pediatric weight management trial," Pediatrics, vol. 128, no. 1, pp. e33–e39, 2011.
31. K. Elfhag and S. Rössner, "Who succeeds in maintaining weight loss? A conceptual review of factors associated with weight loss maintenance and weight regain," Obesity Reviews, vol. 6, no. 1, pp. 67–85, 2005.
32. T. Reinehr, K. Brylak, U. Alexy, M. Kersting, and W. Andler, "Predictors to success in outpatient training in obese children and adolescents," International Journal of Obesity, vol. 27, no. 9, pp. 1087–1092, 2003.
33. J. J. Reilly, J. Armstrong, A. R. Dorosty et al., "Early life risk factors for obesity in childhood: cohort study," BMJ, vol. 330, no. 7504, article 1357, 2005.
34. H. Fiore, S. Travis, A. Whalen, P. Auinger, and S. Ryan, "Potentially protective factors associated with healthful body mass index in adolescents with obese and non-obese parents: a secondary data analysis of the third national health and nutrition examination survey, 1988–1994," Journal of the American Dietetic Association, vol. 106, no. 1, pp. 55–64, 2006.
35. D. S. Ludwig, K. E. Peterson, and S. L. Gortmaker, "Relation between consumption of sugar–sweetened drinks and childhood obesity: a prospective, observational analysis," The Lancet, vol. 357, no. 9255, pp. 505–508, 2001.
36. K. E. Rhee, C. W. DeLago, T. Arscott–Mills, S. D. Mehta, and R. K. Davis, "Factors associated with parental readiness to make changes for overweight children," Pediatrics, vol. 116, no. 1, pp. e94–e101, 2005.
37. A. B. Goldschmidt, M. M. Sinton, V. P. Aspen et al., "Psychosocial and familial impairment among overweight youth with social problems," International Journal of Pediatric Obesity, vol. 5, no. 5, pp. 428–435, 2010.

38. L. H. Epstein, R. A. Paluch, J. N. Roemmich, and M. D. Beecher, "Family–based obesity treatment, then and now: twenty–five years of pediatric obesity treatment," Health Psychology, vol. 26, no. 4, pp. 381–391, 2007.
39. F. G. Huffman, S. Kanikireddy, and M. Patel, "Parenthood–a contributing factor to childhood obesity," International Journal of Environmental Research and Public Health, vol. 7, no. 7, pp. 2800–2810, 2010.

CHAPTER 13

INTERVENING TO REDUCE SEDENTARY BEHAVIORS AND CHILDHOOD OBESITY AMONG SCHOOL–AGE YOUTH: A SYSTEMATIC REVIEW OF RANDOMIZED TRIALS

MAY MAY LEUNG, ALEN AGARONOV, KATERYNA GRYTSENKO, and MING–CHIN YEH

13.1 INTRODUCTION

Childhood obesity has long been recognized as a worldwide growing health concern [1–3]. In the past 2 decades, rates of obesity in the US rose among children aged 6 to 11 years from 11.3% to 19.6%, as well as from 10.5% to 18.1% among adolescents aged 12 to 19 years [4, 5]. Similarly, Great Britain has experienced a threefold increase of overweight in children between 1984 and 2002 [6], and prevalence of obesity among younger children in China has increased from 1.5% to 12.6% between 1989 and 1997 [7]. Early consequences of childhood obesity include asthma, hypertension, and early–onset diabetes mellitus [3]. In addition, childhood obesity has been shown to follow into adulthood [8–11] and may lead to cardiovascular disease, cancer, and an increased chance of mortality after the age of 30 years [12, 13].

A majority of previous studies addressing this epidemic have revolved around modifying dietary intake [14, 15] and physical activity (PA) [16–18]. However, sedentary behavior (SB) appears to be a lifestyle behavior that is increasingly contributing to the prevalence of childhood obesity [19] as research has shown that obese children are more sedentary than their nonobese counterparts [20]. Sedentary behavior largely consists of media use; however, other behaviors that do not expend significant energy, such as attending classes or playing a musical instrument, have been explored as SB [21–23]. It is estimated that children spend approximately one–third of their waking hours using media, which includes watching TV/videos, playing video games, and personal computing [24]. These SB may in turn displace PA, decrease metabolic rate, and/or serve as a conditioned stimulus for eating [25].

Lifestyle interventions aimed at reducing SB have potential to make an impact; however, limited knowledge exists as to the effectiveness of such interventions. In addition, aspects related to an intervention's potential for translation to practice are important to consider for such a significant public health issue as childhood obesity. The main objective of this paper is to assess the effectiveness of interventions that focus on reducing SB among school–age youth. A second objective is to identify the elements of the identified interventions related to potential translation to practice settings, such as cost or health disparity implications and sustainability of intervention impact.

13.2 METHODS

13.2.1 LITERATURE SEARCH

Four databases (Medline, PubMed, PsychInfo, Cochrane Library) were searched for the relevant studies published between 1980 and April 2011. For this paper, such keywords as "sedentary behavior," "sedentary lifestyle," "physical inactivity," "television," "video games," "children," "adolescents," and "intervention" were used alone and/or in combination. Relevant references were extracted and examined, compiling the list in the form of titles and abstracts of the selected studies.

13.2.2 INCLUSION CRITERIA

Identified studies included those that used an intervention aimed at decreasing SB, separately or in combination with body mass index (BMI) or other anthropometric changes, such as waist circumference or triceps skinfold thickness, among children and adolescents, 6 to 19 years of age. We focused on studies that described randomized trials, conducted in the community, school, home, or clinic setting, which lasted at least 12 weeks, and included such strategies as educational, health promotion, behavioral therapy, counseling, or management strategies at the individual and family levels. Studies whose primary goal was to measure changes in PA levels were included if the change in SB was also measured and specified in the results. Sedentary behavior was defined as media–related behavior (time spent watching TV/videotapes, playing video games), breaks from activity, and activities that do not significantly influence the energy expenditure occurring at rest.

13.2.3 EXCLUSION CRITERIA

Searches were conducted only in the English language. Studies based within a controlled laboratory setting were not considered relevant or generalizable, and therefore, not included in the analysis.

13.2.4 SELECTION PROCESS

The results of the preliminary search were reviewed; relevant titles with abstracts were then retrieved. Bibliographies of some systematic review papers were reviewed to identify additional studies. Full articles of relevant abstracts were retrieved for further review. Two authors independently assessed retrieved studies for inclusion based upon the criteria listed above. Any inconsistencies were resolved by discussions with the other author. Summary tables were composed of the selected studies. The tables included study design, setting in which it was conducted, theory, characteristics of the participants, duration of the intervention and followup, brief description of intervention, definition of control group, measures of SB and additional outcomes, key findings, demographic disparities information, and limitations.

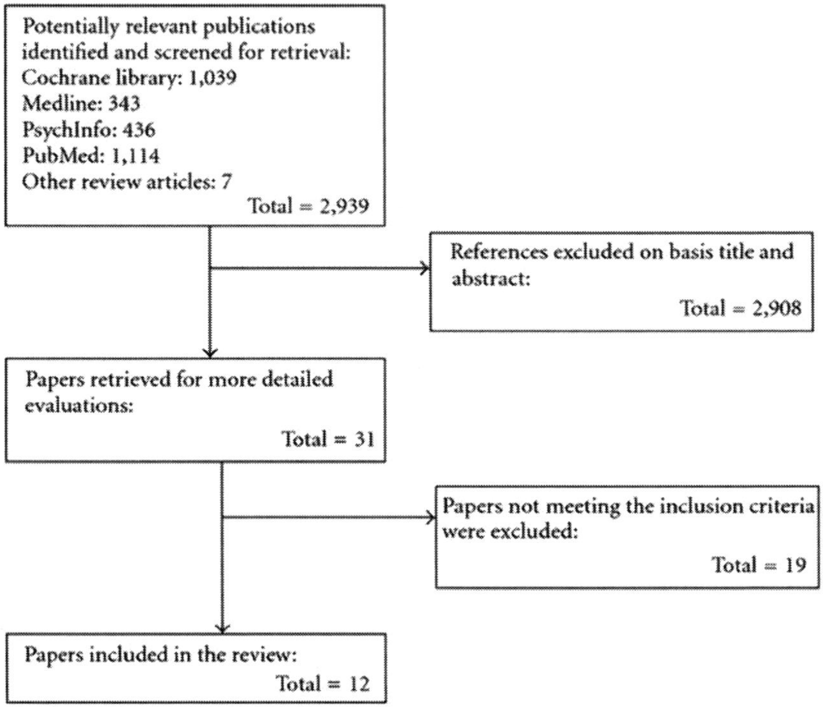

FIGURE 1: Flow chart of the search process.

13.3 RESULTS

A total of 2.939 abstracts were identified through the initial search process. Upon review, 31 full papers were retrieved for further review by two investigators. Of those 31 papers, 12 studies met the inclusion criteria. Figure 1 outlines the flow of the search process and the number of articles that were identified at each stage of the process.

Three studies [22, 23, 26] focused only on SB, 1 study was a PA intervention [27], 6 studies [20, 21, 28–31] were combined SB and PA interventions, and 2 studies [32, 33] targeted SB, PA, and diet. Of the 12 studies, 8 were conducted in the US, 3 in Europe (including the UK, France, and the Netherlands), and 1 in Australia. The majority (7 out of 12) of the studies were conducted in a school setting, while 2 were conducted in a clinic, 1 in community centers, 1 conducted in both community centers and schools, and 1 other was carried out in convenient locations, which included clinics, libraries, and schools.

The definition of SB varied across the studies. Listed here are all the forms of SB that were measured: time spent watching TV and videotapes, playing video games, doing homework, reading, listening to music, using a computer, playing a musical instrument, doing artwork or crafts, talking with parents, playing quiet games indoors, and attending classes or club meetings. Due to the diversity in study design, study duration, setting, population, and measurement outcomes of the interventions, a quantitative synthesis of the evidence was not possible. Therefore, a qualitative assessment of the current evidence stratified by targeted behaviors is presented.

Three studies [22, 23, 26] focused on reducing SB in school–aged children. Escobar–Chaves et al. [26] aimed to reduce TV and other media consumption in families with children of ages 6 to 9 years in Houston, Tex, US. One hundred one families were randomized to either the 6–month intervention, which included a 2–hour workshop and 6 bimonthly newsletters, or a control group. The parents and children also worked together to develop a plan in which alternative activities could be done by the child and family in place of SB. At 6–month followup, there was a trend toward reducing media consumption in the intervention group; however, these results were not statistically significant. The intervention did find a positive impact on proxy behaviors hypothesized to lead to media use reductions, which are also recommended by the American Academy of Pediatrics, such as not having a TV in the child's bedroom.

Robinson [22] randomly assigned 3rd and 4th graders in 1 of 2 public elementary schools in San Jose, Calif, US to receive an 18–lesson, 6–month classroom curriculum to reduce TV, videotape, and video game use. The curriculum, which was taught by the regular classroom teachers, included self–monitoring and self–reporting of media use, followed by a TV turnoff, in which children were challenged not to use media for 10 days. After the turnoff challenge, the children were encouraged to follow a 7–hour/week budget of media use. Each household also received an electronic TV time manager, which monitored and bugeted TV/video use for each household member. Newsletters designed to motivate parents to help their children maintain their TV watching limits were also distributed. At the end of the intervention, children in the intervention group had decreases in multiple anthropometric measures, which included BMI, triceps skinfold thickness, waist circumference, and waist to hip ratio $P < 0.002$, compared to the control group. In addition, reported TV use was lower in the intervention group (8.80 versus 14.46 hours/week; $P < 0.001$); however, no significant changes were reported in video tape and video game use.

Another study conducted by Robinson and borzekowski [23] consisted of a randomized controlled trial among 3rd and 4th graders in San Jose, Calif, US in 2 public elementary schools (n = 181). The intervention was an 18–lesson classroom curriculum focused on reducing screen media exposure. Components of the intervention included children becoming aware of the role TV, videotapes, and video games play in their lives, a TV turnoff in which children attempted to watch no TV/videotapes or play video games for 10 days, children learning how to budget their media use, and participants helping their peers at another school to reduce their media use. Newsletters were also distributed to the parents. Children in the intervention school significantly decreased their weekday TV viewing (1.14 versus 1.96 hours/day; $P < 0.001$) and weekday (0.19 versus 0.52 hours/day; $P < 0.05$) and Saturday video game playing (0.31 versus 0.9 hours/day; $P < 0.05$) compared to controls. Greater effects were found among boys ($P = 0.05$) and more adult–supervised children ($P = 0.03$).

13.3.2 PHYSICAL ACTIVITY STUDY

One study that focused solely on PA in school–aged children was identified. Slootmaker et al. [27] randomized 87 13–to–17 year olds in Amsterdam, The Netherlands to receive either a single brochure with PA recommendations or an accelerometer and access to web–based tailored PA advice for 3 months. When a user logged into the website and uploaded his/her PA score, the website provided individualized PA feedback based on the current PA score and personally adapted suggestions to promote daily PA. At 5–month followup, time spent doing SB was significantly reduced in boys (−1,801 minutes/week; $P = 0.04$). No SB changes were observed in girls.

13.3.3 SEDENTARY BEHAVIOR AND PHYSICAL ACTIVITY STUDIES

Six studies [20, 21, 28–31] that targeted both SB and PA were identified. Epstein et al. [21] randomized obese children of ages 8 to 12 years from 61 families to 1 of 3 treatment groups: (1) increasing exercise (Exercise), (2) decreasing SB (Sedentary), or (3) both increasing exercise and decreasing SB (Combined). All groups received similar information (distributed through manuals) about the benefits of increased PA and the negative effects of SB; however, the groups differed in the types of activities that were reinforced. The Sedentary group was reinforced for decreasing the amount of time they engaged in certain SB; these SBs included media use, imaginative play, talking on the phone, and playing board games. Participants in the Exercise group were reinforced for increasing PA, while those in the Combined group were reinforced for both decreasing SB and increasing PA. Weekly treatment meetings were also conducted for both the parent and child. At 6–month followup, the Sedentary group had greater decrease in percentage overweight than did the Exercise or Combined groups (−18.7 versus −10.3 versus −8.7; $P = 0.026$) and greater decrease in percentage of body fat (−4.7 versus −1.3; $P = 0.037$).

Another study by Epstein et al. 2001 [20] randomly assigned 67 families with an obese child between ages of 8 to 12 years to 1 of 2 treatment groups: (1) increasing PA (Increase) or (2) reducing SB and increasing PA (Combined). The treatment program consisted of 16 weekly meetings, followed by 2 biweekly meetings and 2 monthly meetings during a 6–month intensive program. At 6–month followup, boys showed significantly better percentage of overweight changes in the Combined group than girls (−15.8% versus −1.0%; $P < 0.001$), with no significant differences in the Increase group for boys or girls (−9.3% versus −7.6%). Boys also adhered to the treatment better than girls ($P < 0.01$).

Jones et al. [28] recruited 12 middle schools in central Texas to participate in a 1.5–year randomized clinical trial focused on improving bone health mainly through promoting the increase of PA. A total of 718 6th grade girls participated in the intervention, which consisted of a 16–session health curriculum to promote increased weight–bearing PA and consumption of calcium–rich foods. A physical education component was also included, which consisted of high–impact activities. Relative to the girls in the control group, the intervention group significantly reduced daily TV and video minutes (−12.11 minutes/day;). Total daily minutes of sedentary activity were significantly lower for intervention students relative to controls (mean difference between groups = −17 minutes; $P = 0.04$).

Robinson et al. [29] conducted a randomized controlled trial with 61 8–to–10–year–old African–American (AA) girls and their parents. The 12–week intervention consisted of after–school dance classes and a 5–lesson family–based intervention delivered in participants' homes to reduce media use. At followup, the girls in the intervention group had trends towards lower BMI (adjusted difference = −0.32 kg/m2; 95% CI −0.77 to 0.12) and waist circumference (adjusted difference = −0.63 cm; 95% CI −1.92 to 0.67) and reduced TV, videotape, and video game use (adjusted difference = −4.96 hours/week; 95% CI −11.41 to 1.49).

Salmon et al. [30] randomized, by class, 311 children from 3 government schools in low socioeconomic areas of Melbourne, Australia into one of four conditions: (1) behavioral modification (BM); (2) fundamental movement skills (FMS); (3) combined BM and FMS (BM/FMS); (4) control (usual curriculum). Each of the intervention conditions consisted of 19 lessons promoting PA and decreasing SB. The BM lessons were delivered in the classroom, while the FMS lessons were delivered in PA facilities, which focused on teaching participants physical skills while emphasizing enjoyment and fun. The combined group received both the BM and FMS lessons. There was a significant intervention effect from baseline to postintervention on BMI in the BM/FMS group compared to the control group ($-1.88\,kg/m^2$; $P < 0.01$), which was maintained at 6– and 12–month followup ($-1.53\,kg/m^2$; $P < 0.05$). The BM group reported highest levels of TV viewing compared to the other groups (239.9 minutes/week; $P < 0.05$).

Simon et al. [31] conducted a 4–year randomized controlled trial with a cohort of 954 middle–school adolescents in eastern France. The multilevel intervention focused on influencing intrapersonal, social, and environmental determinants of PA and SB through informational sessions, social support by parents, peers, teachers, and PA instructors and by providing environmental conditions for PA to encourage students to apply the knowledge and skills they learned. The study is currently on going; thus, data reported here were collected 6 months into the intervention. After 6 months of the intervention, high SB (<3 hours/day) was reduced in both girls and boys (OR = 0.54 and 0.52; $P < 0.001$) in the intervention group compared to the control.

13.3.4 SEDENTARY BEHAVIOR, PHYSICAL ACTIVITY, AND DIET STUDIES

Two studies [32, 33] focused on modifying SB, PA, and diet. Gortmaker et al. [32] randomized 5 out of 10 middle schools in Massachusetts to receive an interdisciplinary intervention over the course of 2 school years. The intervention, Planet Health, was included in the existing school curriculum of 4 subjects and physical education classes. The sessions focused on decreasing TV viewing, decreasing consumption of high–fat foods, increasing fruit and vegetable intake, and increasing moderate to vigorous PA. Over the 2–year intervention period, obesity prevalence among girls in the intervention schools decreased compared to controls (OR = 47; $P =$ 0.03), while no differences were observed in boys. The number of hours of TV/video use was reduced in both boys and girls in the intervention group compared to the control group (adjusted difference between groups for boys and girls = −0.40 and −0.58 hours/day; $P < 0.001$).

Sacher et al. [33] recruited 116 obese children in the UK to be randomly assigned to receive the Mind, Exercise, Nutrition, Do it (MEND) program, a multicomponent community–based intervention. This intervention consisted of 18 2–hour group educational and PA sessions held twice weekly in sports centers and schools, in which both parents and children attended. These sessions were followed by a 12–week free family swimming pass. At 6 months, participants in the intervention group had a reduced waist circumference *z*–score (−0.37; $P < 0.0001$) and BMI *z*–score (−0.24; $P <$ 0.0001) compared to controls. Significant differences in SB were observed between the intervention and control groups (15.9 versus 21.7 hours/week; $P = 0.01$). The significant decreases in waist circumference and BMI in the intervention group were sustained up to 9 months after participants completed the educational and PA sessions.

13.4 DISCUSSION

Overall, interventions that focused on decreasing SB, whether alone or in combination with other strategies, such as increasing PA and improving diet, were associated with reduction in time spent on SB and/or improvements in anthropometric measurements related to childhood obesity.

13.4.1 STUDY DESIGN

While the results of the majority of the studies were positive, it is not possible to make any conclusions as to the degree of impact each strategy had on the outcomes due to the variability in study design and outcome measurements. There were only 3 studies [22, 23, 26] that focused solely on the reduction of SB, and only 1 of those studies [22] collected anthropometric measures. The other 9 studies combined other strategies, such as exercise and healthy eating. Similar results in relation to anthropometric measures and SB were observed in these studies compared to the studies solely focused on reducing SB.

Another aspect of the study designs that made it challenging to interpret any further than a qualitative summary is the variation in how SB was defined. Some studies examined SB as only media use, while others collected additional measurements, which included behaviors such as talking with parents, playing quiet indoor games, and attending clubs, in addition to media use [21–23]. More consistent measures of different types of SB across studies would assist in determining their relative impact on childhood obesity.

The intensity and dose of the interventions received by participants also varied between interventions. The duration of study periods ranged from 12 weeks to 4 academic years. In addition, some interventions consisted of a workshop and newsletters, while other interventions consisted of multiple lessons and face–to–face encounters with the participants across similar time periods.

Another challenge when assessing impact, particularly when considering potential for translation into practice, was the limited measures of long–term sustainability of the interventions impact. Only 5 out of the 12 identified studies incorporated postintervention follow–up measures [20, 21, 26, 27, 30, 33], which ranged from 5 to 12 months. Overall, a positive long–term impact was observed in either behavioral or anthropometric outcomes in those 5 studies [20, 21, 26, 27, 30, 33]. However, this highlights the challenges in interpreting impact of the interventions and their potential for translation into real–world settings.

13.4.2 COMMON COMPONENTS

One intervention component that appeared to be repeated in several of the designs was the involvement of family. Whether it was for a clinic–based treatment for obese children or promoting positive behaviors to prevent childhood obesity, parents were engaged to varying degrees. In some interventions, the parents were mailed newsletters to reiterate health messages that were presented to children in school [22, 23], while other interventions included having the parent attend workshops/meetings with their children and share in planning healthy events [20, 21, 26, 29, 31, 33]. In one study [23], children who had greater adult supervision were more likely to respond better to the intervention than less supervised children. These study designs and results highlight the importance of having a supportive family environment to promote the positive behaviors that are being targeted.

Another component that was repeated throughout many of the intervention designs was that the children were provided with tangible ideas and appealing alternatives to sedentary activities, and some had the opportunity to choose how to allocate their time [21, 26, 29–31]. When children are provided choice among alternative activities, they may perceive increased control over their activity options, so the reduction in SB observed in the studies could be partly explained by the provision of suggestions, ideas, and options for students.

13.4.3 DEMOGRAPHIC DISPARITIES

Four of the studies [20, 23, 27, 32] reported differential effects of the intervention between genders while measuring SB outcomes or anthropometric changes. The impact was inconsistent across the studies. The gender differences were observed in both a family–based weight control treatment [20] and school–based interventions to prevent obesity [20, 23, 27, 32] targeting a range of ages. Two of the studies [23, 27] observed a greater effect on boys' SB, while the Epstein et al. [20] study resulted in greater changes in % overweight in boys compared to girls. On the other hand, the Gortmaker et al. [32] study observed BMI changes in girls, but not boys.

There is no clear explanation as to the differential effects by gender and also why the results were inconsistent across intervention, especially since the interventions were originally designed to reach both males and females. Some suggest that gender differences may vary or become more obvious as children become adolescents, with hormonal and environmental differences between sexes emerging at that challenging stage of development [34–36]; however, gender differences were observed with children as young as 8 years. The differential results may suggest that mediators for SB or anthropometric changes may be different between males and females; thus, future interventions may need to be tailored specific to gender.

Obesity rates are disproportionate across the ethnicities and socioeconomic status (SES) groups. Reducing such inequalities in childhood obesity is imperative. Some of the studies did address such disparities by either specifically designing interventions to reach certain at–risk populations, such as AA girls or schools in low–SES areas [29, 30], or by evaluating results across race/ethnicity or SES groups. However, such study designs and data analysis were limited, warranting further interventions to focus on specifically addressing such inequalities.

13.4.4 COSTS

Understanding the costs related to recruitment and implementation of an intervention and its potential cost effectiveness are important aspects to consider when a health practitioner must determine how best to utilize the often–limited resources that are available in community or school settings. In this systematic paper, we aimed to collect any evidence related to cost of the interventions. While there is a need to understand cost–related issues of interventions, unfortunately, as reported in other publications [37], data on cost of the interventions identified for this paper were very limited. Measuring costs related to the different stages of the research process should be incorporated into study designs, and such data should be included when reporting intervention effects.

13.4.5 LIMITATIONS

There were several limitations to the paper. Similar to other papers, this systematic paper is limited by the quantity and quality of the studies that were identified. A qualitative analysis of the evidence was warranted due to the variations in study design and characteristics, including intervention and follow–up duration, strategies used, population, and measurement outcomes. Measurements of SB were mainly self–report; however, to minimize this potential bias, some studies did use measures with high validity and reliability. In addition, the majority of the studies were conducted in the US, which may limit the generalizability to other countries, where cultural values and behavioral patterns of SB may differ.

13.5 IMPLICATIONS

13.5.1 FOR FUTURE RESEARCH

This systematic paper highlights the need for future research to further explore the reduction of SB in relation to preventing and treating childhood obesity. More comprehensive study designs, which include postintervention follow–up measures, are warranted to better understand the impact and potential sustainability of different strategies on outcomes measures related to SB and anthropometry. Additionally, as SB data were mainly self–report, more valid and reliable measures of SB should be developed. Furthermore, addressing childhood obesity inequalities related to race/ethnicity, SES and gender need to be further explored and should be incorporated into the design of future interventions. In addition, a review on cost of the interventions was not possible due to the paucity of available data, thus collecting data related to cost would provide more comprehensive data for public health practitioners to allow them to determine which interventions may be most effective in their settings.

13.5.2 FOR PUBLIC HEALTH

Many of these interventions, while comprehensive, were designed to be incorporated into the regular school classroom with teachers delivering the lessons. Others were designed to be implemented in convenient locations within communities, and sessions could be led by those without extensive health training or education. One study [27] specifically mentioned that the intervention was designed to make it easily applicable to real–life settings. These study designs point to the important consideration of the often–challenging aspect of feasibility when implementing interventions in real–world settings and highlight interventions that may have a "true public health impact" [38] as behavioral science research must be "contextual" and "practical" [39].

A very limited number of the studies focused on interventions that modified school policies and the physical environment in ways that support improved dietary practices and regular PA. Often such interventions are not candidates for reviews because of their limited outcome measures on specific behaviors or weight–related outcomes. However, such strategies are gaining support and have the potential to make a significant and sustainable impact [40].

In conclusion, interventions aimed at reducing SB appear to be effective in decreasing SB and improvements in anthropometric measures of childhood obesity. In addition, several of the studies did consider elements of feasibility and applicability in real–world settings to increase potential translation of research interventions into practice settings. Childhood obesity is a complex epidemic with various contributing factors at multiple levels. To make an impact on reversing the trends, a combined effort of strategies that address multiple determinants, including SB, across multiple settings, such as the school, community, clinic, and household is needed.

REFERENCES

1. K. Silventoinen, S. Sans, H. Tolonen et al., "Trends in obesity and energy supply in the WHO MONICA Project," International Journal of Obesity, vol. 28, no. 5, pp. 710–718, 2004.
2. Y. Wang, C. Monteiro, and B. M. Popkin, "Trends of obesity and underweight in older children and adolescents in the United States, Brazil, China, and Russia," American Journal of Clinical Nutrition, vol. 75, no. 6, pp. 971–977, 2002.
3. C. Gonzalez–Suarez, A. Worley, K. Grimmer–Somers, and V. Dones, "School–based interventions on childhood obesity: a meta–analysis," American Journal of Preventive Medicine, vol. 37, no. 5, pp. 418–427, 2009.
4. C. L. Ogden, M. D. Carroll, L. R. Curtin, M. M. Lamb, and K. M. Flegal, "Prevalence of high body mass index in US children and adolescents, 2007–2008," JAMA: Journal of the American Medical Association, vol. 303, no. 3, pp. 242–249, 2010.
5. C. L. Ogden, K. M. Flegal, M. D. Carroll, and C. L. Johnson, "Prevalence and trends in overweight among US children and adolescents, 1999–2000," Journal of the American Medical Association, vol. 288, no. 14, pp. 1728–1732, 2002.
6. K. L. Rennie and S. A. Jebb, "National prevalence of obesity: prevalence of obesity in Great Britain," Obesity Reviews, vol. 6, no. 1, pp. 11–12, 2005.
7. J. Luo and F. B. Hu, "Time trends of obesity in pre–school children in China from 1989 to 1997," International Journal of Obesity, vol. 26, no. 4, pp. 553–558, 2002.
8. S. S. Guo and W. C. Chumlea, "Tracking of body mass index in children in relation to overweight in adulthood," American Journal of Clinical Nutrition, vol. 70, no. 1, pp. 145S–148S, 1999.
9. M. K. Serdula, D. Ivery, R. J. Coates, D. S. Freedman, D. F. Williamson, and T. Byers, "Do obese children become obese adults? A review of the literature," Preventive Medicine, vol. 22, no. 2, pp. 167–177, 1993.
10. T. J. Parsons, C. Power, S. Logan, and C. D. Summerbell, "Childhood predictors of adult obesity: a systematic review," International Journal of Obesity, vol. 23, supplement 8, pp. S1–S107, 1999.
11. L. E. Thorpe, D. G. List, T. Marx, L. May, S. D. Helgerson, and T. R. Frieden, "Childhood obesity in New York City elementary school students," American Journal of Public Health, vol. 94, no. 9, pp. 1496–1500, 2004.

12. National Institutes of Health, "Clinical Guidelines on the identification, evaluation, and treatment of overweight and obesity in adults—the evidence report," Obesity Research, vol. 6, supplement 2, pp. 51S–209S, 1998.
13. V. Burke, L. J. Beilin, D. Dunbar, and M. Kevan, "Associations between blood pressure and overweight defined by new standards for body mass index in childhood," Preventive Medicine, vol. 38, no. 5, pp. 558–564, 2004.
14. S. Amaro, A. Viggiano, A. Di Costanzo et al., "Kalèdo, a new educational board–game, gives nutritional rudiments and encourages healthy eating in children: a pilot cluster randomized trial," European Journal of Pediatrics, vol. 165, no. 9, pp. 630–635, 2006.
15. A. S. Ask, S. Hernes, I. Aarek, G. Johannessen, and M. Haugen, "Changes in dietary pattern in 15 year old adolescents following a 4 month dietary intervention with school breakfast—a pilot study," Nutrition Journal, vol. 5, no. 1, article 33, 2006.
16. A. L. Carrel, R. R. Clark, S. E. Peterson, B. A. Nemeth, J. Sullivan, and D. B. Allen, "Improvement of fitness, body composition, and insulin sensitivity in overweight children in a school–based exercise program: a randomized, controlled study," Archives of Pediatrics and Adolescent Medicine, vol. 159, no. 10, pp. 963–968, 2005.
17. N. Lazaar, J. Aucouturier, S. Ratel, M. Rance, M. Meyer, and P. Duché, "Effect of physical activity intervention on body composition in young children: influence of body mass index status and gender," Acta Paediatrica, vol. 96, no. 9, pp. 1315–1320, 2007.
18. A. L. Liu, X. Q. Hu, G. S. Ma et al., "Report on childhood obesity in China (6) evaluation of a classroom–based physical activity promotion program," Biomedical and Environmental Sciences, vol. 20, no. 1, pp. 19–23, 2007.
19. J. N. Roemmich, C. M. Gurgol, and L. H. Epstein, "Open–Loop Feedback Increases Physical Activity of Youth," Medicine and Science in Sports and Exercise, vol. 36, no. 4, pp. 668–673, 2004.
20. L. H. Epstein, R. A. Paluch, and H. A. Raynor, "Sex differences in obese children and siblings in family–based obesity treatment," Obesity Research, vol. 9, no. 12, pp. 746–753, 2001.
21. L. H. Epstein, A. M. Valoski, L. S. Vara et al., "Effects of decreasing sedentary behavior and increasing activity on weight change in obese children," Health Psychology, vol. 14, no. 2, pp. 109–115, 1995.
22. T. N. Robinson, "Reducing children's television viewing to prevent obesity: a randomized controlled trial," Journal of the American Medical Association, vol. 282, no. 16, pp. 1561–1567, 1999.
23. T. N. Robinson and D. L. G. Borzekowski, "Effects of the SMART classroom curriculum to reduce child and family screen time," Journal of Communication, vol. 56, no. 1, pp. 1–26, 2006.
24. S. L. Escobar–Chaves and C. A. Anderson, "Media and risky behaviors," Future of Children, vol. 18, no. 1, pp. 147–180, 2008.
25. L. DeMattia, L. Lemont, and L. Meurer, "Do interventions to limit sedentary behaviours change behaviour and reduce childhood obesity? A critical review of the literature," Obesity Reviews, vol. 8, no. 1, pp. 69–81, 2007.

26. S. L. Escobar–Chaves, C. M. Markham, R. C. Addy, A. Greisinger, N. G. Murray, and B. Brehm, "The Fun Families study: intervention to reduce children's TV viewing," Obesity, vol. 18, no. 1, pp. S99–S101, 2010.
27. S. M. Slootmaker, M. J. M. Chinapaw, J. C. Seidell, W. van Mechelen, and A. J. Schuit, "Accelerometers and Internet for physical activity promotion in youth? Feasibility and effectiveness of a minimal intervention [ISRCTN93896459]," Preventive Medicine, vol. 51, no. 1, pp. 31–36, 2010.
28. D. Jones, D. M. Hoelscher, S. H. Kelder, A. Hergenroeder, and S. V. Sharma, "Increasing physical activity and decreasing sedentary activity in adolescent girls—the Incorporating More Physical Activity and Calcium in Teens (IMPACT) study," International Journal of Behavioral Nutrition and Physical Activity, vol. 5, article 42, 2008.
29. T. N. Robinson, J. D. Killen, H. C. Kraemer et al., "Dance and reducing television viewing to prevent weight gain in African–American girls: the Stanford GEMS pilot study," Ethnicity and Disease, vol. 13, no. 1, pp. S65–S77, 2003.
30. J. Salmon, K. Ball, C. Hume, M. Booth, and D. Crawford, "Outcomes of a group–randomized trial to prevent excess weight gain, reduce screen behaviours and promote physical activity in 10–year–old children: Switch–Play," International Journal of Obesity, vol. 32, no. 4, pp. 601–612, 2008.
31. C. Simon, A. Wagner, C. DiVita et al., "Intervention centred on adolescents' physical activity and sedentary behaviour (ICAPS): concept and 6–month results," International Journal of Obesity, vol. 28, supplement 3, pp. S96–S103, 2004.
32. S. L. Gortmaker, K. Peterson, J. Wiecha et al., "Reducing obesity via a school–based interdisciplinary intervention among youth: Planet Health," Archives of Pediatrics and Adolescent Medicine, vol. 153, no. 4, pp. 409–418, 1999.
33. P. M. Sacher, M. Kolotourou, P. M. Chadwick et al., "Randomized controlled trial of the MEND program: a family–based community intervention for childhood obesty," Obesity, vol. 18, supplement 1, pp. S62–S68, 2010.
34. J. N. Roerimich and A. D. Rogol, "Hormonal changes during puberty and their relationship to fat distribution," American Journal of Human Biology, vol. 11, no. 2, pp. 209–224, 1999.
35. J. Brooks–Gunn and J. A. Graber, "Puberty as a biological and social event: implications for research on pharmacology," Journal of Adolescent Health, vol. 15, no. 8, pp. 663–671, 1994.
36. M. P. Warren and J. Brooks–Gunn, "Mood and behavior at adolescence: evidence for hormonal factors," Journal of Clinical Endocrinology and Metabolism, vol. 69, no. 1, pp. 77–83, 1989.
37. C. D. Summerbell, E. Waters, L. D. Edmunds, S. Kelly, T. Brown, and K. J. Campbell, "Interventions for preventing obesity in children," Cochrane Database of Systematic Reviews, no. 3, Article ID CD001871, 2005.
38. P. A. Estabrooks, E. B. Fisher, and L. L. Hayman, "What is needed to reverse the trends in childhood obesity? A call to action," Annals of Behavioral Medicine, vol. 36, no. 3, pp. 209–216, 2008.

39. R. E. Glasgow, "What types of evidence are most needed to advance behavioral medicine?" Annals of Behavioral Medicine, vol. 35, no. 1, pp. 19–25, 2008.
40. B. Swinburn, "Obesity Prevention in Children and Adolescents," Child and Adolescent Psychiatric Clinics of North America, vol. 18, no. 1, pp. 209–223, 2009.

CHAPTER 14

MULTI–INSTITUTIONAL SHARING OF ELECTRONIC HEALTH RECORD DATA TO ASSESS CHILDHOOD OBESITY

L. CHARLES BAILEY, DAVID E. MILOV, KELLY KELLEHER, MICHAEL G. KAHN, MARK DEL BECCARO, FELICIANO YU, THOMAS RICHARDS, and CHRISTOPHER B. FORRE

14.1 INTRODUCTION

Assessing the health and healthcare of the nation's children depends critically on data that are timely, relevant, and accurate. Currently, surveillance and policy decisions rely heavily on labor–intensive periodic and ad hoc surveys using cross–sectional designs [1]. However, electronic health records (EHRs) are rapidly coming into wider use as the medium in which information about care of patients is recorded [2]; in the United States, the 'Meaningful Use' initiative for EHR systems [3] [4] is providing a significant stimulus for this process. The shift of health information into electronic forms more amenable to analysis and exchange creates opportunities to improve healthcare quality, develop new methods for clinical research, and follow the health of patient populations using EHR–derived data [5]. Several policy analysts have suggested that multi–institutional sharing of EHR data presents a new paradigm for advancing population health [6] [7] [8]. In particular, greater facility with EHR data is critical to achieving a learning health system [9] [10], in which information derived from clinical care continuously supports advances in medical understanding and delivery of health care. These advances will be of particular value to child health research, much of which relies on longitudinal changes in growth, health, and development.

Realizing the potential of EHR–derived data requires addressing important questions, including differences in representation of information, variability in data capture, and governance issues [11]. Early efforts include the eMERGE Consortium, which used site–specific combinations of discrete data and natural language processing to identify patients with particular diagnoses that could be pooled for genotype–phenotype studies [12]. The HMO Research Network [13] has piloted a system that allows queries against member sites' records; the Shared Health Research Information Network's (SHRINE) query framework operates across sites that have implemented compatible infrastructure using i2b2 [14]. These tools provide for addressing aspects of heterogeneity between EHRs to support federated case ascertainment for research, but require significant resources to implement the required infrastructure. In parallel, mechanisms for patient–centered health information exchange supporting clinical operations and continuity of care are rapidly evolving [15], with significant involvement in the United States from the Office of the National Coordinator for Health IT [16]. Processes that benefit from frequent iteration, such as quality improvement and public health surveillance, may benefit from hybrid strategies that incorporate limited start–up cost and well–defined data models.

Groups of institutions having unified EHRs, such as the Kaiser Permanente system [17], the Harvard Vanguard Medical Associates [18], and the Nemours foundation [19] have demonstrated the utility of aggregating data to examine larger populations of patients in pediatric as well as adult health care, and analyses based on data from a single EHR are becoming more common. At the same time, we are better defining potential limitations of EHR–derived data [20] and developing ways to address them [21] [22]. Sharing of data from disparate EHR systems to enable population–based research [9] is a logical and widely anticipated extension. As we evolve toward this goal, it will be important to study not only the mechanisms for data sharing, but also the ability of data aggregated through various means to support valid and meaningful conclusions about population health.

America's children, like the population as a whole, are experiencing alarming levels of obesity, although rates appear to have stabilized at 17–19% of children and adolescents [23] [18]. National estimates of childhood obesity are generated by the National Health and Nutrition Examination Survey (NHANES) [24]. Data are combined over two–year intervals to accrue sufficient subjects to generate precise estimates. Sharing of EHR–derived anthropometrics can achieve very large sample sizes to generate interval assessments more rapidly, provide multiple assessments per child to permit longitudinal assessment, and link with other relevant clinical data. In these ways, it can serve as a valuable complement to more in–depth but resource–intensive structured population surveys such as NHANES. This sort of population health surveillance is a key goal of the Meaningful Use initiative, and early efforts to examine feasibility are underway [15], but there are few examples to date in pediatrics.

We report here a study testing the feasibility, validity, and utility of multi–institutional EHR data sharing for monitoring and investigating childhood obesity. The study was expressly designed to minimize resources required by contributing institutions, using processes similar to those used for quality improvement, or for the health information exchange envisioned in the Meaningful Use initiative as a part of routine EHR interoperation. Six pediatric academic health systems from different regions of the United States participated, sharing data from 2007–2008, to examine both the logistical requirements for data sharing and the characteristics of EHR–derived data. The interval matches a data–reporting period for NHANES, to better compare the two approaches. To further assess the utility of EHR–derived data, we examined the association of measured obesity with other clinical data, including the diagnosis of obesity, detection of co–morbidities, and assessment of healthcare utilization. Our goal was to explore the unique potential of EHR–derived data to provide an integrated clinical picture over time.

14.2 METHODS

14.2.1 ETHICS STATEMENT

The Institutional Review Boards at the Children's Hospital of Philadelphia, Nemours Children's Hospital, Nationwide Children's Hospital, Children's Hospital of Colorado, Seattle Children's Hospital, and St. Louis

Children's Hospital approved this study protocol, and granted waivers of individual consent based on absence of individually identifying data. Individual subject identifiers from the EHR were replaced, dates of all visits for a given subject were shifted by a random offset, and subjects' age at each visit was recorded in months.

14.2.2 DATA ACQUISITION

The study was conducted in the Pediatric EHR Data Sharing Network (PEDSNet), a consortium formed in late 2009 in response to the Institute of Medicine's call for development of real–world examples of learning health systems [9] [10]. Each site extracted from their EHR information from all outpatient physician visits in 2007–2008, excluding emergency department and surgical center visits, for all patients with age <18 years. Data included subject sex and age, visit date and department specialty, subject's measured weight and height, and all diagnoses recorded for the visit. Four of the six institutions use the EpicCare EHR [25], one uses Cerner Millennium [26], and one uses Allscripts [27]. Since the Allscripts EHR does not associate diagnoses with a specific visit, that site reported all diagnoses listed as active on the date of the visit. Data were transmitted to the coordinating center, where analytic databases were constructed and quality control tests were run. Ambiguities or apparent errors were corrected by communication with the submitting site and resubmission. Sites reported the feasibility of capturing requested elements, as well as required resources for regulatory review, query definition, and data extraction and de–identification.

For analyses using the NHANES 2007–2008 samples, DEMO and BMX datasets were retrieved from the NHANES web site [24] and age, sex, weight, stature, and MEC sample weight were used for analyses.

14.2.3 DETERMINATION OF OVERWEIGHT AND OBESITY

Body mass index (BMI) was calculated in kg/m^2. Visits for which any of age, sex, or weight were missing were excluded. If measured height was not available for the visit, but values were available for the prior and subsequent visits, height was imputed using linear interpolation. If two or more height values were available only before or after the current visit and yielded consistent percentiles on the NHANES 2000 height–sex–age growth curves, a value for the current visit was imputed based on the corresponding percentile at the current visit. The subject's age, sex, and BMI were then used to calculate a percentile based on the NHANES 2000 curves, using the SAS algorithm published by the CDC. [28] BMI values flagged as outliers (z–score<–4 or >5) using the CDC's modified z–score algorithm [29] were excluded, in keeping with accepted norms. A subject was considered obese if the percentile was 95 or greater, and overweight if the percentile was at least 85 but less than 95.

A diagnosis of obesity was noted if any of the ICD–9–CM codes 278, 278.0, 278.00, 278.01, 278.1, 759.81, 783.1, or V85.54 were present. For specialty–specific analyses, only visits in the same specialty were considered for scoring both BMI and diagnoses, though imputation of height was allowed using data from other visit types, as these data would be accessible in the EHR. For person–level analyses, BMI criteria or diagnostic criteria could be met independently at any eligible visit during the study period.

14.2.4 DETECTION OF OBESITY–RELATED CO–MORBIDITY

For children with evaluable BMI, recorded diagnoses from all visits were clustered into clinically homogeneous categories using Expanded Diagnostic Clusters (EDCs) from the Adjusted Clinical Group (ACG) System [30]. Within each category, we calculated a standardized morbidity ratio of prevalence in children with obesity versus that in the cohort as a whole. Rates were standardized by age and sex.

14.2.5 DATA ANALYSIS

Data management was done using Perl 5.12–5.16 [31] and MySQL 5.5 [32]. Analyses were done using R 2.14 or 2.15 [33] and SAS 9.2 or 9.3 software [34].

For EHR–derived data, estimates of prevalence were computed as unweighted proportions of the indicated population; for NHANES data, MEC sample weights were used. At the recommendation of NHANES staff, no BMI values were excluded as outliers, in order to better match the methods used in their published analyses. A two–tailed Student's t–test was used to assess significance of continuous variables, and $\chi 2$ testing was used for categorical variables. Curves of average BMI values by age were fitted in R using cubic polynomial regression. To assess the reliability over time of BMI, per–subject intraclass correlation coefficients (ICC2) were computed [35]. Multiple linear regression incorporating age, sex, obesity, and comorbidity burden as encoded by ACG Relative Utilization Bands [30] was performed using R.

14.3 RESULTS

14.3.1 DATASET CONSTRUCTION

Sites reported between 5 and 40 person–hours required for retrieval of data elements from their EHR, with the majority of effort being in query construction and regulatory review. Overall, sites required 0–2 revisions to their extraction process, after review by the data coordinating center, to resolve data quality issues. In two cases, sites had systematic errors in their initial data involving calculation of age at visit; of note, both were detectible as outliers in the resulting BMI distributions by comparison to the other sites, without reference to external standards. Age and sex were consistent within subjects across >99% of visits. Anthropometric data were also internally consistent, with <0.35% of visits recording apparent English–metric unit errors based on prior or subsequent visits.

14.3.2 POPULATION

The EHR dataset included 2,491,015 outpatient visits involving 699,767 children 2–17 years of age. Of these, 1,398,655 visits (56%) made by

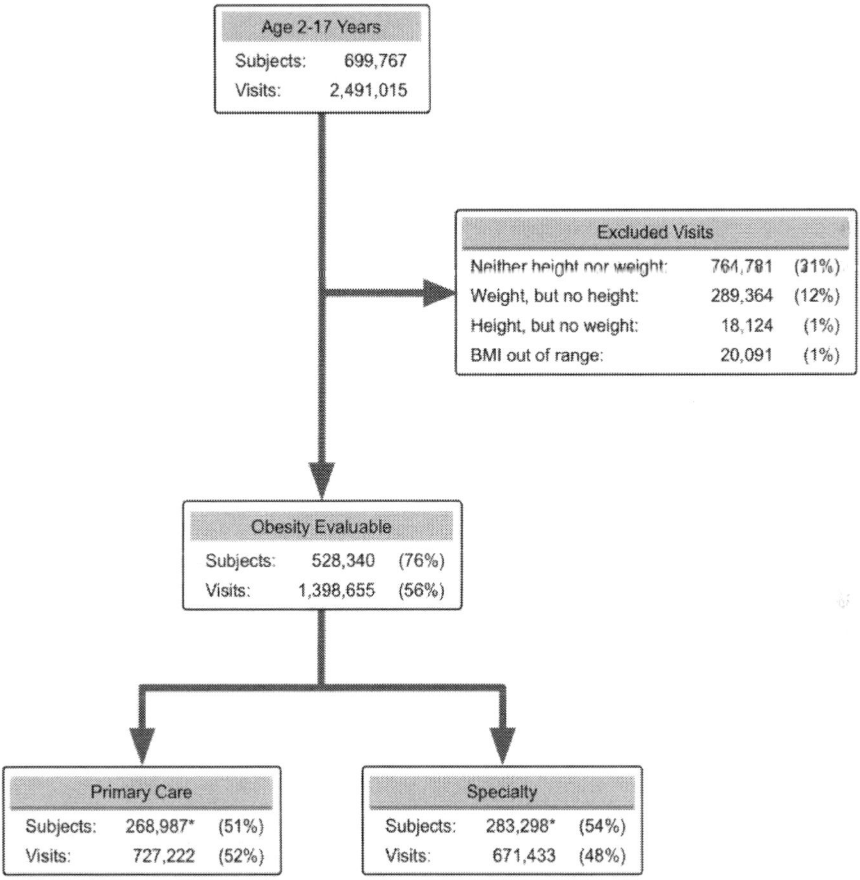

FIGURE 1: Evaluable Population for Obesity Analyses. Development of the dataset for obesity–related analyses, showing the number of evaluable children and visits at each step. Percentages at each step are calculated relative to totals in the prior step. Since patients may have both primary care and specialty visits, subject counts at this step do not sum to the prior total; these values are marked with an asterisk.

528,340 children (76%) had sufficient data to compute a BMI (Figure 1), with a mean of 2.6 (range 1–141; inter–quartile range 1–3) BMI assessments/child. Height was imputed for 21% of these visits. For every month of age from 2–15 years, the dataset contained over 6,000 BMI measurements., Counts decreased steadily for adolescents ages 16–18, to a low of 1674 observations for children 215 months old, likely representing transition of older adolescents out of pediatric care.

Fifty–two percent of subjects were male. Of total visits, 51% were at primary care sites and 49% at specialty clinics; 28% were made by children 2–4 years of age, 39% 5–10 years of age, and 34% 10–17 years of age. Contributions from a single site ranged from 3 to 35% of subjects and 2 to 43% of visits. All proportions were comparable for evaluable visits.

14.3.3 MEASUREMENT OF OBESITY AND OVERWEIGHT

Figure 2 shows a comparison of BMI measurement in the clinical data from the EHR dataset to the U.S. benchmark NHANES survey for the same period. Mean BMI values for each month of age were highly similar in the EHR and NHANES datasets. However, there was substantially higher precision in the EHR–derived data, particularly among adolescents.

BMI measurements were used to estimate the prevalence estimates of obesity and overweight in different age groups, as shown in Table 1. The estimates produced using EHR–derived data were 18% for obesity and 35% obesity plus overweight; these figures align closely with the 18% and 34% estimates, respectively, derived from the NHANES surveys. The differences between EHR–based and NHANES estimates were slightly greater for 2–4 year old children, but they did not reach significance.

Because the dataset contained 101,897 obese or overweight children with multiple visits, we were able to assess the stability over time of EHR–based BMI measurements by calculating the per–child intraclass correlation coefficient (ICC). For obese children, the ICC was 0.90, for overweight but non–obese children 0.81, and for all children 0.97, demonstrating that clinical BMI assessment was a highly reliable process. Among children who were obese at any visit, 85% remained obese or overweight at all visits during the study period.

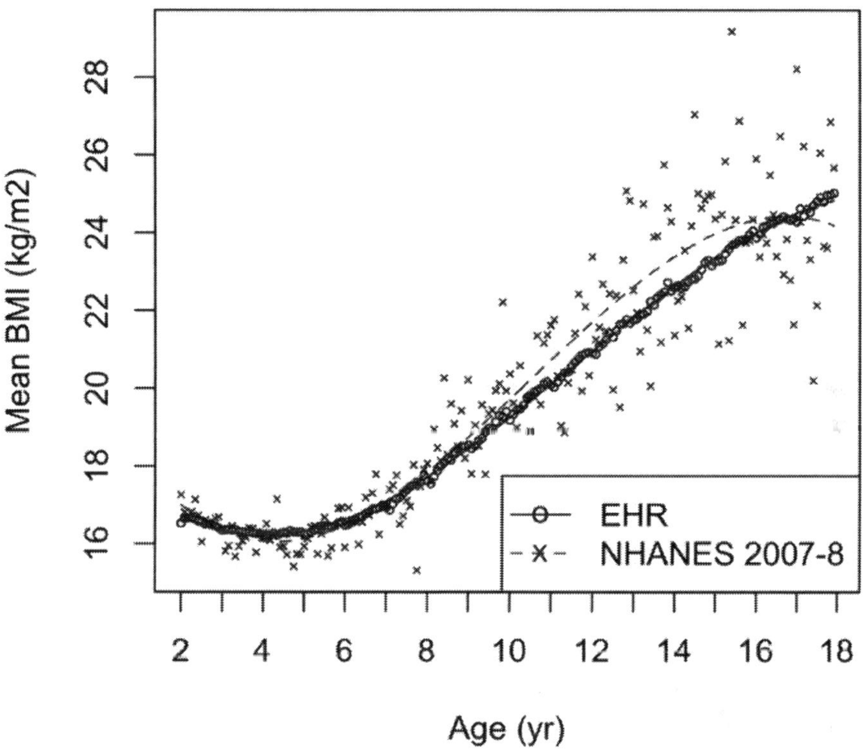

FIGURE 2: Comparison of EHR and NHANES 2007–8 Cohorts. Average measured BMIs for children of both sexes at each month of age from 2–17 years in the multi–institutional EHR cohort and in the NHANES 2007–8 cohort. In addition to individual points, curves fitted to each dataset by cubic polynomial regression are shown.

14.3.4 CORRELATION WITH CLINICAL PRACTICE

Figure 3 presents rates at which clinicians in different specialties made a diagnosis of obesity for children with elevated BMI. Overall, only 20% of children with one or more BMI measurements above the 95th percentile had a diagnosis recorded at any visit. When the analysis was restricted to primary care visits, the rate rose to just 29%; considering only well child checks did not alter this result. The only contexts in which diagnosis rates exceeded 30% were endocrinology and weight management clinics. At the visit level, just 14% of all visits with measured obesity had a diagnosis of obesity recorded.

TABLE 1: Prevalence of Obesity and Overweight in EHR–Derived Data and NHANES Data.

	Fraction of Sample[b]	% Obese	% Overweight, never obese
NHANES 2007–2008[a]			
2–17 years	1.000	18	16
2–4 years	0.194	11	12
5–10 years	0.349	19	15
11–17 years	0.457	20	17
Multi–site EHR data			
2–17 years	1.000	18	17
2–4 years	0.280[c]	14	16
5–10 years	0.418[c]	18	17
11–17 years	0.374[c]	20	17

[a]*All proportions for NHANES data were calculated using MEC sample weights; no BMI outliers were excluded in prevalence estimates following NHANES standard practice.*
[b]*Total raw samples sizes were 3032 for NHANES and 528,340 for multi–site EHR data.*
[c]*Different visits for a given child may appear in different age subgroups, due to the longitudinal nature of the EHR dataset. Therefore, the fractions of children from each age subgroup do not sum to 1000.*
EHR:Electronic Health Record. NHANES: National Health and Nutrition Examination Survey

We used the EHR dataset to detect groups of conditions that most commonly co–occur with obesity (Figure 4). Several of these conditions are known comorbidities of obesity, such as hypertension and hyperlipidemia. However, we were also able to detect associations between obesity and rare disorders such as acute leukemia, multiple sclerosis, and chromosomal anomalies.

In addition, we observed an overall increase for obese children in both primary care visits (ever obese: 4.8±4.0 vs. never obese: 4.0±3.4; $p<0.001$) and specialty visits (3.7±6.6 vs. 2.7±4.0; $p<0.001$). After adjustment for age, sex, and site, 52% of this difference in outpatient utilization was attributable to diagnosed comorbidities, as assessed by ACG Resource Utilization Bands.

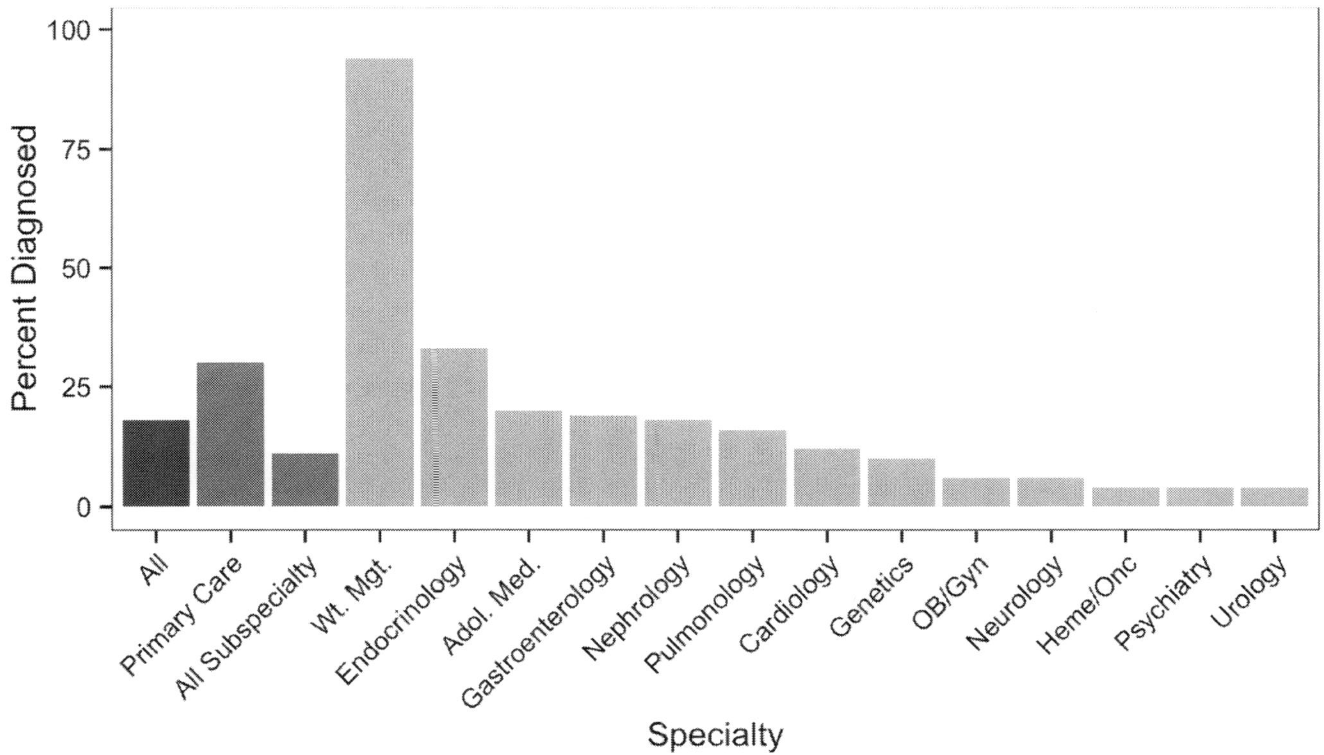

FIGURE 3: Diagnosis of Obesity at Outpatient Visits. Percentages of children who were obese at any time during the study period, and diagnosed as obese at any visit to the indicated specialty. All specialties with a diagnosis rate ≥4% are included.

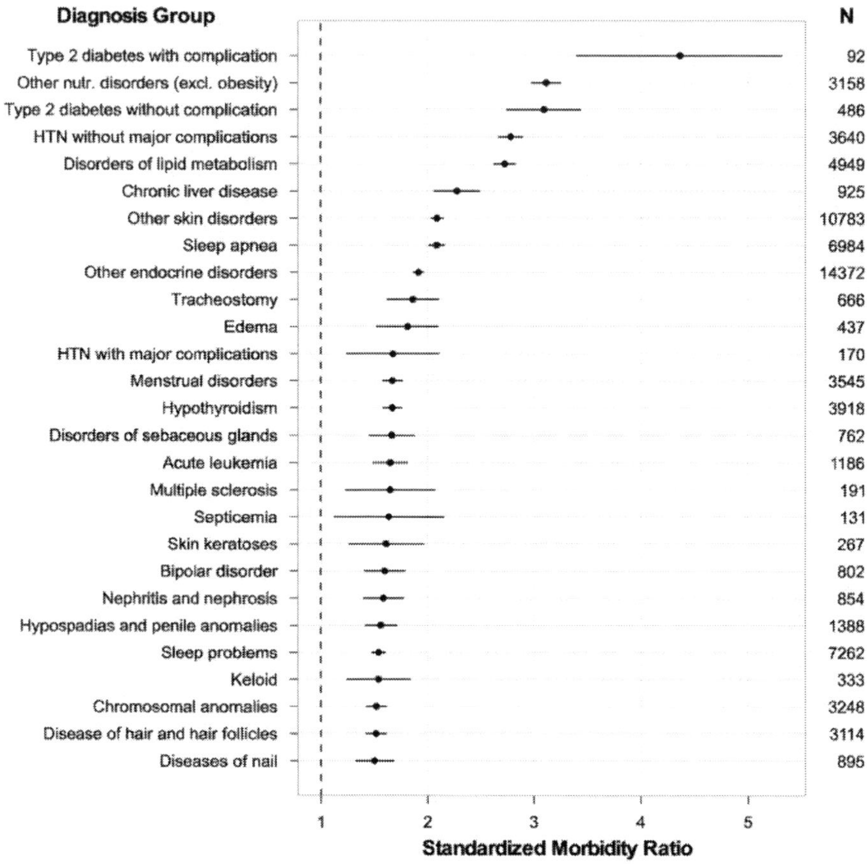

FIGURE 4: Obesity–Related Co–Morbidities. Standardized morbidity ratios (observed prevalence in obese children/expected prevalence from entire cohort) with 95% confidence intervals for diagnostic groups (EDCs) having SMR >1.5 and CI95>1.0 among children with measured obesity. N = total number of children in cohort with diagnosis in that EDC.

14.4 DISCUSSION

This study demonstrates the feasibility and validity of sharing EHR–derived data for assessing obesity in large populations of children. The effort required to retrieve the data was nominal, and largely for query development and validation, a one–time cost that would not apply to refreshing data for ongoing surveillance. The scale of EHR–derived data is significant: this sample from six pediatric centers produced 6,000 BMI assessments per month of age for most of childhood.

Aggregation of data across sites and EHR types represents an important test of principle. This study was designed with the intent to isolate technical and procedural factors that might affect the feasibility of data exchange. To that end, we focused on data types, such as anthropometric measurements, where the meaning of a value, as distinct from the method used to obtain it, was unambiguous. For clinical diagnoses, we used the International Classification of Disease, 9th Edition, Clinical Modification (ICD9–CM) [36], the current standard diagnostic terminology in use in the United States. This provided a common vocabulary, though as noted below, usage of specific codes differed across institutions. In the general case, the problem of semantic interoperability [37], or accounting for the ways in which a common concept is represented in different contexts, remains a barrier to sharing of clinical information. Addressing this problem will require additional work in a number of areas, including development of more robust terminologies and standards for data interchange [37] [38], better understanding of ways in which clinicians interact with the EHR [39] and data are captured [40], and further studies of the operating characteristics of EHR–derived data.

Because most data in the EHR are obtained as part of routine clinical care, their structure will also reflect practice patterns that must be accounted for in secondary analyses. In our case, consistent with expected outpatient practice, patient heights were measured less often than patient weights. Where no height or weight data were available for a child, we considered them inevaluable for obesity, as it would be error–prone, and risk circularity, to impute these directly from a population distribution to an individual. However, unlike weight, for a given individual height velocity is a relatively stable physiologic quantity over the interval considered in this study. Therefore, when at least two measured heights were available, we used a conservative imputation strategy to derive height, and hence BMI, values for visits where it was not directly measured. Although this added no subjects to the dataset, it increased by 21% the number of evaluable visits available for longitudinal and practice type analyses with low risk to validity of data. It also reflects an anticipated, if not yet widely realized, benefit of EHR adoption: information from one site becomes more widely available for use at other sites sharing the EHR.

Secondary use of clinical data for research has also generated concern about the potential consequences of increased variability across measurements. Although the carefully controlled NHANES methodology likely does produce more precise individual measurements, the survey yields an average sample of 16 subjects per month of age. In the EHR data, the potential effects of individual measurement error are damped by the size of the sample and repeated measurement, resulting in highly stable population estimates of BMI and obesity prevalence compatible with NHANES results. This difference was particularly apparent in adolescents, where the EHR–derived data did not display the increased variance seen in NHANES measurements, which was likely due to the differences between individual children in timing of pubertal growth. Furthermore, at the individual level, we found high reliability for BMI assessments over time, suggesting that any error introduced by variation in assessment technique is small.

The cohort size achievable with EHR–derived data permits detection of clinically relevant associations not possible with survey data, such as the known association between acute lymphoblastic leukemia and secondary obesity [41]. EHRs also facilitate construction of cohorts with linked clinical data, to assess the impact of obesity on children with rare primary disorders. In these ways, EHR–based population surveillance can provide an important complement to in–depth but resource–intensive surveys such as NHANES.

It is important to note potential limitations of our EHR–based study. First, data are derived from six centers, leaving gaps in national geographic coverage and underrepresentation of rural areas. In particular, this may contribute to the difference from NHANES in prevalence of obesity in younger children; the higher proportion of measurements from children aged 2–10 in the EHR–derived data than in the NHANES sample is also in keeping with expected patterns of clinical utilization. In most respects, however, our estimates closely match the results of the NHANES stratified sampling model. A fortuitous combination of contributing sites is possible, though participating centers were not selected based on obesity prevalence. The means of 2.6 evaluable visits/child and 1.9 diagnoses/visit also indicate that our results were not likely to have been heavily biased by a subpopulation of children with complex medical conditions affecting their growth. Moreover, the low cost of EHR queries suggests that as the nation's healthcare system becomes increasingly digitized, it will become possible to readily combine data from additional geographic areas and clinical settings, and increase the generalizability of results based on data sharing.

Second, the high data quality observed in the EHR dataset may in part reflect the selection of anthropometric and demographic data, which are semantically unambiguous and directly measured as a matter of routine in pediatrics, as the source of the primary outcome measured. Our results do provide significant reassurance against the concern that clinical data are generally too unreliable for use in research. However, the quality of other types of information, particularly subjective findings or clinical decisions, will depend on different sets of semantic and pragmatic considerations. Further study will be required to assess the fitness of these and other types of EHR–derived data for population–level analyses [42] [43].

Third, rates of obesity diagnosis in the EHR were remarkably low, even in primary care settings. Although unsurprising [44] [45] [46], this is a significant problem, since growth monitoring is a core function of pediatrics. We considered the possibility that our obesity–related ICD9–CM diagnosis cluster did not sufficiently comprise codes in common use. Examination of the most common diagnoses for obese children does not indicate that an alternative code(s), including those for overweight or specific BMI ranges, was used frequently (data not shown). Of note, the inclusion of 783.1 ("abnormal weight gain") in our cluster is the result of this analysis demonstrating that it was the most common weight–related diagnosis given to obese children at one site. It is also possible that obesity is missing because multiple other diagnoses are recorded. However, obese children had on average 1.9 diagnoses/visit, a number unlikely to preclude adding a diagnosis of obesity. The low rates of diagnosis in primary care and well–child visits also argue that addressing more acute problems is not a major factor preventing diagnosis of obesity. We did observe stronger association between a diagnosis of obesity and many comorbid diagnoses than between measured obesity itself and these diagnoses (data not shown), suggesting that the presence of a comorbid condition such as hypertension or diabetes may "prompt" a diagnosis of obesity. Adding entries from the

EHR's problem list at the largest site increased the documentation rate by just 3%, showing that this is not a major alternative method of recording recognition of obesity. It is more likely that pediatricians are not recording obesity diagnoses for reasons other than lack of opportunity, such as a belief that obesity is best addressed by "non–medical" interventions, non–reimbursement of obesity diagnoses, or concern for stigmatization of patients. If we are to improve the quality of care for obese children, we will need to better document the problem in the medical record, where it serves not only as a cognitive marker, but as a trigger for additional decision support around appropriate screening and treatment.

EHRs provide access to primary clinical data, rather than specifically coded or prompted responses as on a case report form. This can bias ascertainment of a datum if it does not reflect a common element of clinical care. However, it can also be valuable, if it permits ascertainment of affected status directly, rather than relying on diagnoses or similar administrative data. As we demonstrate, using administrative data to identify obese children misses over 75% of affected individuals. It is possible that similar biases affect diagnoses of comorbid conditions used in our analyses. The strong associations seen between obesity and several known comorbidities are reassuring in this respect. However, further analyses using EHR–derived data can provide opportunities for direct assessment of other conditions, to more reliably establish association with obesity, and potentially to allow us to better identify subsets of children at higher risk for specific complications of obesity.

Using EHR data to monitor other aspects of population health will benefit increasingly from structured data in the EHR, such as diagnoses, vital signs, medications, and diagnostic results. Free text (e.g. clinical assessments and instructions) will require greater, though not necessarily prohibitive [21], effort to derive useful population–level information. In addition to data type, it will be important to understand operating characteristics of EHR–derived data, since the potential for selection and reporting biases will be different from other survey methods.

Using clinical information from the EHR, we demonstrate robust associations between measured obesity and diagnosed comorbidities such as diabetes and other endocrinopathies [47] [48] [49], hypertension [50] [51], dyslipidemia [52] [4], liver disease [53], and sleep apnea [54]. Obese children had increased overall healthcare utilization as well, about half of which is explained by excess diagnosed comorbidities. Both findings highlight public health implications of the high prevalence of obesity for children today and adults tomorrow. Further study to identify appropriate markers in the medical record of screening for and treatment of obesity–associated morbidity will help to define strategies for addressing these problems.

This study also suggests several opportunities for quality improvement. Overall, 44% of visits did not include sufficient data to assess BMI and 24% of subjects had no assessments over the two–year study period, which is recommended at least annually by the American Academy of Pediatrics as a universal practice [55], and is a core objective of the stage 1 meaningful use measures [56]. Only one in five obese children have the diagnosis recorded, an important step in the medical management of any condition. Moreover, the low opportunity cost of EHR–derived monitoring, potentially coupled with geocoding or other public health data, makes it possible to assess the impact of medical and community–based interventions on obesity in a variety of geographic and demographic settings. Methods validated using EHR–derived data can also provide direct input into design of decision support systems to improve quality at the point of care.

14.5 CONCLUSION

We are still early in the process of incorporating the EHR into clinical and public health practice. This study demonstrates the potential for integrating EHR–derived data from multiple sources to monitor childhood obesity and its correlates. Given the breadth of information collected in EHRs, we believe this potential extends to many areas of population health management; utility for specific conditions will depend on the degree to which critical data are consistently captured and can be meaningfully recovered from the EHR. Further, the close linkage of source data to patient care may allow systems that incorporate EHR–derived data to more effectively translate results into clinical practice. Developing a nationwide cross–institutional data sharing system holds the potential for population health surveillance, quality improvement, and ultimately formation of the digital infrastructure of a transformative, learning health system for the nation [5]. Both health information exchanges and clinical research networks such as HMORN and PEDSNet will contribute to understanding the logistical and scientific requirements for effective use of clinical data in this process.

REFERENCES

1. National Research Council, Institute of Medicine (2004) Children's Health, the Nation's Wealth: Assessing and Improving Child Health. Washington, DC: National Academies Press.
2. Hsiao CJ, Hing E (2012) Use and characteristics of electronic health record systems among office–based physician practices: United States, 2001–2012. NCHS data brief: 1–8.
3. Centers for Medicare and Medicaid Services H (2010) Medicare and Medicaid programs; electronic health record incentive program. Final rule. Federal register 75: 44313–44588.
4. Blumenthal D, Tavenner M (2010) The "meaningful use" regulation for electronic health records. The New England journal of medicine 363: 501–504. doi: 10.1056/nejmp1006114.
5. Olsen L, Aisner D, McGinnis JM (2007) The Learning Healthcare System. Washington, DC: National Academies Press.
6. Diamond CC, Mostashari F, Shirky C (2009) Collecting and sharing data for population health: a new paradigm. Health affairs 28: 454–466. doi: 10.1377/hlthaff.28.2.454.
7. Etheredge LM (2007) A rapid–learning health system. Health affairs 26: w107–118. doi: 10.1377/hlthaff.26.2.w107.
8. Slutsky JR (2007) Moving closer to a rapid–learning health care system. Health affairs 26: w122–124. doi: 10.1377/hlthaff.26.2.w122.
9. Grossman C, Goolsby WA, Olsen L, McGinnis JM (2011) Clinical Data as the Basic Staple of Health Learning: Creating and Protecting a Public Good. Washington, D.C.: Institute of Medicine.
10. Olsen L, Aisner D, McGinnis JM (2007) The Learning Healthcare System. Washington, D.C.: Institute of Medicine.
11. Hripcsak G, Albers DJ (2013) Next–generation phenotyping of electronic health records. Journal of the American Medical Informatics Association : JAMIA 20: 117–121. doi: 10.1136/amiajnl–2012–001145.
12. Kho AN, Pacheco JA, Peissig PL, Rasmussen L, Newton KM, et al. (2011) Electronic medical records for genetic research: results of the eMERGE consortium. Science translational medicine 3: 79re71. doi: 10.1126/scitranslmed.3001807.
13. Lieu TA, Hinrichsen VL, Moreira A, Platt R (2011) Collaborations in population–based health research: the 17th annual HMO Research Network Conference, March

23–25, 2011, Boston, Massachusetts, USA. Clinical medicine & research 9: 137–140. doi: 10.3121/cmr.2011.1025.
14. Weber GM, Murphy SN, McMurry AJ, Macfadden D, Nigrin DJ, et al. (2009) The Shared Health Research Information Network (SHRINE): a prototype federated query tool for clinical data repositories. Journal of the American Medical Informatics Association : JAMIA 16: 624–630. doi: 10.1197/jamia.m3191.
15. Maxson ER, Jain SH, McKethan AN, Brammer C, Buntin MB, et al. (2010) Beacon communities aim to use health information technology to transform the delivery of care. Health affairs 29: 1671–1677. doi: 10.1377/hlthaff.2010.0577.
16. Williams C, Mostashari F, Mertz K, Hogin E, Atwal P (2012) From the Office of the National Coordinator: the strategy for advancing the exchange of health information. Health affairs 31: 527–536. doi: 10.1377/hlthaff.2011.1314.
17. Koebnick C, Smith N, Black MH, Porter AH, Richie BA, et al.. (2012) Pediatric Obesity And Gallstone Disease: Results From A Cross–Sectional Study of Over 510,000 Youth. Journal of pediatric gastroenterology and nutrition.
18. Wen X, Gillman MW, Rifas–Shiman SL, Sherry B, Kleinman K, et al. (2012) Decreasing prevalence of obesity among young children in Massachusetts from 2004 to 2008. Pediatrics 129: 823–831. doi: 10.1542/peds.2011–1833.
19. Falkner B, Gidding SS, Ramirez–Garnica G, Wiltrout SA, West D, et al. (2006) The relationship of body mass index and blood pressure in primary care pediatric patients. The Journal of Pediatrics 148: 195–200. doi: 10.1016/j.jpeds.2005.10.030.
20. Hripcsak G, Knirsch C, Zhou L, Wilcox A, Melton G (2011) Bias associated with mining electronic health records. Journal of biomedical discovery and collaboration 6: 48–52. doi: 10.5210/disco.v6i0.3581.
21. Nadkarni PM, Ohno–Machado L, Chapman WW (2011) Natural language processing: an introduction. Journal of the American Medical Informatics Association : JAMIA 18: 544–551. doi: 10.1136/amiajnl–2011–000464.
22. Rea S, Pathak J, Savova G, Oniki TA, Westberg L, et al.. (2012) Building a robust, scalable and standards–driven infrastructure for secondary use of EHR data: The SHARPn project. Journal of biomedical informatics.
23. Flegal KM, Carroll MD, Ogden CL, Curtin LR (2010) Prevalence and trends in obesity among US adults, 1999–2008. JAMA : the journal of the American Medical Association 303: 235–241. doi: 10.1001/jama.2009.2014.
24. National Center for Health Statistics, Centers for Disease Control and Prevention (2007) National Health and Nutrition Examination Survey Data 2007–2008. In: National Center for Health Statistics, Centers for Disease Control and Prevention., editor. Hyattsville, MS.
25. Epic Systems Corporation (2011) EpicCare Electronic Medical Record.
26. Cerner Corporation (2011) Cerner Millenium. Kansas City, MO.
27. Allscripts Corporation (2011) Allscripts.
28. Centers for Disease Prevention and Control (2011) A SAS Program for the CDC Growth Charts. Centers for Disease Control and Prevention.
29. Centers for Disease Prevention and Control (2011) Cut–offs to define outliers in the 2000 CDC Growth Charts.

30. Weiner JP, Abrams C (2009) The Johns Hopkins ACG System Technical Reference Guide. 9.0 ed. Baltimore, MD.
31. The Perl Foundation (2010) Perl. 5.12 ed.
32. MySQL MySQL Database Management System.
33. R Development Core Team (2011) R: A Language and Environment for Statistical Computing.
34. SAS Institute Inc. (2008) SAS System for Microsoft Windows. 9.2, 9.3 ed.
35. Bartko JJ (1976) On Various Intraclass Correlation Reliability Coefficients. Psychological Bulletin 83: 762–765. doi: 10.1037//0033–2909.83.5.762.
36. National Center for Health Statistics CfDCaP (2012) International Classification of Diseases, Ninth Revision, Clinical Modification (ICD–9–CM). Centers for Disease Control and Prevention. pp. The International Classification of Diseases, Ninth Revision, Clinical Modification (ICD–9–CM) is based on the World Health Organization's Ninth Revision, International Classification of Diseases (ICD–9). ICD–9–CM is the official system of assigning codes to diagnoses and procedures associated with hospital utilization in the United States.
37. Dolin RH, Alschuler L (2011) Approaching semantic interoperability in Health Level Seven. Journal of the American Medical Informatics Association : JAMIA 18: 99–103. doi: 10.1136/jamia.2010.007864.
38. Lim Choi Keung SN, Zhao L, Tyler E, Taweel A, Delaney B, et al. (2012) Cohort identification for clinical research: querying federated electronic healthcare records using controlled vocabularies and semantic types. AMIA Summits on Translational Science proceedings AMIA Summit on Translational Science 2012: 9.
39. Borycki EM, Kushniruk AW, Kuwata S, Kannry J (2011) Engineering the electronic health record for safety: a multi–level video–based approach to diagnosing and preventing technology–induced error arising from usability problems. Stud Health Technol Inform 166: 197–205.
40. Lin MC, Vreeman DJ, McDonald CJ, Huff SM (2012) Auditing consistency and usefulness of LOINC use among three large institutions – using version spaces for grouping LOINC codes. Journal of biomedical informatics 45: 658–666. doi: 10.1016/j.jbi.2012.01.008.
41. Collins L, Zarzabal LA, Nayiager T, Pollock BH, Barr RD (2010) Growth in children with acute lymphoblastic leukemia during treatment. Journal of pediatric hematology/oncology 32: e304–307. doi: 10.1097/mph.0b013e3181ece2bb.
42. Kahn MG, Raebel MA, Glanz JM, Riedlinger K, Steiner JF (2012) A pragmatic framework for single–site and multisite data quality assessment in electronic health record–based clinical research. Medical care 50 Suppl: S21–29 doi: 10.1097/mlr.0b013e318257dd67.
43. Holve E, Segal C, Hamilton Lopez M (2012) Opportunities and challenges for comparative effectiveness research (CER) with Electronic Clinical Data: a perspective from the EDM forum. Medical care 50 Suppl: S11–18 doi: 10.1097/mlr.0b013e318258530f.
44. Lemay CA, Cashman S, Savageau J, Fletcher K, Kinney R, et al. (2003) Underdiagnosis of obesity at a community health center. The Journal of the American Board of Family Practice/American Board of Family Practice 16: 14–21. doi: 10.3122/jabfm.16.1.14.

45. Ruser CB, Sanders L, Brescia GR, Talbot M, Hartman K, et al. (2005) Identification and management of overweight and obesity by internal medicine residents. Journal of general internal medicine 20: 1139–1141. doi: 10.1111/j.1525–1497.2005.0263.x.
46. Patel AI, Madsen KA, Maselli JH, Cabana MD, Stafford RS, et al. (2010) Underdiagnosis of pediatric obesity during outpatient preventive care visits. Academic pediatrics 10: 405–409. doi: 10.1016/j.acap.2010.09.004.
47. Franks S (2008) Polycystic ovary syndrome in adolescents. International journal of obesity 32: 1035–1041. doi: 10.1038/ijo.2008.61.
48. Abrams P, Levitt Katz LE (2011) Metabolic effects of obesity causing disease in childhood. Current opinion in endocrinology, diabetes, and obesity 18: 23–27. doi: 10.1097/med.0b013e3283424b37.
49. Huang RC, de Klerk NH, Smith A, Kendall GE, Landau LI, et al. (2011) Lifecourse childhood adiposity trajectories associated with adolescent insulin resistance. Diabetes care 34: 1019–1025. doi: 10.2337/dc10–1809.
50. National High Blood Pressure Education Program Working Group on High Blood Pressure in Children and Adolescents (2004) The fourth report on the diagnosis, evaluation, and treatment of high blood pressure in children and adolescents. Pediatrics 114: 555–576. doi: 10.1542/peds.114.2.s2.555.
51. Juonala M, Magnussen CG, Berenson GS, Venn A, Burns TL, et al. (2011) Childhood adiposity, adult adiposity, and cardiovascular risk factors. The New England journal of medicine 365: 1876–1885. doi: 10.1056/nejmoa1010112.
52. Cook S, Kavey RE (2011) Dyslipidemia and pediatric obesity. Pediatric clinics of North America 58: 1363 1373, ix.
53. Volovelsky O, Weiss R (2011) Fatty liver disease in obese children–relation to other metabolic risk factors. International journal of pediatric obesity : IJPO : an official journal of the International Association for the Study of Obesity 6 Suppl 159–64. doi: 10.3109/17477166.2011.583661.
54. Tauman R, Gozal D (2011) Obstructive sleep apnea syndrome in children. Expert review of respiratory medicine 5: 425–440. doi: 10.1586/ers.11.7.
55. Barlow SE (2007) Expert committee recommendations regarding the prevention, assessment, and treatment of child and adolescent overweight and obesity: summary report. Pediatrics 120 Suppl 4S164–192. doi: 10.1542/peds.2007–2329c.
56. Centers for Medicare and Medicaid Services (2011) Eligible Professional Meaningful Use Table of Contents Core and Menu Set Objectives.

CHAPTER 15

CHILDHOOD OBESITY, PREVALENCE AND PREVENTION

MAHSHID DEHGHAN, NOORI AKHTAR-DANESH,
and ANWAR T. MERCHANT

15.1 INTRODUCTION

Childhood obesity has reached epidemic levels in developed countries. Twenty five percent of children in the US are overweight and 11% are obese. About 70% of obese adolescents grow up to become obese adults [1–3]. The prevalence of childhood obesity is in increasing since 1971 in developed countries (Table 1). In some European countries such as the Scandinavian countries the prevalence of childhood obesity is lower as compared with Mediterranean countries, nonetheless, the proportion of obese children is rising in both cases [4]. The highest prevalence rates of childhood obesity have been observed in developed countries, however, its prevalence is increasing in developing countries as well. The prevalence of childhood obesity is high in the Middle East, Central and Eastern Europe [5]. For instance, in 1998, The World Health Organization project monitoring of cardiovascular diseases (MONICA) reported Iran as one of the seven countries with the highest prevalence of childhood obesity. The prevalence of BMI (in percentage) between 85th and 95th percentile in girls was significantly higher than that in boys (10.7, SD = 1.1 vs. 7.4, SD = 0.9). The same pattern was seen for the prevalence of BMI > 95th percentile (2.9, SD = 0.1 vs. 1.9, SD = 0.1) [6]. In Saudi Arabia, one in every six children aged 6 to 18 years old is obese [7]. Furthermore, in both developed and developing countries there are proportionately more girls overweight than boys, particularly among adolescent [6,8,9].

TABLE 1: Changes in the prevalence of overweight and obesity in some developed countries

Country/Year	Age/yr	Study (author)	Change in obesity
USA			
1973–1994	5–24	Bogalusa [67]	Mean level increased 0.2 kg/yr, twofold increase in prevalence of obesity
1971–1974	6–19	NHANES I [68]	Relatively stable
1976–1980	6–19	NHANES II [68]	Relatively stable
1988–1994	6–19	NHANES III [68]	Doubled to 11%
1999–2000	6–19	NHANES IV [68]	Increased by 4%
Japan			
1974–1993	6–14	Kotani [69]	Doubled (5% to 10%)
UK			
1984–1998	7–11	Lobstein [70]	Changed from 8% to 20%
Spain			
1985/6–1995/6	6–7	Moreno [71]	Changed from 23% to 35%
France			
1992–1996	5–12	Rolland–Cachera [72]	Changed from 10% to 14%
Greece			
1984–2000	6–12	Krassas [73]	Increased by 7%

Overweight and obesity in childhood have significant impact on both physical and psychological health; for example, overweight and obesity are associated with Hyperlipidaemia, hypertension, abnormal glucose tolerance, and infertility. In addition, psychological disorders such as depression occur with increased frequency in obese children [10]. Overweight children followed up for 40 [11] and 55 years [12] were more likely to have cardiovascular and digestive diseases, and die from any cause as compared with those who were lean.

15.2 DEFINITION OF CHILDHOOD OBESITY

Although definition of obesity and overweight has changed over time [13,14], it can be defined as an excess of Body Fat (BF). There is no consensus on a cutoff point for excess fatness of overweight or obesity in children and adolescents. Williams et al. [15] measured skin fold thickness of 3320 children aged 5–18 years and classified children as fat if their percentage of body fat was at least 25% and 30%, respectively, for males and females. The Center for Disease Control and Prevention defined overweight as at or above the 95th percentile of BMI for age and "at risk for overweight" as between 85th to 95th percentile of BMI for age [16,17]. European researchers classified overweight as at or above 85th percentile and obesity as at or above 95th percentile of BMI [18].

There are also several methods to measure the percentage of body fat. In research, techniques include underwater weighing (densitometry), multi–frequency bioelectrical impedance analysis (BIA) and magnetic resonance imaging (MRI). In the clinical environment, techniques such as body mass index (BMI), waist circumference, and skin fold thickness have been used extensively. Although, these methods are less accurate than research methods, they are satisfactory to identify risk. While BMI seems appropriate for differentiating adults, it may not be as useful in children because of their changing body shape as they progress through normal growth. In addition, BMI fails to distinguish between fat and fat–free mass (muscle and bone) and may exaggerate obesity in large muscular children. Furthermore, maturation pattern differs between genders and different ethnic groups. Studies that used BMI to identify overweight and obese children based on percentage of body fat have found high specificity (95–100%), but low sensitivity (36–66%) for this system of classification [19]. While health consequences of obesity are related to excess fatness, the ideal method of classification should be based on direct measurement of fatness. Although methods such as densitometry can be used in research practice, they are not feasible for clinical settings. For large population–based studies and clinical situations, bioelectrical impedance analysis (BIA) is widely used. Cross–sectional studies have shown that BIA predicts total body water (TBW), fat–free mass (FFM), and fat mass or percentage of body fat (%BF) among children [20–23]. Also, it has been shown that BIA provides accurate estimation of changes on %BF and FFM over time [24]. Waist circumference, as a surrogate marker of visceral obesity, has been added to refine the measure of obesity related risks [25]. Waist circumference seems to be more accurate for children because it targets central obesity, which is a risk factor for type II diabetes and coronary heart disease. To the best of our knowledge there is no publication on specific cut off points for waist circumference, but there are some ongoing studies.

15.3 CAUSES OF OBESITY

Although the mechanism of obesity development is not fully understood, it is confirmed that obesity occurs when energy intake exceeds energy expenditure. There are multiple etiologies for this imbalance, hence, and the rising prevalence of obesity cannot be addressed by a single etiology. Genetic factors influence the susceptibility of a given child to an obesity–conducive environment. However, environmental factors, lifestyle preferences, and cultural environment seem to play major roles in the rising prevalence of obesity worldwide [26–29]. In a small number of cases, childhood obesity is due to genes such as leptin deficiency or medical causes such as hypothyroidism and growth hormone deficiency or side effects due to drugs (e.g. – steroids) [30]. Most of the time, however, personal lifestyle choices and cultural environment significantly influence obesity.

15.4 BEHAVIORAL AND SOCIAL FACTORS

15.4.1 DIET

Over the last decades, food has become more affordable to larger numbers of people as the price of food has decreased substantially relative to income and the concept of 'food' has changed from a means of nourishment to a marker of lifestyle and a source of pleasure. Clearly, increases in physical activity are not likely to offset an energy rich, poor nutritive diet. It takes between 1–2 hours of extremely vigorous activity to counteract a single large–sized (i.e., >=785 kcal) children's meal at a fast food restaurant. Frequent consumption of such a diet can hardly be counteracted by the average child or adult [31].

15.4.1.1 CALORIE INTAKE

Although overweight and obesity are mostly assumed to be results of increase in caloric intake, there is not enough supporting evidence for such phenomenon. Food frequency methods measure usual diet, but estimate caloric intake poorly [32]. Other methods such as 24–hour recall or food diaries evaluate caloric intakes more accurately, however, estimate short–term not long–term intake [32]. Total energy intake is difficult to measure accurately at a population level. However, a small caloric imbalance (within the margin of error of estimation methods) is sufficient over a long period of time to lead to obesity. With concurrent rise in childhood obesity prevalence in the USA, the National Health and Nutrition Examination Survey (NHANES) noted only subtle change in calorie intake among US children from the 1970s to 1988–1994. For this period, NHANES III found an increase calorie intake only among white and black adolescent females. The same pattern was observed by the latest NHANES (1999–2000). The Bogalusa study which has been following the health and nutrition of children since 1973 in Bogalusa (Louisiana), reported that total calorie intake of 10–year old children remained unchanged during 1973–1988 and a slight but significant decrease was observed when energy intake was expressed per kilogram body weight [33]. The result of a survey carried out during the past few decades in the UK suggested that average energy intakes, for all age groups, are lower than they used to be [34]. Some small studies also found similar energy intake among obese children and their lean counterparts [6,35–37].

15.4.1.2 FAT INTAKE

While for many years it has been claimed that the increase in pediatric obesity has happened because of an increase in high fat intake, contradictory results have been obtained by cross–sectional and longitudinal studies. Result of NHANES has shown that fat consumption of American children has fallen over the last three decades. For instance; mean dietary fat consumption in males aged 12–19 years fell from 37.0% (SD = 0.29%) of total caloric intake in 1971–1974 to 32.0% (SD = 0.42%) in 1999–2000. The pattern was the same for females, whose fat consumption fell from 36.7% (SD = 0.27%) to 32.1% (SD = 0.61%) [38,39]. Gregory et al. [40] reported that the average fat intake of children aged 4–18 years in the UK is close to the government recommendation of 35% energy. On the other hand, some cross–sectional studies have found a positive relationship between fat intake and adiposity in children even after controlling for confounding factors [41,42]. The main objection to the notion that dietary fat is responsible for the accelerated pediatric obesity epidemic is the fact that at the same time the prevalence of childhood obesity was increasing, the consumption of dietary fat in different populations was decreasing. Although fat eaten in excess leads to obesity, there is not strong enough evidence that fat intake is the chief reason for the ascending trend of childhood obesity.

15.4.1.3 OTHER DIETARY FACTORS

There is a growing body of evidence suggesting that increasing dairy intake by about two servings per day could reduce the risk of overweight by up to 70% [43]. In addition, calcium intake was associated with 21% reduced risk of development of insulin resistance among overweight younger adults and may reduce diabetes risk [44]. Higher calcium intake and more dairy servings per day were associated with reduced adiposity in children studied longitudinally [45,46]. There are few data reporting the relation between calcium or dairy intake and obesity among children.

Between 1970 and 1997, the United State Department of Agriculture (USDA) surveys indicated an increase of 118% of per capita consumption of carbonated drinks, and a decline of 23% for beverage milk [47]. Soft drink intake has been associated with the epidemic of obesity [48] and type II diabetes [49] among children. While it is possible that drinking soda instead of milk would result in higher intake of total energy, it cannot be concluded definitively that sugar containing soft drinks promote weight gain because they displace dairy products.

15.4.2 PHYSICAL ACTIVITY

It has been hypothesized that a steady decline in physical activity among all age groups has heavily contributed to rising rates of obesity all around the world. Physical activity strongly influenced weight gain in a study of monozygotic twins [50]. Numerous studies have shown that sedentary behaviors like watching television and playing computer games are associated with increased prevalence of obesity [51,52]. Furthermore, parents report that they prefer having their children watch television at home rather than play outside unattended because parents are then able to complete their chores while keeping an eye on their children [53]. In addition, increased proportions of children who are being driven to school and low participation rates in sports and physical education, particularly among adolescent girls [51], are also associated with increased obesity prevalence. Since both parental and children's choices fashion these behaviors, it is not surprising that overweight children tend to have overweight parents and are themselves more likely to grow into overweight adults than normal weight children [54]. In response to the significant impact that the cultural environment of a child has on his/her daily choices, promoting a more active lifestyle has wide ranging health benefits and minimal risk, making it a promising public health recommendation.

15.5 PREVENTION

Almost all public health researchers and clinicians agree that prevention could be the key strategy for controlling the current epidemic of obesity [55]. Prevention may include primary prevention of overweight or obesity itself, secondary prevention or avoidance of weight regains following weight loss, and prevention of further weight increases in obese individuals unable to lose weight. Until now, most approaches have focused on changing the behavior of individuals on diet and exercise and it seems that these strategies have had little impact on the growing increase of the obesity epidemic.

15.5.1 WHAT AGE GROUP IS THE PRIORITY FOR STARTING PREVENTION?

Children are often considered the priority population for intervention strategies because, firstly, weight loss in adulthood is difficult and there are a greater number of potential interventions for children than for adults. Schools are a natural setting for influencing the food and physical activity environments of children. Other settings such as preschool institutions and after–school care services will have similar opportunities for action. Secondly, it is difficult to reduce excessive weight in adults once it becomes established. Therefore it would be more sensible to initiate prevention and treatment of obesity during childhood. Prevention may be achieved through a variety of interventions targeting built environment, physical activity and diet.

15.5.2 BUILT ENVIRONMENT

The challenge ahead is to identify obesogenic environments and influence them so that healthier choices are more available, easier to access, and widely promoted to a large proportion of the community (Table 2). The neighborhood is a key setting that can be used for intervention. It encompasses the walking network (footpaths and trails, etc.), the cycling network (roads and cycle paths), public open spaces (parks) and recreation facilities (recreation centers, etc.). While increasing the amount of public open space might be difficult within an existing built environment, protecting the loss of such spaces requires strong support within the community. Although the local environment, both school and the wider community, plays an important role in shaping children's physical activity, the smaller scale of the home environment is also very important in relation to shaping children's eating behaviors and physical activity patterns. Surprisingly, we know very little about specific home influences and as a setting, it is difficult to influence because of the total numbers and heterogeneity of homes and the limited options for access [56]. Of all aspects of behavior in the home environment, however, television viewing has been researched in greatest detail [57–59].

TABLE 2: Some interventions strategies that could be considered for prevention of childhood obesity

I. Built environment
  1. Walking network
    a. Footpaths (designated safe walking paths)
    b. Trails (increasing safety on trails)
  2. The cycling network
    a. Roads (designated cycling routes)
    b. Cycle paths
  3. Publis open spaces (parks)
  4. Recreation facilities (providing safe and inexpensive recreation centers)
II. Physical activity
  1. Increasing sports participation
  2. Improving and increasing physical education time
  3. Using school report cards to make the parents aware of their children's weight problem
  4. Enhancing active modes of transport to and from school
    a. Walking (e.g. walking bus)
    b. Cycling
    c. Public transport
III. TV watching
  1. Restricting television viewing
  2. Reducing eating in front of the television
  3. Ban or restriction on television advertising to children
IV. Food sector
  1. Applying a small tax on high–volume foods of low nutritional value (e.g. soft drinks, confectionary, and snack foods)
  2. Food labeling and nutrition "signposts" (e.g. logos for nutritious foods)
  3. Implementing standards for product formulation

15.5.3 PHYSICAL ACTIVITY

Stone et al. [60] reviewed the impact of 14 school–based interventions on physical activity knowledge and behavior. Most of the outcome variables showed significant improvements for the intervention. One interdisciplinary intervention program in the USA featured a curriculum–based approach to influence eating patterns, reduce sedentary behaviors (with a strong emphasis on television viewing), and promote higher activity levels among children of school grades 6 to 8. Evaluation at two years showed a reduction in obesity prevalence in girls (OR = 0.47; 95%CI: 0.24 – 0.93), but not in boys (OR = 0.85; 95%CI: 0.52 – 1.39) compared to controls. The reduction in television viewing (by approximately 30 min/day) was highly significant for both boys and girls. Increases in sports participation and/or physical education time would need policy–based changes at both school and education sector levels [61]. Similarly, increases in active modes of transport to and from school (walking, cycling, and public transport) would require policy changes at the school and local government levels, as well as support from parents and the community. In some communities a variety of such programs have been implemented e.g. road crossings, 'walking bus', and designated safe walking and cycling routes [51].

15.5.4 EFFECTS OF DIETARY PATTERN AND TV WATCHING

It appears that gains can be made in obesity prevention through restricting television viewing. Although, it seems that reduced eating in front of the television is at least as important as increasing activity [58]. Fast foods are one of the most advertised products on television and children are often the targeted market. Reducing the huge volume of marketing of energy–dense foods and drinks and fast–food restaurants to young children, particularly through the powerful media of television, is a potential strategy that has been advocated. Television advertising to children under 12 years of age has not been permitted in Sweden since commercial television began over a decade ago, although children's television programs from other countries, and through satellite television, probably dilute the impact of the ban in Sweden. Norway, Denmark, Austria, Ireland, Australia, and Greece also have some restrictions on television advertising to young children [51]. The fact that children would still be seeing some television advertisements during adult programs or other types of marketing, such as billboards, does not contradict the rationale for the control on the television watching of young children.

15.5.5 FOOD SECTOR

Food prices have a marked influence on food–buying behaviour and, consequently, on nutrient intake [62]. A small tax (but large enough to affect sales) on high–volume foods of low nutritional value, such as soft drinks, confectionery, and snack foods, may discourage their use. Such taxes currently applied in some parts of the USA and Canada [63]. In addition, food labeling and nutrition 'signposts' such as logos that indicate that a food meets certain nutrition standards might help consumers make choices of healthy foods. An example is the 'Pick the Tick' symbol program run by the National Heart Foundations in Australia and New Zealand [64]. The 'Pick the Tick' symbols made it easier for consumers to identify healthier food choices and are frequently used by shoppers. In addition, the nutrition criteria for the products serve as 'de facto' standards for product formulation, and many manufacturers will formulate or reformulate products to meet those standards.

15.5.6 EFFECTIVENESS OF THE PREVENTION METHODS

It has been shown that focusing on reducing sedentary behaviour and encouraging free play has been more effective than focusing on forced exercise or reducing food intake in preventing already obese children from gaining more weight [65]. Recent efforts in preventing obesity include the initiative of using school report cards to make the parents aware of their children's weight problem. Health report cards are believed to aid prevention of obesity. In a study in the Boston area, parents who received health and fitness report cards were almost twice as likely to know or acknowledge that their child was actually overweight than those parents who did not get a report card [66]. They also were over twice as likely to plan weight–control activities for their overweight children.

A summary of prevention and intervention strategies is presented in Table 2.

15.6 CONCLUSION

Obesity is a chronic disorder that has multiple causes. Overweight and obesity in childhood have significant impact on both physical and psychological health. In addition, psychological disorders such as depression occur with increased frequency in obese children. Overweight children are more likely to have cardiovascular and digestive diseases in adulthood as compared with those who are lean. It is believed that both over–consumption of calories and reduced physical activity are mainly involved in childhood obesity.

Apparently, primary or secondary prevention could be the key plan for controlling the current epidemic of obesity and these strategies seem to be more effective in children than in adults. A number of potential effective plans can be implemented to target built environment, physical activity, and diet. These strategies can be initiated at home and in preschool institutions, schools or after–school care services as natural setting for influencing the diet and physical activity and at home and work for adults. Both groups can benefit from an appropriate built environment. However, further research needs to examine the most effective strategies of intervention, prevention, and treatment of obesity. These strategies should be culture specific, ethnical, and consider the socio–economical aspects of the targeting population.

REFERENCES

1. Nicklas TA, T. B, K.W. C, G. B: Eating Patterns, Dietary Quality and Obesity. Journal of the American College of Nutrition 2001, 20:599–608.
2. Parsons TJ, Power C, Logan S, Summerbell CD: Childhood predictors of adult obesity: a systematic review. International Journal of Obesity 1999, 23:S1–S107.

3. Whitaker RC, Wright JA, Pepe MS, Seidel KD, Dietz WH: Predicting obesity in young adulthood from childhood and parental obesity. New England Journal of Medicine 1997, 337:869–873.
4. Livingstone MB: Childhood obesity in Europe: a growing concern. Public Health Nutr 2001, 4:109–116.
5. James PT: Obesity: The worldwide epidemic. Clinics in Dermatology 2004, 22:276–280.
6. Kelishadi R, Pour MH, Sarraf–Zadegan N, Sadry GH, Ansari R, Alikhassy H, Bashardoust N: Obesity and associated modifiable environmental factors in Iranian adolescents: Isfahan Healthy Heart Program – Heart Health Promotion from Childhood. Pediatr Int 2003, 45:435–442.
7. AlNuaim AR, Bamgboye EA, AlHerbish A: The pattern of growth and obesity in Saudi Arabian male school children. International Journal of Obesity 1996, 20:1000–1005.
8. McCarthy HD, Ellis SM, Cole TJ: Central overweight and obesity in British youth aged 11–16 years: cross sectional surveys of waist circumference. BMJ 2003, 326:624.
9. Ruxton CH, Reilly JJ, Kirk TR: Body composition of healthy 7–and 8–year–old children and a comparison with the 'reference child'. International Journal of Obesity 1999, 23:1276–1281.
10. Daniels SR, Arnett DK, Eckel RH, Gidding SS, Hayman LL, Kumanyika S, Robinson TN, Scott BJ, St Jeor S, Williams CL: Overweight in children and adolescents: pathophysiology, consequences, prevention, and treatment. Circulation 2005, 111:1999–2012.
11. Mossberg HO: 40–Year Follow–Up of Overweight Children. Lancet 1989, 2:491–493.
12. Must A, Jacques PF, Dallal GE, Bajema CJ, Dietz WH: Long–Term Morbidity and Mortality of Overweight Adolescents – A Follow–Up of the Harvard Growth Study of 1922 to 1935. New England Journal of Medicine 1992, 327:1350–1355.
13. Flegal KM, Carroll MD, Ogden CL, Johnson CL: Prevalence and trends in obesity among US adults, 1999–2000. JAMA 2002, 288:1723–1727.
14. Kuczmarski RJ, Flegal KM: Criteria for definition of overweight in transition: background and recommendations for the United States. Am J Clin Nutr 2000, 72:1074–1081.
15. Williams DP, Going SB, Lohman TG, Harsha DW, Srinivasan SR, Webber LS, Berenson GS: Body Fatness and Risk for Elevated Blood–Pressure, Total Cholesterol, and Serum–Lipoprotein Ratios in Children and Adolescents. American Journal of Public Health 1992, 82:358–363.
16. Flegal KM, Wei R, Ogden C: Weight–for–stature compared with body mass index–for–age growth charts for the United States from the Centers for Disease Control and Prevention. American Journal of Clinical Nutrition 2002, 75:761–766.
17. Himes JH, Dietz WH: Guidelines for Overweight in Adolescent Preventive Services – Recommendations from An Expert Committee. American Journal of Clinical Nutrition 1994, 59:307–316.
18. Flodmark CE, Lissau I, Moreno LA, Pietrobelli A, Widhalm K: New insights into the field of children and adolescents' obesity: the European perspective (vol 28, pg 1189, 2004). International Journal of Obesity 2004., 28: OpenURL

19. Lazarus R, Baur L, Webb K, Blyth F: Body mass index in screening for adiposity in children and adolescents: systematic evaluation using receiver operating characteristic curves. Am J Clin Nutr 1996, 63:500–506.
20. Danford LC, Schoeller DA, Kushner RF: Comparison of two bioelectrical impedance analysis models for total body water measurement in children. Ann Hum Biol 1992, 19:603–607.
21. Deurenberg P, van der KK, Paling A, Withagen P: Assessment of body composition in 8–11 year old children by bioelectrical impedance. Eur J Clin Nutr 1989, 43:623–629.
22. Deurenberg P, Kusters CS, Smit HE: Assessment of body composition by bioelectrical impedance in children and young adults is strongly age–dependent. Eur J Clin Nutr 1990, 44:261–268.
23. Deurenberg P, Pieters JJ, Hautvast JG: The assessment of the body fat percentage by skinfold thickness measurements in childhood and young adolescence. Br J Nutr 1990, 63:293–303.
24. Phillips SM, Bandini LG, Compton DV, Naumova EN, Must A: A longitudinal comparison of body composition by total body water and bioelectrical impedance in adolescent girls. Journal of Nutrition 2003, 133:1419–1425.
25. Stevens J: Obesity, fat patterning and cardiovascular risk. Adv Exp Med Biol 1995, 369:21–27.
26. Hill JO, Peters JC: Environmental contributions to the obesity epidemic. Science 1998, 280:1371–1374.
27. Goodrick GK, Poston WS, Foreyt JP: Methods for voluntary weight loss and control: update 1996. Nutrition 1996, 12:672–676.
28. Eckel RH, Krauss RM: American Heart Association call to action: obesity as a major risk factor for coronary heart disease. AHA Nutrition Committee. Circulation 1998, 97:2099–2100.
29. Grundy SM: Multifactorial causation of obesity: implications for prevention. Am J Clin Nutr 1998, 67:563S–572S.
30. Link K, Moell C, Garwicz S, Cavallin–Stahl E, Bjork J, Thilen U, Ahren B, Erfurth EM: Growth hormone deficiency predicts cardiovascular risk in young adults treated for acute lymphoblastic leukemia in childhood. J Clin Endocrinol Metab 2004, 89:5003–5012.
31. Styne DM: Obesity in childhood: what's activity got to do with it? American Journal of Clinical Nutrition 2005, 81:337–338.
32. Willett W: Food Frequency Methods. In Nutritional Epidemiology. Volume 5. 2nd edition. Oxford University Press; 1998:74. OpenURL
33. Nicklas TA: Dietary Studies of Children – the Bogalusa Heart–Study Experience. Journal of the American Dietetic Association 1995, 95:1127–1133.
34. Prentice AM, Jebb SA: Obesity in Britain – Gluttony Or Sloth. British Medical Journal 1995, 311:437–439.
35. Bellisle F, Rolland–Cachera MF, Deheeger M, Guilloud–Bataille M: Obesity and food intake in children: evidence for a role of metabolic and/or behavioral daily rhythms. Appetite 1988, 11:111–118.
36. Griffiths M, Payne PR: Energy expenditure in small children of obese and non–obese parents. Nature 1976, 260:698–700.

37. Maffeis C, Zaffanello M, Pinelli L, Schutz Y: Total energy expenditure and patterns of activity in 8–10–year–old obese and nonobese children. J Pediatr Gastroenterol Nutr 1996, 23:256–261.
38. Troiano RP, Briefel RR, Carroll MD, Bialostosky K: Energy and fat intakes of children and adolescents in the united states: data from the national health and nutrition examination surveys. Am J Clin Nutr 2000, 72:1343S–1353S.
39. Wright JD, Kennedy–Stephenson J, Wang CY, McDowell MA, Johnson CL: Trends in intake of energy and macronutrients – United States, 1971–2000 (Reprinted from MMWR, vol 53, pg 80–82, 2004). Journal of the American Medical Association 2004, 291:1193–1194.
40. Gregory JW, Lowe S: National Diet and Nutrition Survery: Young People Aged 4 to 18 Years : Report of the Diet and Nutrition Survey. London, The Stationery Office.; 2000.
41. Maffeis C, Pinelli L, Schutz Y: Fat intake and adiposity in 8 to 11 year–old obese children. International Journal of Obesity 1996, 20:170–174
42. Tucker LA, Seljaas GT, Hager RL: Body fat percentage of children varies according to their diet composition. Journal of the American Dietetic Association 1997, 97:981–986.
43. Heaney RP, Davies KM, Barger–Lux MJ: Calcium and weight: clinical studies. J Am Coll Nutr 2002, 21:152S–155S.
44. Pereira MA, Jacobs DRJ, Van Horn L, Slattery ML, Kartashov AI, Ludwig DS: Dairy consumption, obesity, and the insulin resistance syndrome in young adults: the CARDIA Study. JAMA 2002, 287:2081–2089.
45. Carruth BR, Skinner JD: The role of dietary calcium and other nutrients in moderating body fat in preschool children.
46. Int J Obes Relat Metab Disord 2001, 25:559–566.
47. Skinner JD, Bounds W, Carruth BR, Ziegler P: Longitudinal calcium intake is negatively related to children's body fat indexes. J Am Diet Assoc 2003, 103:1626–1631.
48. Putnam JJ, Allshouse JE: Food consumption, prices, and expenditures, 1970–97. Washington,D.C., Food and Consumers Economics Division, Economic Research Service, US Department of Agriculture; 1999.
49. Ludwig DS, Peterson KE, Gortmaker SL: Relation between consumption of sugar–sweetened drinks and childhood obesity: a prospective, observational analysis. Lancet 2001, 357:505–508.
50. Gittelsohn J, Wolever TM, Harris SB, Harris–Giraldo R, Hanley AJ, Zinman B: Specific patterns of food consumption and preparation are associated with diabetes and obesity in a Native Canadian community. J Nutr 1998, 128:541–547.
51. Heitmann BL, Kaprio J, Harris JR, Rissanen A, Korkeila M, Koskenvuo M: Are genetic determinants of weight gain modified by leisure–time physical activity? A prospective study of Finnish twins. American Journal of Clinical Nutrition 1997, 66:672–678.
52. Swinburn B, Egger G: Preventive strategies against weight gain and obesity. Obes Rev 2002, 3:289–301.
53. Tremblay MS, Willms JD: Is the Canadian childhood obesity epidemic related to physical inactivity? Int J Obes Relat Metab Disord 2003, 27:1100–1105.

54. Gordon–Larsen P, Griffiths P, Bentley ME, Ward DS, Kelsey K, Shields K, Ammerman A: Barriers to physical activity: qualitative data on caregiver–daughter perceptions and practices. Am J Prev Med 2004, 27:218–223.
55. Carriere G: Parent and child factors associated with youth obesity. Statistics Canada; 2003.
56. Muller MJ, Mast M, Asbeck I, Langnase K, Grund A: Prevention of obesity—is it possible? Obes Rev 2001, 2:15–28.
57. Campbell K, Crawford D, Jackson M, Cashel K, Worsley A, Gibbons K, Birch LL: Family food environments of 5–6–year–old–children: does socioeconomic status make a difference? Asia Pac J Clin Nutr 2002, 11 Suppl 3:S553–S561.
58. Gortmaker SL, Peterson K, Wiecha J, Sobol AM, Dixit S, Fox MK, Laird N: Reducing obesity via a school–based interdisciplinary intervention among youth: Planet Health. Arch Pediatr Adolesc Med 1999, 153:409–418.
59. Robinson TN: Reducing children's television viewing to prevent obesity: a randomized controlled trial. JAMA 1999, 282:1561–1567.
60. Dietz WH, Gortmaker SL: Preventing obesity in children and adolescents. Annu Rev Public Health 2001, 22:337–353.
61. Stone EJ, McKenzie TL, Welk GJ, Booth ML: Effects of physical activity interventions in youth. Review and synthesis. Am J Prev Med 1998, 15:298–315.
62. Dwyer T, Coonan WE, Leitch DR, Hetzel BS, Baghurst RA: An investigation of the effects of daily physical activity on the health of primary school students in South Australia. Int J Epidemiol 1983, 12:308–313.
63. Guo X, Popkin BM, Mroz TA, Zhai F: Food price policy can favorably alter macronutrient intake in China. J Nutr 1999, 129:994–1001.
64. Jacobson MF, Brownell KD: Small taxes on soft drinks and snack foods to promote health. Am J Public Health 2000, 90:854–857.
65. Young L, Swinburn B: Impact of the Pick the Tick food information programme on the salt content of food in New Zealand. Health Promot Int 2002, 17:13–19.
66. Caterson ID, Gill TP: Obesity: epidemiology and possible prevention. Best Pract Res Clin Endocrinol Metab 2002, 16:595–610.
67. Chomitz VR, Collins J, Kim J, Kramer E, McGowan R: Promoting healthy weight among elementary school children via a health report card approach. Archives of Pediatrics & Adolescent Medicine 2003, 157:765–772.
68. Freedman DS, Srinivasan SR, Valdez RA, Williamson DF, Berenson GS: Secular increases in relative weight and adiposity among children over two decades: the Bogalusa Heart Study. Pediatrics 1997, 99:420–426.
69. Zametkin AJ, Zoon CK, Klein HW, Munson S: Psychiatric aspects of child and adolescent obesity: a review of the past 10 years. J Am Acad Child Adolesc Psychiatry 2004, 43:134–150.
70. Kotani K, Nishida M, Yamashita S, Funahashi T, Fujioka S, Tokunaga K, Ishikawa K, Tarui S, Matsuzawa Y: Two decades of annual medical examinations in Japanese obese children: do obese children grow into obese adults? Int J Obes Relat Metab Disord 1997, 21:912–921.
71. Lobstein TJ, James WP, Cole TJ: Increasing levels of excess weight among children in England. Int J Obes Relat Metab Disord 2003, 27:1136–1138.

72. Moreno LA, Sarria A, Popkin BM: The nutrition transition in Spain: a European Mediterranean country. Eur J Clin Nutr 2002, 56:992–1003.
73. Rolland–Cachera MF, Deheeger M, Thibault H: [Epidemiologic bases of obesity]. Arch Pediatr 2001, 8 Suppl 2:287s–289s.
74. GE K, T T, C T, T K: Prevalence and trends in overweight and obesity among children and adolescents in Thessaloniki, Greece. J Pediatr Endocrinol Metab 2005, 14:1319–1365.

This chapter was originally published under the Creative Commons License. Dehghan, M., Akhtar-Danesh, N., and Merchant, A. T. Childhood Obesity, Prevalence and Prevention. Nutrition Journal 2005, 4:24 doi:10.1186/1475-2891-4-24.

AUTHOR NOTES

CHAPTER 2

Acknowledgements

We thank the twins and families from Canada, Sweden, Denmark, and Australia for their participation. Denmark: We thank secretary Jytte Duerlund for expert technical help during data collection. Australia: We thank Ann Eldridge, Marlene Grace, Kerrie McAloney (sample collection); David Smyth, Harry Beeby, Daniel Park (IT support).

Integrity of the Data & Data Sharing

All authors, external and internal, had full access to all of the data (including statistical reports and tables) in the study and can take responsibility for the integrity of the data and the accuracy of the data analysis. Data sharing: No additional data available.

Author Contributions

Conceived and designed the experiments: LD KOK DP JH AS FR MJW PL NGM. Performed the experiments: KOK DP JH AS FR MJW PL NGM. Analyzed the data: MG. Contributed reagents/materials/analysis tools: KOK DP JH AS FR MJW PL NGM. Wrote the paper: FTT.

CHAPTER 3

Acknowledgments

We thank the GIANT Consortium (Metabolism Initiative and Program in Medical and Population Genetics, Broad Institute, Cambridge, Massachusetts, USA) for letting us access to unpublished obesity associated SNPs that have been genotyped in the current study. We thank Nabila Bouatia–Naji (CNRS–8199–Lille North of France University, Pasteur Institute, Lille, France) for advice on SNPs selection and genotyping and Marion Marchand (CNRS–8199–Lille North of France University, Pasteur Institute, Lille, France) for performing part of genotyping. We thank M. Deweirder and F. Allegaert (CNRS–8199–Lille North of France University, Pasteur Institute, Lille, France) for DNA bank management. We are indebted to all subjects who participated in these studies.

Author Contributions

Conceived and designed the experiments: AM DM PF. Performed the experiments: SL SG VV. Analyzed the data: AM KK SLR–S. Contributed reagents/materials/analysis tools: SL MK SG VV. Wrote the paper: AM. Cohort investigators: M–RJ CM MG KK SLR–S MK AP A–LH JL AR SD AAK. Supervised the study: PE M–RJ PF. Equally contributed as last authors: M–RJ PF. Equal corresponding authors: AM PF.

CHAPTER 4

Competing interests

VLF as Senior Vice President of Nutrition Impact, LLC, performs consulting and database analyses for various food and beverage companies and related entities. EEQ is an employee of Dairy Research Institute/National Dairy Council.

Authors' contributions

VLF designed the study, was primarily responsible for the data analysis, and provided critical input into the manuscript; EEQ helped with data analysis and drafted the manuscript. All authors read and approved the final manuscript.

Sources of support

Dairy Research Institute administered by the National Dairy Council

CHAPTER 5

Competing interests

The authors declare no competing interests.

Authors' contribution

AR performed the literature search and drafted the manuscript. FM revised the manuscript and supported the process of writing the manuscript. AW gave the final approval of the version to be published. All authors read and approved the final manuscript.

Acknowledgements

We would like to express our appreciation to PD Dr. Annegret Mündermann (University of Konstanz) and Dr. Stefanie Everke–Buchanan (University of Konstanz and Zeppelin University Friedrichshafen) for their writing assistance.

Funding sources/trial registrations

The corresponding author received a stipend of the Department of Sport Science, University of Konstanz.

CHAPTER 6

Competing interests

None of the authors have any financial or conflict of interest to declare.

Authors' contributions

Author contributions include the following: AG was responsible for conception and design, analysis and interpretation of data, and drafting the article. LA, LD and HLM were responsible for conception and design of data, and critical revision of paper for important intellectual content. MB and KG were responsible for statistical analyses and interpretation of data, and revising the article critically for important statistical and intellectual content. All authors had full access to all of the data (including statistical reports and tables) in the study, can take responsibility for the integrity of the data and the accuracy of the data analysis, and approved the final version to be published. In collaboration with Statistics Canada, MB was responsible for the design and implementation of the study.

Funding

The 1983 research funding was provided by the Ontario Ministry of Community and Social Services. The follow up in 2001 was funded by a grant from the Canadian Institutes of Health Research (CIHR) awarded to Dr.

Boyle. Dr. Gonzalez was funded by a Canadian Institutes of Health Research (CIHR) Postdoctoral Fellowship from the Institute of Gender and Health and a Lawson Postdoctoral Fellowship. Dr. Boyle received support from a Canada Research Chair from CIHR in Social Determinants of Child Health. Dr. MacMillan received support from the David R. (Dan) Offord Chair in Child Studies.

Neither funding agency had direct involvement in the design and conduct of the study; in collection, management, analysis, and interpretation of the data; or in preparation, review, or approval of the manuscript.

CHAPTER 8

Acknowledgments

This work was supported by an interdisciplinary doctoral grant from the University Research Fund (BOF–ID) of the University of Antwerp to V. M. Conraads, J. Ramet, D. K. Vissers, and L. Bruyndonckx. V. M. Conraads is a senior clinical investigator supported by the Fund for Scientific Research (FWO), Flanders (Belgium). The authors kindly thank Dr. Ilse Van Brussel (Laboratory of Physiopharmacology, Faculty of Medicine and Health Sciences, Antwerp University, Belgium) for providing images of endothelial and smooth muscle cells to prepare Figure 2.

CHAPTER 9

Acknowledgments

D. Gozal is supported by National Institutes of Health Grants no. HL–065270 and HL–086662. R. Bhattacharjee was supported by a fellowship from Jazz Pharmaceuticals. J. Kim and R. Bhattacharjee contributed equally to this paper.

CHAPTER 11

Author Contributions

Conceived and designed the experiments: AG CR RH HR. Performed the experiments: AG. Analyzed the data: AG. Wrote the paper: AG. Provided expert input and advice regarding statistical analyses: CR AW.

CHAPTER 12

Acknowledgment
The authors thank Professor J. P. Thissen for his help in writing this manuscript.

CHAPTER 13

Conflict of Interests
The authors declare that they have no conflict of interests.

CHAPTER 14

Acknowledgments
The authors are grateful to Saira Khan and Peixin Zhang for assistance with data analyses, and to the information systems staff at participating institutions for initial retrieval of data from the EHR.

Author Contributions
Conceived and designed the experiments: LCB CBF. Performed the experiments: LCB DM KK MK MDB FY TR CBF. Analyzed the data: LCB TR CBF. Contributed reagents/materials/analysis tools: LCB DM KK MK MDB FY TR. Wrote the paper: LCB CBF. Reviewed and revised the manuscript: LCB DM KK MK MDB FY TR CBF.

CHAPTER 15

Authors' contributions
All authors had equal contribution in writing this manuscript.

Acknowledgements
We would like to thank Claire Vayalumkal for her helpful comments and careful reading of the final manuscript.

INDEX

G

H

I

J

L

Made in the USA
Las Vegas, NV
21 September 2026

ce1a69f8-a54d-4463-9471-a6b93d0d7b6eR01